Springer Series in Operations Research And Financial Engineering

Series Editors:
Thomas V. Mikosch
Sidney I. Resnick
Stephen M. Robinson

For other titles published in this series, go to
http://www.springer.com/series/3182

John A. Muckstadt
Amar Sapra

Principles of Inventory Management

When You Are Down to Four, Order More

John A. Muckstadt
School of Operations Research
  and Information Engineering
Cornell University
286 Rhodes Hall
Ithaca, NY 14853-3801
USA
JAM61@cornell.edu

Amar Sapra
Department of Quantitative Methods
  and Information Systems
Indian Institute of Management
  Bangalore
Bannerghatta Road
Bangalore 560 076
India
amar.sapra@iimb.ernet.in

Series Editors:
Thomas V. Mikosch
Department of Mathematical Sciences
University of Copenhagen
DK-1017 Copenhagen
Denmark
mikosch@math.ku.dk

Stephen M. Robinson
Department of Industrial and Systems Engineering
University of Wisconsin-Madison
Madision, WI 53706
USA
smrobins@wisc.edu

Sidney I. Resnick
Cornell University
School of Operations Research and
  Information Engineering
Ithaca, NY 14853
USA
sirl@cornell.edu

ISSN 1431-8598
ISBN 978-0-387-24492-1   ISBN 978-0-387-68948-7 (eBook)
DOI 10.1007/978-0-387-68948-7
Springer New York Dordrecht Heidelberg London

Library of Congress Control Number: 2009939430

Mathematics Subject Classification (2010): 90-01, 90B05, 90C39

Printed on acid-free paper

Springer is part of Springer Science+Business Media (www.springer.com)

To my wife, Linda, my parents, my children, and grandchildren, who have supported and inspired me.
—Jack Muckstadt

To my parents and siblings for their continued support and encouragement. My parents have lived a difficult life and have denied themselves most pleasures in life just to make sure that their children were able to obtain the best possible education.
—Amar Sapra

Preface

The importance of managing inventories properly in global supply chains cannot be denied. Each component of these numerous supply chains must function appropriately so that inventories are managed efficiently. To manage efficiently requires the leaders and staffs in each organization to comprehend certain basic principles and laws. The purpose of this book is to discuss these principles.

The contents of this text represent a collection of lecture notes that have been created over the past 33 years at Cornell University. As such, the topics discussed, the sequence in which they are presented, and the level of mathematical sophistication required to understand the contents of this text are based on my interests and the backgrounds of my students. Clearly, not all topics found in the vast literature on quantitative methods used to model and solve inventory management problems can be covered in a one-semester course. Consequently, this book is limited in scope and depth.

The contents of the book are organized in a manner that I have found to be effective in teaching the subject matter. After an introductory chapter in which the fundamental issues pertaining to the management of inventories are discussed, we introduce a variety of models and algorithms. Each such model is developed on the basis of a set of assumptions about the manner in which an operating environment functions.

In Chapter 2 we study the classic economic order quantity problem. This type of problem is based on the assumption that demands occur at a constant, continuous, and known rate over an infinite planning horizon. Furthermore, the cost structure remains constant over this infinite horizon as well. The focus is on managing inventories at a single location.

The material in Chapter 3 extends the topic covered in Chapter 2. Several multi-location or multi-item models are analyzed. These analyses are based on what are called power-of-two policies. Again, the underlying operating environments are assured to be deterministic and unchanging over an infinite horizon.

The assumptions made about the operating environment are altered in Chapter 4. Here the planning horizon is finite in length and divided into periods. Demands and costs are assumed to be known in each period, although they may change from period to period.

In all subsequent chapters, uncertainty is present in the operating environment. In Chapter 5 we study single-period problems in which customer demand is assumed to be described by a random variable. In Chapter 6, the analysis is extended to multiple periods. The discussion largely focuses on establishing properties of optimal policies in finite-horizon settings when demand is described by a non-stationary process through time. Serial systems are also discussed. The objective is to minimize the expected costs of holding inventories and stocking out. Thus, the cost structure in this chapter is limited to the case where there are no fixed ordering costs.

In Chapter 7, we study environments in which demands can occur at any point in time over an infinite planning horizon. Whereas we assumed in Chapter 6 that inventory procurement decisions were made periodically, in this chapter we assume such decisions are made continuously in time. The underlying stochastic processes governing the demand processes are stationary over the infinite planning horizon, as are the costs. As in Chapter 6, we assume there are no fixed ordering costs.

The analysis in Chapter 7 is confined to managing items in a single location. In Chapter 8 we extend the analysis to multi-echelon systems. Thus the underlying system is one in which inventory decisions are made continuously through time, but now in multiple locations. The importance of understanding the interactions of inventory policies between echelons is the main topic of this chapter.

Chapters 9 and 10 contain extensions of the materials in Chapters 7 and 6, respectively. In both chapters, we introduce the impact that fixed ordering costs have on the form of optimal operating policies as well as on the methods used to model and solve the resulting optimization problems. Both exact and approximate models are presented along with appropriate algorithms and heuristics. A proof of the optimality of so-called (s, S) policies is given, too.

As mentioned, the materials contained in this text are ones that have been taught to Cornell students. These students are seniors and first year graduate students. As such, they have studied optimization methods, probability theory (non-measure-theoretic) and stochastic processes in undergraduate level courses prior to taking the inventory management course. In addition to presenting fundamental principles to them, the intent of the course is also to demonstrate the application of the topics they have studied previously.

The text is written so that sections can be read mostly independently. To make this possible, notation is presented in each major section of each chapter. The text could be used in different ways. For example, a half semester course could consist of material in Chapter 2, Section 4.1, Sections 5.1–5.2, Sections 6.2–6.3, most of Sections 9.1–

9.2, Sections 10.1–10.2, and Section 10.5. While we have chosen to examine stochastic lot sizing problems at the end of the text, these materials could easily be studied in a different sequence. For example, Chapter 9 could be studied after Chapter 3, and Chapter 10 could be studied after Chapter 6. Rearranging the sequence in which the text can be read is possible because of the way it has been written.

I have mentioned that the scope of this text is limited. I encourage readers to study other texts to complete their understanding of the basic principles underlying the topic of inventory management. These texts include those authored by Sven Axsäter; Ed Silver and Rein Peterson; Steve Nahmias; Craig Sherbrooke; Paul Zipkin; and Evan Porteus. Each of these authors has made exceptional contributions to the science and practice of inventory management.

Ithaca, NY   *John A. Muckstadt*
May 2009

Acknowledgments

I began my study of inventories while a student at the University of Michigan. My teachers there, Richard C. Wilson and Herbert P. Galliher, taught me the basics of the subject. These two were great teachers and engineers. They prodded and encouraged me during and after my student years. I am deeply indebted to them.

As is often the case in a person's life, an event occurred that altered every professional activity I have undertaken thereafter. This event occurred for me in the early 1970s when I was asked to develop an approach for computing procurement quantities for engines and other repairable items for the U.S. Air Force's F-15 aircraft. At that time I was an active duty Air Force Officer. Suddenly, I had to truly learn and then apply the principles of inventory theory. The people with whom I worked on this project were capable, dedicated, and truly of great character. At the Air Force Logistics Command Headquarters, where I worked, these people included Major General George Rhodes, Colonel Fred Gluck, Major Gene Perkins, Captain Jon Reynolds, Captain Mike Pearson, MSgt Robert Kinsey, Tom Harruff, Vic Presutti, and Perry Stewart. I learned much from my friends and colleagues at the RAND Corporation: Irv Cohen, Gordon Crawford, Steve Drezner, Murray Geisler, Jack Abel, Mort Berman, Lou Miller, Bob Paulson, Hy Shulman, and John Lu. I also benefited greatly from research conducted at RAND by Craig Sherbrooke. Many of the ideas presented in Chapter 8 are directly related to his efforts. Also, I had the distinct privilege of learning about the practice of inventory management from Bernie Rosenman, who headed the Army Inventory Research Office, and his colleagues Karl Kruse and Alan Kaplan.

Since 1974 I have been on the faculty at Cornell and have had the opportunity to work with some of the finest scholars in the field of operations research. Peter Jackson, Bill Maxwell, Paat Rusmevichientong, and Robin Roundy all have greatly influenced my thinking about the principles of inventory management. I have been fortunate to have taught and worked with many gifted students. Almost 1,000 students have been taught inventory management principles at Cornell since 1974. I am especially indebted to many former Ph.D. students, who, without exception, have been wonderful people and a great joy to work with. They include Kripa Shanker, Peter Knepell, Mike Isaac, Jim Rappold, Kathryn Caggiano, Andy Loerch, Bob Sheldon, Ed Chan, Alan Bowman, David Murray, Jong Chow, Eleftherios Iacovou, Susan Alten, Chuck Sox, Howard Singer, Sophia Wang, Juan Pereira, and, most recently, Retsef Levi, Tim Huh, and Ganesh Janakiraman. Major sections of Chapter 10 are due to these latter three. Thanks also to Tim Huh, Retsef Levi, Ganesh Janakiraman, and Joseph Geunes for their early adoption of the book and helpful feedback.

Amar Sapra, my co-author and former Cornell student, urged me for many years to write this book. Without his encouragement and substantial assistance, the book would not have been completed.

I cannot express with mere words how thankful I am to all of these truly exceptional people.

I also appreciate the heroic efforts of June Meyermann, who had to decipher my handwriting as she typed the manuscript. She is a jewel. Kathleen King and Paat Rusmevichientong have provided substantial support in the preparation of this book, as well.

Lastly, and most importantly, my wife Linda has been very supportive of the time I have spent working on the text. The many hours that I have not been available for activities with her are too numerous to count. I deeply appreciate her constant love and support.

Contents

1

Inventories Are Everywhere

This morning I began the day by pouring a glass of orange juice from a half gallon container, filling a bowl with cereal, which was stored in a large box in a kitchen cabinet, taking a banana from a bunch sitting on our kitchen countertop along with many other items, slicing the banana onto the cereal, pouring milk into the bowl from a gallon container, and then sitting at a table to enjoy my breakfast. There are six chairs at my breakfast table, but, of course, I occupy only one. When taking the cereal from the cabinet, I had to choose from six different cereals we have stocked. I could have selected either low or high pulp content orange juice, since we stock both types; I could have chosen either 1% or skim milk to place on my cereal. The kitchen remains full of food items and food preparation materials that will be used at some later time. The remainder of my house contains many other types of items sitting idly, waiting to be used at a future time. My Jaguar convertible will not be used today. It is raining, so I will take the Dodge minivan to the office.

All of the items I have mentioned are examples of inventories that we have around us that support our daily living. But why do we have these inventories? Is it simply convenience or are there economic factors at play as well?

Inventories are obviously prevalent in the commercial world. Retail stores are stocked with an abundance of material. Manufacturing facilities are also filled with inventories of raw materials, work in process, and perhaps finished goods. But they are also stocked with inventories of equipment, machines, spare parts, and people, among other things. Governments stockpile material, too, including items to be used in emergencies, such as vaccines that will be used in the event of a biological attack, salt that will used to keep roads clear in the winter, and military equipment and material, to mention only a few. The Federal Reserve Banks have inventories of money to ensure the smooth execution of commerce in the U.S. economy. Regional blood banks stock large quantities of blood for use in emergencies as well as for meeting day-to-day needs. All

J.A. Muckstadt and A. Sapra, *Principles of Inventory Management: When You Are Down to Four, Order More*, Springer Series in Operations Research and Financial Engineering, DOI 10.1007/978-0-387-68948-7_1, © Springer Science+Business Media, LLC 2010

these and many more types of inventories are evident throughout the world. But again why are these inventories created and maintained at particular levels?

In general, inventories exist because there is an imbalance between the supply of an item at a location and its consumption or sale there. The imbalances are the consequence of many technical, economic, social, and natural forces. Note that inventories are a consequence and not a cause of some policy or action. Hence inventories become a dependent rather than an independent variable.

If I choose to go to the grocery store once or twice a week (this is my policy), then I must carry inventory of food sufficient to satisfy my needs until the next trip to the store. If a manufacturing plant contains equipment that is designed to make components efficiently when operating, but is engineered in a way that requires a lengthy setup time between production runs of different component types, then an economic production run of a particular component type will yield a large number of units, and the production runs will occur infrequently. Inventories are therefore created in each run to meet delivery requirements between successive production runs. These examples illustrate that policy and technology together dictate that inventories must exist.

1.1 The Roles of Inventory

We all recognize the necessity of carrying inventories to sustain operations within an economy. One role of management is to determine policies that create and distribute inventories most effectively. As we have mentioned, there are many forces that affect the choice of a policy that managements might select. These policies, to a major extent, reflect the environment in which a company operates. The environmental factors, in turn, result in the roles inventories play in a corporation's or supply chain's strategy.

Let us consider one way to think about defining types of inventories and the roles they play. While there are other ways to categorize inventories, we think of them as being one of the following types: anticipation stock, cycle stock, safety stock, pipeline stock, and decoupling stock. We will discuss each type.

Anticipation stocks are created by a firm not to meet immediate needs, but to meet requirements in the more distant future. In a manufacturing setting, for example, immediate needs could be current orders that must be fulfilled or those expected to be demanded within a manufacturing lead time. In businesses with seasonal demands (say snow shovels), production may occur throughout the year to build up inventories that will be depleted in a few weeks or months. The buildup occurs because production capacity is incapable of meeting the demand at the time it occurs. Thus the decision to limit production capacity results in creating inventories in anticipation of demand.

In some instances, anticipation stocks are created owing to speculation. If raw material prices are expected to increase, then it may be advantageous to purchase large quantities of them at a point in time in anticipation of the need for their use at a much later point in time.

Agricultural output is dictated by the growing seasons for crops in a particular location. Hence production occurs in anticipation of future demand and market prices. The harvested crops may not be consumed for several years.

These examples illustrate that inventories may be created because of capacity limitations, speculative motives, or seasonal cycles. These are all examples of anticipation stocks.

A second role of inventory is to meet current demand from stock which was created earlier because of the cyclic nature of the incoming supply of inventory. Suppose that a product is ordered from a supplier each month. Then the amount received each month must be adequate to meet demands throughout the ensuing month. Assume that demand occurs at a constant, continuous rate throughout the month. The left portion of the graph in Figure 1.1 illustrates the effect of receiving material at the beginning of the month, or more generally a cycle, and the constant rate at which the inventory level decreases as a consequence of the constant, continuous nature of the demand process. The average inventory carried in stock due to the cycle length, one month in our example, is equal to the average time a unit remains in stock times the demand rate. Since inventory is depleted at a constant rate, the average time a unit of stock remains on hand is one half the cycle length. If the cycle length is altered, the average amount of cycle stock changes in a proportional manner. Reducing cycle lengths reduces cycle stock levels. Replenishment of inventories is not

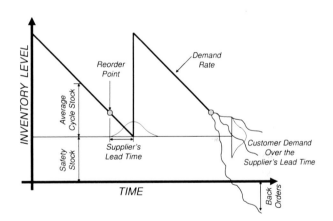

Fig. 1.1. Cycle stock and safety stock.

instantaneous in most instances. The length of time between the placement of an order on a supplier and its receipt is called a lead time. To ensure that adequate stock is on hand, an order is placed to replenish stocks when the inventory level reaches a particular value, which is called the reorder point. The inventory graph in the right-hand portion of Figure 1.1 illustrates possible demand patterns that reduce on-hand stock during a lead time when demand during the lead time is uncertain. Thus total demand may be greater than, less than, or about the same as predicted for this period. Furthermore, as illustrated in Figure 1.1, the lead time is also subject to variation. To protect against uncertain demand over an uncertain lead time, another type of stock is created called safety stock, or demand-driven safety stock.

As we will show later in the book, safety stocks may be necessary because productive capacity of a supplier is limited. When such a capacity limitation on supply exists, lead times are not fixed and hence more inventory may be required to ensure customer service is maintained. When capacity limitations exist, these inventories are called capacity-driven safety stocks.

There is a complex relationship that exists between cycle and safety stocks, which will be discussed in detail in subsequent chapters. The goal of many companies is to keep reducing cycle lengths, thereby reducing cycle stocks. But doing so also increases the number of cycles per year and correspondingly the number of times the company is exposed to the possibility for a stockout to occur. The need to maintain service to customers may thus force the company to increase safety stock levels.

The fourth type of stock that exists in a system is called pipeline stock. This stock exists because of the length of time it takes from the issuing of an order for stock replenishment until it is ready for issue or sale at the receiving location. This time is the replenishment lead time. Pipeline stock is equal to the expected demand over the lead time, which is equal to the expected demand per day times the length of the lead time, measured in days. This is a consequence of the well-known law called Little's law. Again observe that pipeline stock is proportional to the lead time length, so doubling the average lead time doubles the pipeline stock.

Decoupling stock is another type of safety stock. In manufacturing settings, there are successive operations in a plant corresponding to the production of products. Each operation corresponds to the physical transformation of material, basic processing of material, assembly of components, or possibly testing of the product at various production stages. For example, assembling of an automotive engine is normally accomplished through a sequence of tasks. Each task corresponds to adding components to the partially assembled engine. The tasks are performed at work stations. In order to keep the assembly process operating smoothly, inventories are introduced between successive stations. These inventories protect against variation in processing times or machine breakdowns at a station, and are called decoupling stocks. They are given this name because the presence of these stocks essentially permits each station to operate independently of all others. There will be stock available to work on when a task is completed, and there will be a place to temporarily store the output of the task performed at each station. Each station is neither starved for material to work on nor blocked from sending its output to the next station. Hence all operations are essentially decoupled from one another.

Multi-echelon inventory systems operate smoothly when a request from a supplying location can be satisfied with minimal or no delay. To ensure that this smooth flow exists between echelons, safety or decoupling stocks are often created.

While inventories play different roles and are created for different purposes, the question remains as to how much inventory of each type should exist.

1.2 Fundamental Questions

There are four fundamental questions that must be answered pertaining to inventories. The first is: what items should be stocked in a system? The answer depends on the objectives of a business and the strategy employed to achieve the objectives. Walmart and Amazon.com are both retail companies, but they differ in fundamental ways. One way they are different is in the range of stock they offer. Walmart stores may stock many tens of thousands of item types. You can order any one of over 40 million item types from Amazon.com. Thus the breadth of the product offering is a key decision that must be made by a company.

The second question is: where should the item be stocked? Should all stores in a retail chain stock the same item types? Amazon.com does not have retail stores. It supplies its customers from its own as well as supplier warehouses. What items should be stocked in each warehouse? Should all items be stocked everywhere, or should certain items be stocked in only a single location?

Xerox maintains a large inventory of service parts. There are many hundreds of thousands of different part types that are stocked in their multi-echelon resupply system. There are also many thousands of technicians who repair Xerox machines. What part types should they carry in the trunks of their vehicles or in an inventory locker? How should these technicians be resupplied? Where in this complex resupply system should each part be stocked?

The third question is: how much should be ordered when an order is placed? As we will see, the answer to this question will depend on a large number of factors that we will discuss in the next section. These factors will also determine the answer to the fourth question, which is "when should an order be placed?"

The material presented in this book focuses almost entirely on answering the third and fourth questions. The second question is addressed indirectly when we examine multi-echelon systems. To answer these questions we will construct a variety of mathematical models, each built on a different set of assumptions concerning the way the system being studied operates. Thus our goal in this book is to show how to represent a broad range of problem environments mathematically and to show how to answer the third and fourth questions for each such environment.

1.3   Factors Affecting Inventory Policy Decisions

When constructing mathematical models that address the questions raised in the previous section, we must consider several key factors. Models, by their nature, are representations or abstractions of real operating environments. Hence all the factors affecting inventory policy decisions are not always captured or represented in a model. We will examine many models in this text, and each differs in the manner in which the individual factors are expressed mathematically. Before we begin exploring these mathematical models, let us discuss these underlying factors that affect inventory policy decision making.

1.3.1   System Structure

The first factor is the supply chain's structure. The structure indicates the manner in which both material and information flow in a supply chain system. This system may consist of many stages or echelons. If the environment being represented is a service parts system for a high tech company, the system structure will likely look like the one found in Figure 1.2. In such a system, a central warehouse may stock a broad range of item types received from a variety of manufacturing sources. These items are distributed according to some policy to regional locations, which may be located in many countries. Within the United States there may be four or more such regional warehouses. These regional warehouses are responsible for supplying

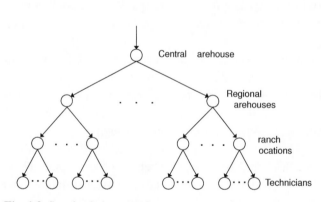

Fig. 1.2. Supply chain example.

a set of locations, which we are calling branches. There may be 75 or more such branches in the United States. Service technicians receive stock needed to repair equipment, found in customer locations, from these branches. In many cases, there are thousands of technicians servicing customer locations.

The central warehouse, regional warehouses, branches, and technician levels in the diagram each represent an echelon in the system in our service parts system example. How material flows and how information flows depends on this echelon structure.

The echelon structures found in complex supply chains are more complicated than the one we portrayed in our example. In an automotive supply chain there are echelons corresponding to raw material suppliers, component suppliers, manufacturing plants, vehicle assembly plants, and car dealers. The suppliers of raw materials and components will likely deliver material to more than one car company and to multiple locations for a given car company.

In this automotive example, and in most retail situations, the various echelons in the supply chain are owned by different economic entities. This fact makes management of inventories within a supply chain much more difficult. Inventory policies as well as information flows need to be coordinated throughout a supply chain, that is, across echelons, to ensure timely and cost-effective delivery of inventories. Coordination of policies, even when the echelons are part of the same company, is often lacking. Poor coordination negatively affects a supply chain's performance.

Since supply chains are increasingly becoming global, their echelon structures are sometimes affected by governmental requirements for local content, tax policies, labor rules, cost, etc. Thus a supply chain's structural complexity is in part the result of national and regional policies. Such policies play a major role in the echelon structure of firms with businesses in the European Union, for example.

1.3.2   The Items

A second factor to be considered is the nature of the items being stocked in the supply chain and at a particular location. The number of items being stocked and their interactions are important when establishing stocking policies. The total amount of space, for example, that is available in a warehouse will limit the amount of inventory held for each item type. The ability to process incoming freight in a warehouse might limit the frequency at which each item can be received.

Clearly Amazon.com has many inventory management issues that it must address daily simply because of the range of item types it stocks. These issues are not the same as those considered by a small retail shop whose owner can manage inventories of a limited number of items with a very simple system. But even in small operations, efficient management of inventories is an essential component of economic success.

Additionally, items differ in their physical attributes. They differ by weight and volume. Storing automotive muffler systems, which have unusual shapes, is different than storing items that are small and are in boxes. Furthermore, if there are only a small number of box sizes, then warehouses can be designed much more efficiently.

Obsolescence is also an issue. This is a major factor in electronics and style goods industries where product life cycles are short.

Products are sometimes perishable. Foods, hospital supplies, and blood are all examples of items that must be managed carefully owing to their limited shelf lives. Certain of these items require refrigeration. The need to refrigerate products affects the design of supply chains.

Some products are not unique in the eyes of a customer. As the number of available products of basically the same type increases (soft drinks, for example), substitutions occur more frequently. Hence substitution of one product for another is common. If a retailer stocks out of one item, the customer may take another in its place. This phenomenon is so prevalent that it makes demand estimation quite difficult to do with a high degree of accuracy. One must analyze inventory and sales data carefully so as to not miss the effect of substitutions.

Market requirements also differ among items. Demand rates and variability in demand differ. Some products will be required by the customer immediately, such as bread in a grocery store; for others, such as a Jaguar, customers are willing to wait to get exactly what they want. More will be said about demand characteristics subsequently.

Another important attribute of items is whether they are a consumable, such as foods, or a repairable item, such as a jet engine. Clearly the management of such items will be different, which implies that supply chains and inventory policies will differ between consumable and repairable items.

Finally, the most obvious way that items differ is their cost. The cost of an automobile engine differs dramatically from that of a toothpick. Hence policies for controlling items will depend to a large extent on the cost of the items and the cost inherent in storing and managing them. The nature of these costs is discussed in what follows.

1.3.3 Market Characteristics

As we discussed, not all item types are the same. A key way in which they differ is their demand rates. In most commercial settings, some items have very high demand rates relative to the vast majority of the items, which have low demand rates. It is not unusual for a small percentage of the items sold to account for 80% to 90% of the units or value of units shipped. This is an example of Pareto's law, or, as it is sometimes called, the "80–20" rule. In the inventory context, this rule implies that approximately 20% of the items account for about 80% of a company's total sales revenue or items sold. Within this context, the item types that yield this 80% of sales are sometimes called the A items. Those items comprising the next 15% of sales are called the B type items, and the remainder are called C type items. The C type items often account for about 50% of the item types sold, although they generate only about 5% of the sales.

In many instances, this ABC type classification also holds for a company's customers. That is, a very large portion of sales of a company goes to a few customers, the A customers. Smaller portions go to the B and C customers, where, as before, 50% of the customers generate only a small fraction of the company's sales.

The two Pareto curves in Figures 1.3 and 1.4 illustrate the nature of these ABC curves for items and customers. The graph in Figure 1.3 represents the cumulative

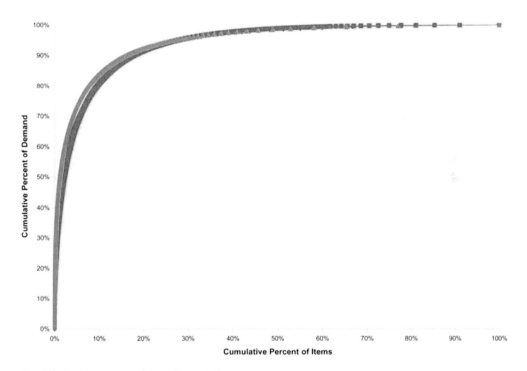

Fig. 1.3. Pareto analysis for on-line retailer.

percentage of demand as a function of the cumulative percentage of items for a major on-line retailer for three product lines. While the classical Pareto curve assumes that 80% of demand would be in 20% of the items, these data show that in the on-line retail sector this assumption does not hold. In fact, about 10% of the items account for over 80% of the demand in the example.

The graph in Figure 1.4 illustrates a Pareto curve for customers versus demand. The graph was constructed from data obtained from a firm that produces components for the automotive, truck, construction and farm equipment sectors. Note that in this case 80% of the total demand over a year arose from 20% of the customers (187 out of 935), an exact example of the classic 80–20 rule.

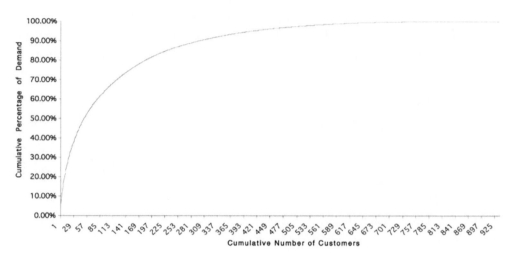

Fig. 1.4. Pareto analysis by customer for an industrial company.

A reason for performing an ABC analysis is to help understand what items to stock at which locations. Furthermore, delivery promises made to customers for these item types should probably differ. Unless margins are large, it is very difficult to achieve high off-the-shelf service and to provide the service profitably. Demand for lower-volume products is often highly variable, thereby resulting in substantial forecast errors. Given the generally high forecast errors for low demand rate items, companies have relatively larger amounts of safety and cycle stock in low demand rate items. It is not unusual for 20% or more of the inventory stocked by a company to be in the C type items. Such inventories are prone to become obsolete, and hence can become a severe financial liability to a company.

The graph in Figure 1.5 illustrates the volatility of the demand. These data correspond to actual demands experienced by a firm with which we have worked.

The models we will develop all make assumptions concerning the way demand processes evolve over time and what we know about these patterns. We will assume in some cases that demand is perfectly predictable, while in others we will assume that forecast errors exist. In the latter case, we will assume some statistical distribution of observed demand relative to the forecast of demand, such as a Poisson distribution or a normal distribution. The inventory optimization models require us to stipulate the underlying mathematical models for these distributions. In this text, as is the case in virtually all texts, these distributions of forecast errors are called the demand distributions. Thus what will be called demand distributions are in fact the distribution of observed demand relative to the forecast.

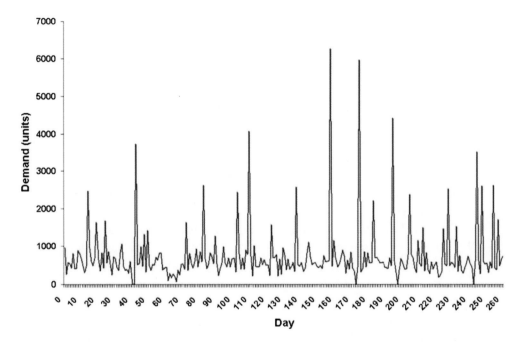

Fig. 1.5. Historical demand by day.

The models that are used to describe demand processes can represent seasonal patterns, high or low degrees of uncertainty (as measured by the coefficient of variation, which is the standard deviation of demand over a period of time divided by the expected demand over that same time period), correlations between items over time, or other factors. As stated, in some instances we will assume that demand processes are known with certainty, while in others we may assume that demand processes are either stationary or non-stationary.

The perception of demand and the demand model that is used will depend in practice on the information systems employed throughout a supply chain. The degree to which collaboration exists throughout a supply chain affects the forecast errors and hence the demand model. While variability may exist when supply chain partners plan operational strategy and tactics collaboratively, uncertainty will be reduced dramatically.

Uncertainty in demand forecasting and the degree of collaboration or information sharing are tightly linked. The greater the degree of collaboration, the lower the coefficient of variation of the demand model.

1.3.4 Lead Times

Lead times exist in supply chains, and greatly affect stock levels. In general, lead times measure the time lag between the placement of an order and its receipt. Normally, we think of lead time length as a measure of the responsiveness of a supplier. The longer the lead time, the more uncertain we are as to the demand over the lead time and, therefore, the requirement for inventory.

When modeling the demand process, we usually are thinking of the demand that arises over a lead time. This is the case since safety stock requirements are directly related to the length of the lead time. When a lead time's length is uncertain, then the demand over a lead time almost surely becomes more uncertain and again safety stock requirements increase.

We will show how lead times affect inventory requirements in many places in this text. We will make assumptions about lead time lengths and uncertainty when we develop decision models. In some cases we will assume lead times are constants and in others that lead times are random variables. We will assume that lead times of successive orders may or may not cross. That is, if an order is placed for an item type at a point in time and another order for the same item type is placed at a later point in time, we may assume that the order placed first must arrive first. This is called the no crossing assumption.

1.3.5 Costs

Cost is one of the key factors in determining an inventory policy. Most models for planning inventory requirements consider costs of various types. Some of the types of costs that are found in practice and captured in models include purchasing, carrying or holding, stockout, and obsolescence costs. Additionally, costs of receiving, ordering, processing, and fulfilling customer orders (including accounting and information acquisition) are considered in some models.

Purchasing costs differ by application. In manufacturing settings, purchasing implies acquisition of components and raw materials as well as obtaining supplies and equipment. In retail environments, purchasing pertains to the acquisition of products offered for sale, normally to customers. Inventory policies used to manage the acquisition of materials, components, or finished goods will depend on the unit costs and the fixed costs incurred when placing orders. Unit costs may depend on the volume purchased; sometimes there are discounts for all units purchased when an order passes a threshold. Other times the incremental cost of purchasing additional units is lower than the per unit purchasing cost of the preceding unit. We have all seen ads that say "buy one at full price and get the second unit for 50% off." Thus variable purchasing costs can be quantity-dependent.

Fixed procurement costs are those costs that when incurred are independent of the quantity purchased. We incur such costs every time we go to a store to purchase food. There is a cost to get to the store that is independent of how much we purchase. Suppose we went to a grocery store to purchase the ingredients needed to make a single sandwich every time we wanted to eat a sandwich. Thus the frequency at which we would visit the grocery store to purchase these ingredients would exactly match the frequency at which we desire to eat a sandwich. An alternative strategy would be to purchase quantities of many different ingredients (ham, roast beef, turkey, tuna fish, various breads, etc.) that can be used to assemble sandwiches that we may desire to eat over the next several days. Thus the frequency of visiting the grocery store would be lower, but there are other potential problems. Perhaps I would not eat all that has been purchased before it spoils. Then there is another cost that is incurred. If I purchased very large quantities of various ingredients I would have also committed money that could have been used for other activities.

Then there are tradeoffs. The frequency of making purchases results in firms expending resources. As the frequency of making purchases increases, the amount of fixed costs incurred increases; however, the cost of holding or carrying inventories for most items for short periods of time is relatively low. When purchases are made infrequently, fixed ordering costs are low, but more inventory is carried in the form of cycle stocks.

The carrying costs for firms consist of the opportunity cost of not applying the capital to other projects; the cost of this capital invested in inventory by the firm; out of pocket costs for insurance, taxes, damage, pilferage, and warehouse operating costs (fixed and variable); and the cost of obsolescence.

Costs of carrying inventory vary greatly by product, stocking location, potential for technological obsolescence, physical storage costs, and so forth. These holding costs are often charged using a simple approximation which may be appropriate in an aggregate sense for a range of investment levels a firm makes in managing its inventories. The approximation is to charge holding costs proportional to the monetary value of the on-hand inventory. The proportionality factor normally ranges from .15 to .25, on an annual basis, while a value of .40 is possible for items that require special storage facilities and are subject to obsolescence. Certain drugs or electronic devices are examples of such items.

Many of the models developed in this text are based on the assumption that we will use this proportionality method for calculating inventory holding costs. In several environments we will examine, the goal will be to minimize the average annual operating costs. In these cases, the holding cost proportionality constant has the dimension $/$ invested/year. In other models, in which time is divided into periods, the holding costs are charged on the basis of the number of units of stock held at a period's end. The holding costs are usually calculated using a proportionality factor, but also by taking the period's length into account when setting the proportionality constant.

Holding costs are incurred when supply exceeds demand; but, on occasion, there may not be enough inventory on hand to meet all the demand at the point in time at which the demand arises. In these cases, some form of stockout cost is incurred. Sometimes customers simply choose to purchase something else, in which a so-called lost sales cost for the item is incurred. This cost certainly includes the cost of the lost profit margin, but also includes the possibility that future sales may be affected as well. In retail settings, a customer can immediately determine whether or not the desired item is available; however, when purchasing from firms such as L.L. Bean or Amazon.com, notifying customers of shortages is not quite as simple. Customers may be informed that goods are not in stock at the time a request for them is made. Perhaps an estimate of when the stock will be available is provided to the customer. Customers may receive several updates about product availability over that time. Providing notification to customers requires the provider of the goods to maintain a complex information infrastructure, which is costly. Once the goods do arrive they may be shipped to the customer using a premium transportation mode, such as air freight, which can increase delivery costs substantially.

In manufacturing settings, stockouts have serious implications for supply chain partners. Production schedules can be disrupted, causing customers to adjust their plans as well. Premium transportation modes are sometimes used. It is not uncommon for low-value items to be shipped via expensive air freight because of a temporary shortage at a component plant. Suppliers to automotive companies do this so that assembly lines will not be halted for a lack of parts.

Anyone who has worked in a material management organization knows that a very substantial amount of management's time is devoted to dealing with inventory shortages. Hence the indirect costs of managing shortfalls internally and externally are not insignificant. While these crises are commonplace, measuring the cost of dealing with them is difficult since there are no accounting measures that directly track these incurred costs.

The models that we will study are divided into either backorder or lost sales models. In the backorder case, we assume customers wait for the inventory to arrive and eventually have their orders satisfied. In the backorder case, we capture shortage costs in either of two ways. First, there may be a penalty cost incurred given that a demand arises and cannot be met from stock within a customer's desired response time. This cost is charged independently of how long a customer must wait before receiving the ordered item. Second, the penalty cost may be charged as a function of the length of time a customer must wait to receive the desired products. Thus stockout costs are charged from the time an order is received (or due) until it is finally satisfied in this case. Some models contain both types of costs.

In the models we will study where sales are lost if inventory is not available on time to meet a customer's request, a penalty cost will be charged in proportion to the number of sales that are lost.

1.4   Measuring Performance

Inventory systems exist to provide customers with products and services. Normally the level of service expected is high, but the level expected depends on the type of product, the customer being served, and the timeliness of both the need and supply that is possible or required. I may be willing to wait for several weeks to get a new Jaguar that meets my exact desires, but I will not wait at all for almost all grocery, book, electronic, or other media products. Suppliers of products cannot possibly meet all possible ranges of demands for all products immediately, because of the cost that would be incurred. Thus, given the nature of the product and customer requirements, supply chains are constructed and inventory strategies are employed to meet demands using the physical and management infrastructures inherent in the design of the supply chains.

The inventory strategies used are based on some form of model. The models may be very simplistic in nature or may be quite complex. Inventory strategies are based on models of the supply chain and demand which could be simplistic or complex. Complexity is a function of model components chosen to describe demand, costs, constraints and the intricacies of the supply chain. Independently of the model's complexity, there is some goal that is to be achieved through the application of the model.

The goal may be to establish a policy that would maximize profit or minimize relevant costs when implemented. Of course the optimal policy will depend on the assumptions made about the way costs are incurred, how demands arise and are satisfied, and other elements of a supply chain's operations. Different assumptions lead to different policies.

Sometimes the system objective is to minimize the cost of achieving some performance goal, such as a customer fill rate target or a maximum expected waiting time to satisfy demands. By fill rates, we mean the fraction of demand that is satisfied from stock on time, that is, without backordering. In practice, fill rate goals are often set for each portion of a supply chain without much understanding of the impacts on cost of these performance targets. Consequently, too much inventory is often present in supply chains consisting of many echelons. Hence, substantial inefficiencies exist in the managing of supply chain inventories. Judicious use of models can improve understanding of policies for managing inventories and information within supply chains.

Our goal in this text is to present a variety of models and algorithms that are useful in setting policies, computing stock levels, and estimating financial and operational performance within supply chains.

2

EOQ Model

The first model we will present is called the economic order quantity (EOQ) model. This model is studied first owing to its simplicity. Simplicity and restrictive modeling assumptions usually go together, and the EOQ model is not an exception. However, the presence of these modeling assumptions does not mean that the model cannot be used in practice. There are many situations in which this model will produce good results. For example, these models have been effectively employed in automotive, pharmaceutical, and retail sectors of the economy for many years. Another advantage is that the model gives the optimal solution in closed form. This allows us to gain insights about the behavior of the inventory system. The closed-form solution is also easy to compute (compared to, for example, an iterative method of computation).

In this chapter, we will develop several models for a single-stage system in which we manage inventory of a single item. The purpose of these models is to determine how much to purchase (order quantity) and when to place the order (the reorder point). The common thread across these models is the assumption that demand occurs continuously at a constant and known rate. We begin with the simple model in which all demand is satisfied on time. In Section 2.2, we develop a model in which some of the demand could be backordered. In Section 2.3, we consider the EOQ model again; however, the unit purchasing cost depends on the order size. In the final section, we briefly discuss how to manage many item types when constraints exist that link the lot size decisions across items.

J.A. Muckstadt and A. Sapra, *Principles of Inventory Management: When You Are Down to Four, Order More*, Springer Series in Operations Research and Financial Engineering, DOI 10.1007/978-0-387-68948-7_2, © Springer Science+Business Media, LLC 2010

2.1 Model Development: Economic Order Quantity (EOQ) Model

We begin with a discussion of various assumptions underlying the model. This discussion is also used to present the notation.

1. Demand arrives continuously at a constant and known rate of λ units per year. Arrival of demand at a continuous rate implies that the optimal order quantity may be non-integer. The fractional nature of the optimal order quantity is not a significant problem so long as the order quantity is not very small; in practice, one simply rounds off the order quantity. Similarly, the assumption that demand arrives at a constant and known rate is rarely satisfied in practice. However, the model produces good results where demand is relatively stable over time.
2. Whenever an order is placed, a fixed cost K is incurred. Each unit of inventory costs $\$I$ to stock per year per dollar invested in inventory. Therefore, if a unit's purchasing cost is C, it will cost $I \cdot C$ to stock one unit of that item for a year.
3. The order arrives τ years after the placement of the order. We assume that τ is deterministic and known.
4. All the model parameters are unchanging over time.
5. The length of the planning horizon is infinite.
6. All the demand is satisfied on time.

Our goal is to determine the order quantity and the reorder interval. Since all the parameters are stationary over time, the order quantity, denoted by Q, also remains stationary. The reorder interval is related to when an order should be placed, since a reorder interval is equal to the time between two successive epochs at which an order is placed and is called the cycle length. A cycle is the time between the placing of two successive orders. The question of when to place an order has a simple answer in this model. Since demand occurs at a deterministic and fixed rate and the order once placed arrives exactly τ years later, we would want the order to arrive exactly when the last unit is being sold. Thus the order should be placed τ years before the depletion of inventory.

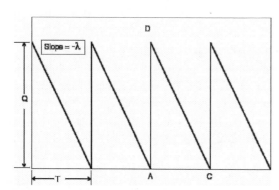

Fig. 2.1. Change in inventory over time for the EOQ model.

The first step in the development of the model is the construction of cost expressions. Since total demand per year is λ, the total purchasing cost for one year is $C\lambda$. Similarly, the number of orders placed per year is equal to λ/Q. Therefore, the total annual average cost of placing orders is $K\lambda/Q$. The derivation of the total holding cost per year is a bit more involved. We will begin by first computing the average inventory per cycle. Since each cycle is identical to any other cycle, the average inventory per year is the same as the average inventory per cycle. The holding cost is equal to the average inventory per year times the cost of holding one unit of inventory for one year. Using Figure 2.1, we find the average inventory per cycle is equal to:

$$\frac{\text{Area of triangle } ADC}{\text{Length of the cycle}} = \frac{\frac{1}{2}QT}{T} = \frac{Q}{2}.$$

The annual cost of holding inventory is thus equal to $ICQ/2$.

Adding the three types of costs together, we get the following objective function, which we want to minimize over Q:

$$\min_{Q \geq 0} Z(Q) = C\lambda + \frac{K\lambda}{Q} + \frac{ICQ}{2}. \tag{2.1}$$

Before we compute the optimal value of Q, let us take a step back and think about what the optimal solution should look like. First, the higher the value of the fixed cost K, the fewer the number of orders that should be placed every year. This means that the quantity ordered per order will be high. Second, if the holding cost rate is high, placing orders more frequently is economical since inventory will on average be lower. A higher frequency of order placement leads to lower amounts ordered per order. Therefore, our intuition tells us that the optimal order quantity should increase as the fixed ordering cost increases and decrease as the holding cost rate increases.

To compute the optimal order quantity, we take the first derivative of $Z(Q)$ with respect to Q and set it equal to zero:

$$\frac{dZ}{dQ} = 0 - K\frac{\lambda}{Q^2} + \frac{IC}{2} = 0$$

$$\text{or} \quad Q^* = \sqrt{\frac{2K\lambda}{IC}}, \tag{2.2}$$

where Q^* is the optimal order quantity. Note that the derivative of the purchasing cost $C\lambda$ is zero since it is independent of Q. The following examples illustrate the computation of the optimal order quantity using (2.2).

In our first example, we assume an office supplies store sees a uniform demand rate of 10 boxes of pencils per week. Each box costs $5. If the fixed cost of placing an order is \$10 and the holding cost rate is .20 per year, let us determine the optimal order quantity using the EOQ model. Assume 52 weeks per year.

In this example $K = 10$, $I = .20$, $C = 5$, and the annual demand rate is $\lambda = (10)(52) = 520$. Substituting these values in (2.2), we get

$$Q^* = \sqrt{\frac{2(520)(10)}{(0.2)(5)}} = 101.98 \approx 102.$$

In our second example, suppose the regional distribution center (RDC) for a major auto manufacturer stocks approximately 20,000 service parts. The RDC fulfills demands of dozens of dealerships in the region. The RDC places orders with the national distribution center (NDC), which is also owned by the auto manufacturer. Given the huge size of these facilities, it is deemed impossible to coordinate the inventory management of the national and regional distribution centers. Accordingly, each RDC manages the inventory on its own regardless of the policies at the NDC.

We consider one part, a tail light, for a specific car model. The demand for this part is almost steady throughout the year at a rate of 100 units per week. The purchasing cost of the tail light paid by the RDC to the NDC is \$10 per unit. In addition, the RDC spends on average \$0.50 per unit in transportation. A breakdown of the different types of costs is as follows:

1. The RDC calculates its interest rate to be 15% per year.
2. The cost of maintaining the warehouse and its depreciation is \$100,000 per year, which is independent of the amount of inventory stored there. In addition, the costs of pilferage and misplacement of inventory are estimated to be 5 cents per dollar of average inventory stocked.
3. The annual cost of a computer-based order management system is \$50,000 and is not dependent on how often orders are placed.
4. The cost of invoice preparation, postage, time, etc. is estimated to be \$100 per order.
5. The cost of unloading every order that arrives is estimated to be \$10 per order.

Let us determine the optimal order quantity. The first task is to determine the cost parameters. The holding cost rate I is equal to the interest rate (.15) plus the cost rate for pilferage and misplacement of inventory (.05). Therefore, $I = .20$. This rate applies to the value of the inventory when it arrives at the RDC. This value includes not only the purchasing cost (\$10) but also the value added through transportation (\$0.5). Therefore, the value of C is \$10.50. Finally, the fixed cost of order placement includes all costs that depend on the order frequency. Thus, it includes the order receiving cost (\$10) and the cost of invoice preparation, etc. (\$100), but not the cost of the order management system. Therefore, $K = 110$. We now substitute these parameters into (2.2) to get the optimal order quantity:

$$Q^* = \sqrt{\frac{2\lambda K}{IC}} = \sqrt{\frac{2(5200)(110)}{(0.2)(10.5)}} = 738.08 \text{ units.}$$

Next, let us determine whether or not the optimal EOQ solution matches our intuition. If the fixed cost K increases, the numerator of (2.2) increases and the optimal order quantity Q^* increases. Similarly, as the holding cost rate I increases, the denominator of (2.2) increases and the optimal order quantity Q^* decreases. Clearly, the solution fulfills our expectations.

To gain more insights, let us explore additional properties the optimal solution possesses. Figure 2.2 shows the plot of the average annual fixed order cost $K\lambda/Q$ and the annual holding cost $ICQ/2$ as functions of Q. The average annual fixed order cost decreases as Q increases because fewer orders are placed. On the other hand, the average annual holding cost increases as Q increases since units remain in inventory longer. Thus the order quantity affects the two types of costs in opposite ways. The annual fixed ordering cost is minimized by making Q as large as possible, but the holding cost is minimized by having Q as small as possible. The two curves intersect at $Q = Q_1$. By definition of Q_1,

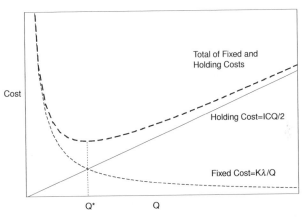

Fig. 2.2. Fixed and holding costs as functions of the order quantity.

$$K\frac{\lambda}{Q_1} = \frac{ICQ_1}{2} \Rightarrow Q_1 = \sqrt{\frac{2K\lambda}{IC}}$$

and, in this case, $Q_1 = Q^*$. The exact balance of the holding and setup costs yields the optimal order quantity. In other words, the optimal solution is the *best compromise* between the two types of costs. (As we will see throughout this book, inventory models are based on finding the best compromise between opposing costs.) Since the annual holding cost $ICQ^*/2$ and the fixed cost $K\lambda/Q^*$ are equal in the optimal solution, the optimal average annual total cost is equal to

$$Z(Q^*) = C\lambda + \frac{ICQ^*}{2} + K\frac{\lambda}{Q^*} = C\lambda + 2K\frac{\lambda}{Q^*}$$

$$= C\lambda + 2K\frac{\lambda}{\sqrt{\frac{2\lambda K}{IC}}} = C\lambda + \sqrt{2\lambda KIC}, \tag{2.3}$$

where we substitute for Q^* using (2.2).

Let us compute the optimal average annual cost for the office supplies example. Using (2.3), the optimal cost is equal to

$$C\lambda + \sqrt{2\lambda KIC} = (5)(520) + \sqrt{2(520)(10)(5)(0.2)} = \$2701.98.$$

The purchasing cost is equal to $(\$5)(520) = \$2600$, and the holding and order placement costs account for the remaining cost of $101.98.

2.1.1 Robustness of the EOQ Model

In the real world, it is often difficult to estimate the model parameters accurately. The cost and demand parameter values used in models are at best an approximation to their actual values. The policy computed using the approximated parameters, henceforth referred to as approximated policy, cannot be optimal. The optimal policy cannot be computed without knowing the true values of the model's parameters. Clearly, if another policy is used, the realized cost will be greater than the cost of the true optimal policy. The following example illustrates this point.

Suppose in the office supplies example that the fixed cost of order placement is estimated to be \$4 and the holding cost rate is estimated to be .15. Let us calculate the alternative policy and the cost difference between employing this policy and the optimal policy. Recall that the average annual cost incurred when following the optimal policy is \$2701.98. To compute the alternative policy, we substitute the estimated parameter values into (2.2):

$$Q^* = \sqrt{\frac{2(520)(4)}{(0.15)(5)}} = 74.48.$$

The realized average annual cost if this policy is used when $K = 10$ is

$$Z(Q^*) = C\lambda + \frac{K\lambda}{Q^*} + \frac{ICQ^*}{2}$$

$$= (5)(520) + \frac{(10)(520)}{74.48} + \frac{(0.2)(5)(74.48)}{2} = \$2707.06.$$

Thus the cost difference between the alternative and optimal policies is $2707.06 − $2701.98 = $5.08. Note that the cost of implementing the alternative policy is calculated using the actual cost parameters.

Let us now derive an upper bound on the realized average annual cost of using the approximate policy relative to the optimal cost. Suppose the *actual* order quantity is denoted by Q_a^*. This is the answer we would get from (2.2) if we could use the true cost and demand parameters. Let the true fixed cost and holding cost rate be denoted by K_a and I_a, respectively. We assume that the purchasing cost C and the demand rate λ have been estimated accurately. The estimates of the fixed cost and holding cost rate are denoted by K and I, respectively. The estimated order quantity is denoted by Q^*. Let $Q^*/Q_a^* = \alpha$ or $Q^* = \alpha Q_a^*$. Thus

$$Q_a^* = \sqrt{\frac{2\lambda K_a}{I_a C}}$$

$$Q^* = \sqrt{\frac{2\lambda K}{IC}} = \alpha\sqrt{\frac{2\lambda K_a}{I_a C}}$$

$$\Rightarrow \alpha = \sqrt{\left(\frac{K}{I}\right)\left(\frac{I_a}{K_a}\right)}.$$

Since the purchasing cost $C\lambda$ is not influenced by the order quantity, we do not include it in the comparison of costs. The true average annual operating cost (the sum of the holding and order placement costs) is equal to

$$Z(Q_a^*) = \sqrt{2K_a\lambda I_a C}.$$

The *actual* incurred average annual cost (sum of the holding and order placement costs) corresponding to the estimated order quantity Q^* is equal to

$$
\begin{aligned}
Z(Q^*) &= \frac{K_a\lambda}{Q^*} + \frac{I_a C Q^*}{2}\\
&= \frac{K_a\lambda}{\alpha Q_a^*} + \frac{I_a C(\alpha Q_a^*)}{2}\\
&= \frac{K_a\lambda}{\alpha\sqrt{\frac{2\lambda K_a}{I_a C}}} + \frac{I_a C\alpha\sqrt{\frac{2\lambda K_a}{I_a C}}}{2}\\
&= \frac{1}{2}\left(\alpha + \frac{1}{\alpha}\right)\sqrt{2K_a\lambda I_a C} = \frac{1}{2}\left(\alpha + \frac{1}{\alpha}\right)Z(Q_a^*).
\end{aligned}
$$

Therefore, if the estimated order quantity is α times the optimal order quantity, the average annual cost corresponding to the estimated order quantity is $\frac{1}{2}\left(\alpha + \frac{1}{\alpha}\right)$ times the optimal cost. For example, if $\alpha = 2$ (or $\frac{1}{2}$), that is, the estimated order quantity is 100% greater (or 50% lower) than the optimal order quantity, then the cost corresponding to the estimated order quantity is 1.25 times the optimal cost. Similarly, when $\alpha = 3$ (or $\frac{1}{3}$), the cost corresponding to the estimated order quantity is approximately 1.67 times the optimal cost.

Two observations can be made. First, and importantly, even for significant inaccuracies in the order quantity, the cost increase is modest. As we showed, the cost increase is only 25% for a 100% increase in the estimated order quantity from the optimal order quantity. The moderate effect of inaccuracies in the cost parameters on the actual incurred average annual cost is very profound. Second, the cost increase is symmetric around $\alpha = 1$ in a multiplicative sense. That is, the cost increase is the same for $Q^*/Q_a^* = \alpha$ or $Q^*/Q_a^* = \frac{1}{\alpha}$. This observation will be useful in the discussion presented in the following chapter.

How do we estimate α? Clearly, if we could estimate α precisely, then we could compute Q_a^* precisely as well and there would be no need to use the estimated order quantity. Since we cannot ascertain its value with certainty, perhaps we can estimate upper and lower bounds for α. These bounds can give us bounds on the cost of using the estimated order quantity relative to the optimal cost. The following example illustrates this notion in more detail.

Suppose in the office supplies example that the retailer is confident that his *actual* fixed cost is at most 120% but no less than 80% of the estimated fixed cost. Similarly, he is sure that his actual holding cost rate is at most 110% but no less than 90% of the estimated holding cost rate. Let us determine the maximum deviation from the optimal cost by implementing a policy obtained on the basis of the estimated parameter values.

We are given that

$$0.8 \leq \frac{K_a}{K} \leq 1.2,$$

and that

$$0.9 \leq \frac{I_a}{I} \leq 1.1.$$

The cost increases when the estimated order quantity is either less than or more than the optimal order quantity. Our approach will involve computing the lower and upper bounds on α and then computing the cost bounds corresponding to these values of α. The maximum of these cost bounds will be the maximum possible deviation of the cost of using the estimated order quantity relative to the optimal cost.

Since $\alpha = \sqrt{\left(\frac{K}{I}\right)\left(\frac{I_a}{K_a}\right)}$, we use the upper bound on I_a/I and the lower bound on K_a/K to get an upper bound on α. Thus,

$$\alpha \le \sqrt{1.1 \times \frac{1}{0.8}} = 1.17.$$

The cost of using the estimated order quantity corresponding to $\alpha = 1.17$ is $\frac{1}{2}(1.17 + \frac{1}{1.17}) = 1.013$ times the optimal cost.

Similarly, to get a lower bound on α, we use the lower bound on I_a/I and the upper bound on K_a/K. Thus,

$$\alpha \ge \sqrt{0.9 \times \frac{1}{1.2}} = 0.87.$$

The cost of using the estimated order quantity corresponding to $\alpha = 0.87$ is $\frac{1}{2}(0.87 + \frac{1}{0.87}) = 1.010$ times the optimal cost. The upper bound on the cost is the maximum of 1.01 and 1.013 times the optimal cost. Therefore, the cost of using the estimated order quantity is at most 1.3% higher than would be obtained if the optimal policy were implemented.

2.1.2   Reorder Point and Reorder Interval

In the EOQ model, the demand rate and lead time are known with certainty. Therefore, an order is placed such that the inventory arrives exactly when it is needed. This means that if the inventory is going to be depleted at time t and the lead time is τ, then an order should be placed at time $t - \tau$. If we place the order before time $t - \tau$, then the order will arrive before time t. Clearly holding costs can be eliminated by having the order arrive at time t. On the other hand, delaying the placement of an order so that it arrives after time t is not permissible since a backorder will occur.

How should we determine the reorder point in terms of the inventory remaining on the shelf? There are two cases depending upon whether the lead time is less than or greater than the reorder interval, that is, whether $\tau \le T$ or $\tau > T$. We discuss the first case here; the details for the second case are left as an exercise. Since the on-hand inventory at the time an order arrives is zero, the inventory at time $t - \tau$ should be equal to the total demand realized during the time interval $(t - \tau, t]$, which is equal to $\lambda \tau$. Therefore, the reorder point when $\tau \le T$ is equal to

$$r^* = \lambda \tau. \tag{2.4}$$

In other words, whenever the inventory drops to the level $\lambda \tau$, an order must be placed. Observe that r^* does not depend on the optimal order quantity.

On the other hand, when $\tau > T$, the reorder point is equal to

$$r^* = \lambda \tau_1, \tag{2.5}$$

where τ_1 is the remainder when τ is divided by T. That is, $\tau = mT + \tau_1$, where m is a positive integer.

The time between the placement of two successive orders, T, is equal to the time between the receipt of two successive order deliveries, since the lead time is a known constant. Since orders are received when the inventory level is zero, the quantity received, Q^*, is consumed entirely at the demand rate λ by the time the next order is received. Therefore, if the optimal reorder interval is denoted by T^*, $Q^* = \lambda T^*$. Hence

$$T^* = \frac{Q^*}{\lambda} = \sqrt{\frac{2K}{\lambda IC}}. \tag{2.6}$$

Suppose in our office supplies example that the lead time is equal to 2 weeks. In this case, the reorder interval T^* is equal to

$$T^* = \frac{Q^*}{\lambda} = \frac{101.98}{520} = 0.196 \text{ year} = 10.196 \text{ weeks}.$$

Since $\tau = \frac{2}{52}$ years, which is less than the reorder interval, we can use (2.4) to find the reorder point. The reorder point r^* is equal to

$$r^* = \lambda \tau = (520)(2/52) = 20.$$

2.2 EOQ Model with Backordering Allowed

In this section we will relax one of the assumptions we have made about satisfying all demand on time. We will now allow some of the demand to be backordered, but there will be a cost penalty incurred. The rest of the modeling assumptions remain unaltered. As a result, the cost function now consists of four components: the purchasing cost, the fixed cost of order placement, the inventory holding cost, and the backlog penalty cost. The system dynamics are shown in Figure 2.3.

Each order cycle is comprised of two sub-cycles. The first sub-cycle (ADC) is T_1 years long and is characterized by positive on-hand inventory which decreases at the demand rate of λ. The second sub-cycle (CEF) is T_2 years long, during which demand is backordered. Hence there is no on-hand inventory during this time period. Since no demand is satisfied in this latter period, the backlog increases at the demand rate λ. The total length of a cycle is $T = T_1 + T_2$.

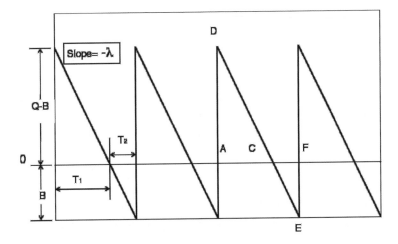

Fig. 2.3. Change in inventory over time for the EOQ model with backordering allowed.

There are two decisions to be made: How much to order whenever an order is placed, and how large the maximum backlog level should be in each cycle. The order quantity is denoted by Q as before, and we use B to denote the maximum amount of backlog allowed. When an order arrives, all the backordered demand is satisfied immediately. Thus, the remaining $Q - B$ units of on-hand inventory satisfies demand in the first sub-cycle. Since this on-hand inventory decreases at rate λ and becomes zero in T_1 years,

$$Q - B = \lambda T_1. \tag{2.7}$$

In the second sub-cycle, the number of backorders increases from 0 to B at rate λ over a period of length T_2 years. Thus

$$B = \lambda T_2 \tag{2.8}$$

and

$$Q = \lambda (T_1 + T_2) = \lambda T. \tag{2.9}$$

The cost expressions for the purchasing cost and annual fixed ordering cost remain the same as for the EOQ model and are equal to $C\lambda$ and $K\lambda/Q$, respectively. The expression for the average annual holding cost is different. We first compute the average inventory per cycle and then multiply the result by the holding cost IC to get the annual holding cost. The average inventory per cycle is equal to

$$\frac{\text{Area of triangle ADC}}{T} = \frac{\frac{(Q-B)T_1}{2}}{T}.$$

We next substitute for T_1 and T. This results in an expression which is a function only of Q and B, our decision variables.

$$\text{Average Inventory per Cycle} = \frac{\frac{(Q-B)^2}{\lambda}}{2\frac{Q}{\lambda}} = \frac{(Q-B)^2}{2Q}.$$

The computation of the average annual backordering cost is similar. We let π be the cost of backordering a unit for one year. The first step is to compute the average number of backorders per cycle. Since all cycles are alike, this means that the average number of outstanding backorders per year is the same as the average per cycle. To get the average backorder cost per year, we multiply the average backorder quantities per year by the backorder cost rate. The average number of backorders per cycle is equal to the area of triangle CEF in Figure 2.3 divided by the length of the cycle T:

$$\frac{\text{Area of triangle CEF}}{T} = \frac{\frac{BT_2}{2}}{T} = \frac{\frac{B^2}{2\lambda}}{\frac{Q}{\lambda}} = \frac{B^2}{2Q}.$$

In the last equality, we have used the relationships $T_2 = B/\lambda$ and $T = Q/\lambda$. Therefore, the average annual backorder cost is equal to $\pi\frac{B^2}{2Q}$.

We now combine all the cost components and express the average annual cost of managing inventory as

$$Z(B,Q) = C\lambda + K\frac{\lambda}{Q} + IC\frac{(Q-B)^2}{2Q} + \pi\frac{B^2}{2Q}. \tag{2.10}$$

Before we obtain the optimal solution, let us anticipate what properties we expect the optimal solution to possess. As before, if the fixed order cost K increases, fewer orders will be placed, which will increase the order quantity. An increase in the holding cost rate should drive the order quantity to lower values. The effect of the backorder cost on the maximum possible number of units backordered should be as follows: the higher the backorder cost, the lower the maximum desirable number of backorders. We will state additional insights after deriving the optimal solution.

To obtain the optimal solution, we take the first partial derivatives of $Z(B,Q)$ in (2.10) with respect to Q and B and set them equal to zero. This yields two simultaneous equations in Q and B:

$$\frac{\partial Z(B,Q)}{\partial Q} = -K\frac{\lambda}{Q^2} + IC\frac{(Q-B)}{Q} - IC\frac{(Q-B)^2}{2Q^2} - \pi\frac{B^2}{2Q^2}, \tag{2.11}$$

$$\frac{\partial Z(B,Q)}{\partial B} = -IC\frac{(Q-B)}{Q} + \pi\frac{B}{Q}. \tag{2.12}$$

Equation (2.12) appears simpler, so let us set it to zero first. This results in

$$B = Q\frac{IC}{IC+\pi}. \tag{2.13}$$

Substituting for B in (2.11) results in

$$\begin{aligned}
\frac{\partial Z(B,Q)}{\partial Q} &= -K\frac{\lambda}{Q^2} + IC\frac{\pi}{IC+\pi} - IC\frac{(\frac{\pi}{IC+\pi})^2}{2} - \pi\frac{(\frac{IC}{IC+\pi})^2}{2} \\
&= -K\frac{\lambda}{Q^2} + IC\frac{\pi}{IC+\pi} - IC\frac{\pi}{2(IC+\pi)} \\
&= -K\frac{\lambda}{Q^2} + IC\frac{\pi}{2(IC+\pi)} \\
\Rightarrow Q^* &= \sqrt{\frac{2K\lambda(IC+\pi)}{IC\pi}} = \sqrt{\frac{(IC+\pi)}{\pi}}\sqrt{\frac{2K\lambda}{IC}} = Q_E\sqrt{\frac{IC+\pi}{\pi}},
\end{aligned} \tag{2.14}$$

where Q_E is the optimal solution to the EOQ model. Also,

$$B^* = Q^*\frac{IC}{IC+\pi} = \sqrt{\frac{2K\lambda IC}{(IC+\pi)\pi}}, \tag{2.15}$$

$$T^* = \frac{Q^*}{\lambda} = \sqrt{\frac{2K(IC+\pi)}{\lambda IC\pi}}.$$

Again, let us return to our office supplies example. Assume now that the pencils sold by the office-supplies retailer are somewhat exotic and that they are not available anywhere else in the town. This means that the customers wait if the retailer runs out of the pencil boxes. Sensing this, the retailer now wants to allow backordering of demand when determining the inventory replenishment policy. Suppose that the cost to backorder a unit per year is \$10. Let us determine the optimal order quantity and the maximum quantity of backorders that will be permitted to accumulate. Since $IC = (0.2)(5) = 1$, $\pi = 10$, and $Q_E = 101.98$, we can use (2.14) to determine

$$Q^* = Q_E\sqrt{\frac{IC+\pi}{\pi}} = (101.98)\sqrt{\frac{1+10}{10}} = 106.96.$$

Next, we use (2.15) to determine the maximum level of backlog

$$B^* = Q^* \frac{IC}{IC + \pi} = (106.96) \frac{1}{1 + 10} = 9.72.$$

Several observations can now be made.

1. Our pre-derivation intuition holds. As K increases, Q^* increases, implying a decrease in the number of orders placed per year. As the holding cost rate I increases, $\frac{IC+\pi}{IC} = 1 + \frac{\pi}{IC}$ decreases, and Q^* decreases. Finally, as π increases, the denominator of (2.15) increases, and B^* decreases.
2. The maximum number of backorders per cycle, B^*, cannot be more than the order quantity per cycle Q^* since $\frac{B^*}{Q^*} = \frac{IC}{IC+\pi} \le 1$.
3. Q^* is never smaller than Q_E, the optimal order quantity for the EOQ model without backordering. Immediately after the arrival of an order, part of the order is used to fulfill the backordered demand. This saves the holding cost on that part of the received order and allows the placement of a bigger order. As π increases, backordering becomes more expensive and B^* decreases. As a result, the component of Q^* that is used to satisfy the backlog decreases and Q^* comes closer to Q_E.
4. As the holding cost rate I increases, $\frac{IC}{IC+\pi}$ increases and B^* increases.

To improve our understanding even further, let us compare the fixed cost to the sum of the holding and backordering costs at $Q = Q^*$ and $B = B^*$.

From (2.15), $\frac{B^*}{Q^*} = \frac{IC}{IC+\pi}$, which means that

$$\frac{Q^* - B^*}{Q^*} = 1 - \frac{B^*}{Q^*} = 1 - \frac{IC}{IC + \pi} = \frac{\pi}{IC + \pi}.$$

Now, the holding cost at $Q = Q^*$ and $B = B^*$ is equal to

$$\frac{IC(Q^* - B^*)^2}{2Q^*} = \frac{ICQ^*}{2} \left(\frac{Q^* - B^*}{Q^*} \right)^2 = \frac{ICQ^*}{2} \left(\frac{\pi}{IC + \pi} \right)^2.$$

Similarly, the backordering cost at $Q = Q^*$ and $B = B^*$ is equal to

$$\frac{\pi B^{*2}}{2Q^*} = \frac{\pi}{2} \left(\frac{IC}{IC + \pi} \right)^2 Q^*.$$

Therefore, the sum of the holding and backordering costs is equal to

$$\frac{Q^*}{2} \left(\frac{\pi^2 IC + (IC)^2 \pi}{(IC + \pi)^2} \right) = \frac{Q^*}{2} \left(\frac{IC\pi}{IC + \pi} \right),$$

which, after substitution of Q^*, is equal to $\sqrt{\frac{\lambda KIC\pi}{2(IC+\pi)}}$.

On the other hand, the fixed cost of order placement at $Q = Q^*$ is equal to

$$\frac{\lambda K}{Q^*} = \sqrt{\frac{\lambda KIC\pi}{2(IC+\pi)}},$$

where we have substituted for Q^*. The expressions for the average annual fixed cost and the sum of the holding and backordering costs are equal. Once again we note that in an optimal solution the costs are balanced.

2.2.1   The Optimal Cost

Since the optimal average annual cost is equal to the purchasing cost plus two times the average annual fixed cost,

$$Z(B^*, Q^*) = C\lambda + K\frac{\lambda}{Q^*} + IC\frac{(Q^*-B^*)^2}{2Q^*} + \pi\frac{(B^*)^2}{2Q^*} = C\lambda + 2K\frac{\lambda}{Q^*}$$

$$= C\lambda + \sqrt{\frac{2K\lambda IC\pi}{IC+\pi}}.$$

Let us compute the cost of following the optimal policy in our backordering example. Recall that $K = 10$, $\lambda = 520$, $I = .20$, $C = \$5$, and $\pi = 10$. Thus the optimal cost is

$$Z(B^*, Q^*) = C\lambda + \sqrt{\frac{2K\lambda IC\pi}{IC+\pi}} = (5)(520) + \sqrt{\frac{2(10)(520)(0.2)(5)(10)}{(5)(0.2)+10}} = 2697.23.$$

The purchasing cost is equal to $2600 and the remaining cost of $97.23 arises from inventory/backlog management. Notice that the optimal cost is lower than the cost obtained without backordering.

2.3   Quantity Discount Model

In this section, we will study inventory management when the unit purchasing cost decreases with the order quantity Q. In other words, a discount is given by the seller if the buyer purchases a large number of units. Our objective is to determine the optimal ordering policy for the buyer in the presence of such incentives. The remainder of the model is the same as stated in Section 2.1.

We will discuss two types of quantity discount contracts: all units discounts and incremental quantity discounts. An example of all units discounts is as follows.

Table 2.1. Example data for all units discount.

Quantity Purchased	Per-Unit Price
0–100	$5.00
101–250	$4.50
251 and higher	$4.00

If the buyer purchases 100 or fewer units, he pays $5 for each unit purchased. If he purchases at least 101 but no more than 250 units, he pays $4.50 for each unit purchased. Thus the cost of purchasing 150 units in a single order is $(150)(4.5) = \$675$. Similarly, his purchasing cost comes down to only $4.00 per unit if he purchases at least 251 units. Therefore, in the all units discounts model, as the order quantity increases, the unit purchasing cost decreases for *every* unit purchased. Figure 2.4 shows the total purchasing cost function for the all units discounts.

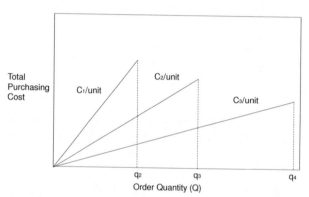

Fig. 2.4. Total purchasing cost for all units discount.

The incremental quantity discount is an alternative type of discount. Let us consider an example. Here the unit purchasing price is $1.00 for every unit up to 200 units. If

Table 2.2. Example data for incremental quantity discount.

Quantity Purchased	Per-Unit Price
0–200	$1.00
201–500	$0.98
501 and higher	$0.95

the order quantity is between 201 and 500, the unit purchasing price drops to $0.98 but *only* for units numbered 201 through 500. The total purchasing cost of 300 units will be $(200)(1) + (300 - 200)(0.98) = \$298$. Similarly, if at least 501 units are purchased, the unit purchasing price decreases to \$0.95 for units numbered 501, 502, etc. Therefore, in the incremental quantity discounts case, the unit purchasing cost decreases only for units beyond a certain threshold and not for every unit as in the all units discounts case. Figure 2.5 illustrates the total purchasing cost function.

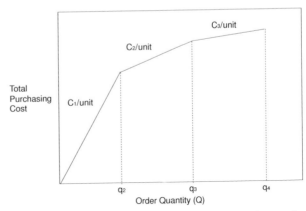

Fig. 2.5. Total purchasing cost for incremental quantity discount.

Why are these discounts offered by suppliers? The reason is to make the customers purchase more per order. As we learned before, large orders result in high inventory holding costs because an average unit spends a longer time in the system before it gets sold. Thus the seller's price discount subsidizes the buyer's inventory holding cost.

We will begin with an analysis of the all units discounts contract. We want to find the order quantity that minimizes the average annual sum of the purchasing, holding, and fixed ordering costs.

2.3.1 All Units Discount

Similarly to the EOQ model, we assume that backordering is not allowed. Let m be the number of discount possibilities. In the example in Table 2.1, there are three discount levels, so $m = 3$. Let $q_1 = 0, q_2, q_3, \ldots, q_j, q_{j+1}, \ldots, q_m$ be the order quantities at which the purchasing cost changes. Even though we gave an example in which units were discrete, for simplicity in our analysis we will assume from now on that units are infinitely divisible. The unit purchasing cost is the same for all Q in $[q_j, q_{j+1})$; let the corresponding unit purchasing cost be denoted by C_j. Thus, the jth lowest unit purchasing cost is denoted by C_j; C_m is the lowest possible purchasing cost with C_1 being the highest unit purchasing cost. Going back to the example in Table 2.1, $q_1 = 0$, $q_2 = 101$, $q_3 = 251$, $C_1 = \$5$, $C_2 = \$4.50$, and $C_3 = \$4$.

The expression for the average annual cost is similar to (2.1) with one difference. The purchasing cost now depends on the value of Q. We rewrite the average annual cost to take this difference into account:

$$Z_j(Q) = C_j\lambda + K\frac{\lambda}{Q} + \frac{IC_jQ}{2}, \quad q_j \leq Q < q_{j+1}. \tag{2.16}$$

Thus we have a family of cost functions indexed by the subscript j. The jth cost function is defined for only those values of Q that lie in $[q_j, q_{j+1})$. These functions are shown in Figure 2.6. The solid portion of each curve corresponds to the interval in which it is defined. The average annual cost function is thus a combination of these solid portions.

Two observations can be made. First, the average annual cost function is not continuous. It is segmented such that each segment is defined over a discount interval $[q_j, q_{j+1})$. The segmented nature of the average annual cost function makes solving the problem slightly more difficult since we now cannot simply take a derivative and set it to zero to find the optimal solution.

It is the second observation that paves the way for an approach to find the optimal solution. Different curves in Figure 2.6 are arranged in an order of decreasing unit purchasing cost. The curve at the top corresponds to the highest per-unit purchasing cost C_1, and the lowest curve corresponds to the lowest per-unit purchasing cost C_m. Also, the curves do not cross each other. To obtain the optimal solution, we will start from the bottom-most curve which is defined for Q in $[q_m, \infty)$, and compute the lowest possible cost in this interval. Then, in the second iteration, we will consider the second-lowest curve and check to see if we could do better in terms of the cost. The algorithm continues as long as the cost keeps decreasing. The steps of the algorithm are outlined in the following subsection.

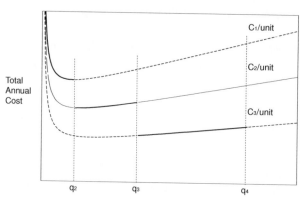

Fig. 2.6. Total cost for all units discount.

2.3.2 An Algorithm to Determine the Optimal Order Quantity for the All Units Discount Case

Step 1: Set $j = m$. Compute the optimal EOQ for the mth cost curve, which we denote by Q_m^*:

$$Q_m^* = \sqrt{\frac{2K\lambda}{IC_m}}.$$

Step 2: Is $Q_m^* \geq q_m$? If yes, Q_m^* is the optimal order quantity and we are done. If not, the minimum cost occurs at $Q = q_m$ owing to the convexity of the cost function. Since the minimum point $Q_m^* < q_m$, the cost function for the mth discount level is increasing on the right of q_m. Consequently, among all the feasible order quantities, that is, for order quantities greater than or equal to q_m, the minimum cost occurs at $Q = q_m$.

Compute the cost corresponding to $Q = q_m$. Let this cost be denoted by Z_{min} and $Q_{min} = q_m$ and go to Step 3.

Step 3: Set $j = j - 1$. Compute the optimal EOQ for the jth cost curve:

$$Q_j^* = \sqrt{\frac{2K\lambda}{IC_j}}.$$

Step 4: Is Q_j^* in $[q_j, q_{j+1})$? If yes, compute $Z(Q_j^*)$ and compare with Z_{min}. If $Z(Q_j^*) < Z_{min}$, Q_j^* is the optimal order quantity. Otherwise, Q_{min} is the optimal order quantity. In either case, we are done.

Otherwise, if Q_j^* is not in $[q_j, q_{j+1})$, then the minimum cost for the jth curve occurs at $Q = q_j$ owing to the convexity of the cost function. Compute this cost $Z(q_j)$. If $Z(q_j) < Z_{min}$, then set $Q_{min} = q_j$ and $Z_{min} = Z(q_j)$. If $j \geq 2$, go to Step 3; otherwise, stop.

We now demonstrate the execution of the above algorithm with the following example.

Let us revisit the office supplies company once again. Suppose now that the retailer gets an all units discount of 5% per box of pencils if he purchases at least 110 pencil boxes in a single order. The deal is further sweetened if the retailer purchases at least 150 boxes in which case he gets a 10% discount. Should the retailer change the order quantity?

We apply the algorithm to compute the optimal order quantity. There are three discount categories, so $m = 3$. Also, $C_1 = 5$, $C_2 = 5(1 - 0.05) = 4.75$, $C_3 = 5(1 - 0.1) = 4.50$, $q_1 = 0$, $q_2 = 110$, and $q_3 = 150$.

Iteration 1: Initialize $j = 3$.

Step 1: We compute $Q_3^* = \sqrt{\frac{2(520)(10)}{(0.2)(4.50)}} = 107.5$.

Step 2: Since Q_3^* is not greater than $q_3 = 150$, the minimum cost occurs at $Q = 150$. This cost is equal to

$$Z_{min} = (4.50)(520) + \frac{(10)(520)}{150} + \frac{(0.2)(4.50)(150)}{2} = \$2442.17.$$

Step 3: Now, we set $j = 2$. We compute $Q_2^* = \sqrt{\frac{2(520)(10)}{(0.2)(4.75)}} = 104.63$.

Step 4: Once again, Q_2^* is not feasible since it does not lie in the interval $[110, 150)$. The minimum feasible cost thus occurs at $Q = q_2 = 110$ and is equal to

$$Z(q_2) = (4.75)(520) + \frac{(10)(520)}{110} + \frac{(0.2)(4.75)(110)}{2} = \$2569.52,$$

which is higher than Z_{min} and so the value of Z_{min} remains unchanged. Now we proceed to Iteration 2.

Iteration 2:

Step 3: Now, we set $j = 1$. Recall that the value of Q_1^* is equal to 101.98.

Step 4: Q_1^* is feasible since it lies in the interval $[0, 110)$. The corresponding cost is equal to \$2701.98, which is also higher than Z_{min}.

Thus the optimal solution is to order 150 units and the corresponding average annual cost of purchasing and managing inventory is equal to \$2442.17.

Observe that the algorithm stops as soon as we find a discount for which Q_j^* is feasible. This does not mean that the optimal solution is equal to Q_j^*; the optimal solution can still correspond to the $(j+1)$st or higher-indexed discount. That is, the optimal solution cannot be less than Q_j^*.

2.3.3 Incremental Quantity Discounts

The incremental quantity discount case differs from the all units discount case. In this situation, as the quantity per order increases, the unit purchasing cost declines incrementally on additional units purchased as opposed to on all the units purchased. Let $q_1 = 0, q_2, \ldots, q_j, q_{j+1}, \ldots, q_m$ be the order quantities at which the unit purchasing cost changes. The number of discount levels is m. In the example, we assumed that the units are discrete; for analysis we will assume that the units are infinitely divisible and the purchasing quantity can assume any real value. The unit purchasing cost is the same for all values of Q in $[q_j, q_{j+1})$, and we denote this cost by C_j. By definition, $C_1 > C_2 > \cdots > C_j > C_{j+1} > \cdots > C_m$. In the above example, $m = 3$, $q_2 = 201$, $q_3 = 501$ (q_1 by definition is 0), $C_1 = \$1$, $C_2 = \$0.98$, $C_3 = \$0.95$.

Our goal is to determine the optimal number of units to be ordered. We first write an expression for the average annual purchasing cost if Q units are ordered. Let Q be in the jth discount interval, that is, Q lies between q_j and q_{j+1}. The purchasing cost for Q in this interval is equal to

$$C(Q) = C_1(q_2 - q_1) + C_2(q_3 - q_2) + \cdots + C_{j-1}(q_j - q_{j-1}) + C_j(Q - q_j).$$

Note that only the last term on the right-hand side of the above expression depends on Q. Let R_j be used to denote the sum of the terms that are independent of Q. That is,

$$R_j = C_1(q_2 - q_1) + C_2(q_3 - q_2) + \cdots + C_{j-1}(q_j - q_{j-1}), \quad j \geq 2,$$

with $R_1 = 0$. Therefore,

$$C(Q) = R_j + C_j(Q - q_j).$$

Now the average unit purchasing cost for Q units is equal to $\frac{C(Q)}{Q}$, which is equal to

$$\frac{C(Q)}{Q} = \frac{R_j}{Q} + C_j - C_j \frac{q_j}{Q}.$$

The average annual purchasing cost when each order consists of Q units is equal to $\frac{C(Q)}{Q} \lambda$. The fixed order cost and holding cost terms remain similar to those in the basic EOQ model except that the average annual holding cost per unit is now equal to $I \frac{C(Q)}{Q}$. This is not surprising since the holding cost per unit depends on the *cost* of each unit, which is a function of the order size given the cost structure here. The average annual cost of managing inventory is thus equal to

$$Z(Q) = \frac{C(Q)}{Q} \lambda + K \frac{\lambda}{Q} + \frac{I \frac{C(Q)}{Q} Q}{2}$$

$$= \left(\frac{R_j}{Q} + C_j - C_j \frac{q_j}{Q} \right) \lambda + K \frac{\lambda}{Q} + \frac{I(R_j + C_j(Q - q_j))}{2}.$$

Realigning terms yields

$$Z(Q) = C_j \lambda + (R_j - C_j q_j + K) \frac{\lambda}{Q} + \frac{IC_j Q}{2} + \frac{I(R_j - C_j q_j)}{2},$$

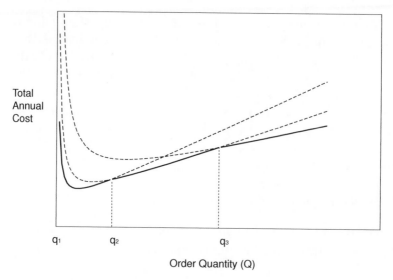

Total
Annual
Cost

q_1 q_2 q_3

Order Quantity (Q)

Fig. 2.7. Total cost for incremental quantity discount case.

which is valid for $q_j \le Q < q_{j+1}$.

Figure 2.7 shows the $Z(Q)$ function. Once again, we have a family of curves, each of which is valid for a given interval. The valid portion is shown using a solid line. Thus the curve with the solid line constitutes the $Z(Q)$ function. Unlike the all units discount cost function, $Z(Q)$ is a continuous function. It turns out that the optimal solution Q^* cannot be equal to any one of $\{q_1, q_2, \ldots, q_m, q_{m+1}\}$. We use this fact to construct an algorithm to determine the optimal order quantity.

2.3.4 An Algorithm to Determine the Optimal Order Quantity for the Incremental Quantity Discount Case

The algorithm for this case is as follows.

Step 1: Compute the order quantity that minimizes $Z_j(Q)$ for each j, which is denoted by Q_j^* and is obtained by setting $\frac{dZ_j(Q)}{dQ} = 0$:

$$\frac{dZ_j(Q)}{dQ} = -(R_j - C_j q_j + K)\frac{\lambda}{Q^2} + \frac{IC_j}{2}$$

$$\Rightarrow Q_j^* = \sqrt{\frac{2(R_j - C_j q_j + K)\lambda}{IC_j}}.$$

This step gives us a total of m possible order quantities.

Step 2: In this step we check which one of these potential values for Q^* is feasible, that is, $q_j \le Q_j^* < q_{j+1}$. Disregard the ones that do not satisfy this inequality.

Step 3: Calculate the cost $Z_j(Q_j^*)$ corresponding to each remaining Q_j^*. The order quantity Q_j^* that produces the least cost is the optimal order quantity.

Let us now illustrate the execution of this algorithm. Suppose in the preceding example the retailer is offered an incremental quantity discount. Let us determine the optimal order quantity.

Step 1: $R_1 = 0$, $R_2 = C_1(q_2 - q_1) = (5)(110 - 0) = 550$, and $R_3 = C_1(q_2 - q_1) + C_2(q_3 - q_2) = R_2 + (4.75)(150 - 110) = 740$. We compute

$$Q_1^* = \sqrt{\frac{2(R_1 - C_1 q_1 + K)\lambda}{IC_1}} = \sqrt{\frac{2(0 - 0 + 10)(520)}{(0.2)(5)}} = 101.98,$$

$$Q_2^* = \sqrt{\frac{2(R_2 - C_2 q_2 + K)\lambda}{IC_2}} = \sqrt{\frac{2(550 - (4.75)(110) + 10)(520)}{(0.2)(4.75)}} = 202.61,$$

$$Q_3^* = \sqrt{\frac{2(R_3 - C_3 q_3 + K)\lambda}{IC_3}} = \sqrt{\frac{2(740 - (4.5)(150) + 10)(520)}{(0.2)(4.5)}} = 294.39.$$

Step 2: We disregard Q_2^* since it does not lie in the interval $[110, 150)$. Q_1^* and Q_3^* are feasible.

Step 3: We now compute the costs corresponding to Q_1^* and Q_3^*:

$$Z(Q_1^*) = C_1 \lambda + (R_1 - C_1 q_1 + K)\frac{\lambda}{Q_1^*} + \frac{IC_1 Q_1^*}{2} + \frac{I(R_1 - C_1 q_1)}{2}$$

$$= (5)(520) + (0 - 0 + 10)\frac{520}{101.98} + \frac{(0.2)(5)(101.98)}{2} + \frac{(0.2)(0 - 0)}{2}$$

$$= 2701.98,$$

$$Z(Q_3^*) = C_3 \lambda + (R_3 - C_3 q_3 + K)\frac{\lambda}{Q_3^*} + \frac{IC_3 Q_3^*}{2} + \frac{I(R_3 - C_3 q_3)}{2}$$

$$= (4.5)(520) + (740 - (4.5)(150) + 10)\frac{520}{294.39} + \frac{(0.2)(4.5)(294.39)}{2}$$

$$+ \frac{(0.2)(740 - (4.5)(150))}{2}$$

$$= 2611.45.$$

Thus, the optimal solution is to order 294.39 units and the corresponding average annual cost of purchasing and managing inventory is equal to $2611.45.

2.4 Lot Sizing When Constraints Exist

In the preceding portions of this chapter, we focused on determining the optimal ordering policy for a single item. In many if not most real-world situations, decisions are not made for each item independently. There may be limitations on space to store items in warehouses; there may be constraints on the number of orders that can be received per year; there may be monetary limitations on the value of inventories that are stocked. Each of these situations requires stocking decisions to be made jointly among the many items managed at a location. We will illustrate how lot sizing decisions can be made for a group of items.

Holding costs are often set to limit the amount of space or investment consumed as a result of the lot sizing decisions. Rather than assuming a holding cost rate is used to calculate the lot sizes, suppose a constraint is placed on the average amount of money invested in inventory. Thus a budget constraint is imposed that limits investment across items. Let Q_i be the procurement lot size for item i and C_i the per-unit purchasing cost for item i, $i = 1, \ldots, n$, where n is the number of items being managed. The sum

$$\sum_{i=1}^{n} C_i \frac{Q_i}{2}$$

measures the average amount invested in inventory over time. Let b be the maximum amount that can be invested in inventory on average. Furthermore, suppose our goal is to minimize the average annual total fixed procurement cost over all item types while adhering to the budget constraint. Let λ_i and K_i represent the average annual demand rate and fixed order cost for item i, respectively. Recall that $\lambda_i K_i / Q_i$ measures the average annual fixed order cost incurred for item i given Q_i is the lot size for item i. Define $F_i(Q_i) = \lambda_i K_i / Q_i$.

This procurement problem can be stated as follows:

$$\text{minimize} \sum_{i=1}^{n} \frac{\lambda_i K_i}{Q_i} = \sum_{i=1}^{n} F_i(Q_i)$$

subject to

$$\sum_{i=1}^{n} C_i \frac{Q_i}{2} \leq b,$$

$$Q_i \geq 0.$$

We will obtain the solution for this problem by constructing the corresponding Karush–Kuhn–Tucker conditions. There are four such conditions. The parameter θ is the Lagrange multiplier associated with the budget constraint.

(i) $$\frac{dF_i(Q_i)}{dQ_i} + \frac{\theta C_i}{2} \geq 0$$

(ii) $$\theta \left(\sum_{i=1}^{n} C_i \frac{Q_i}{2} - b \right) = 0$$

(iii) $$Q_i \left(\frac{dF_i(Q_i)}{dQ_i} + \frac{\theta C_i}{2} \right) = 0$$

(iv) $$\begin{cases} \sum_{i=1}^{n} C_i \frac{Q_i}{2} \leq b \\ Q_i \qquad \geq 0. \end{cases}$$

The function $F_i(Q_i)$ is a strictly decreasing function for all i. Hence if there were no limit on investment in inventory, $Q_i^* = \infty$. Thus in an optimal solution to the problem the budget constraint must hold as a strict equality. That is,

$$\sum_{i=1}^{n} C_i \frac{Q_i}{2} = b$$

in any optimal solution.

Observe that $Q_i > 0$ in any optimal solution. As a consequence of (iii),

$$\frac{dF_i(Q_i)}{dQ_i} + \frac{\theta C_i}{2} = 0.$$

This observation results in

$$Q_i^* = \sqrt{\frac{2\lambda_i K_i}{\theta C_i}}.$$

Since $\sum_{i=1}^{n} C_i \frac{Q_i}{2} = b$,

$$\sum_{i=1}^{n} \frac{C_i}{2} \sqrt{\frac{2\lambda_i K_i}{\theta C_i}} = \frac{1}{\sqrt{\theta}} \sum_{i=1}^{n} \sqrt{\frac{\lambda_i C_i K_i}{2}} = b$$

and

$$\theta = \left[\frac{1}{b} \sum_{i=1}^{n} \sqrt{\frac{\lambda_i C_i K_i}{2}} \right]^2.$$

Observe that the expression for Q_i^* is the same as the EOQ solution with I equal to θ. Thus θ is the imputed holding cost rate given the budget limitation, b. Note also how θ is related to the value of b: θ is proportional to the square of the inverse of b.

Let us examine an example problem. Suppose $n = 2$ and $b = 7000$. Additional data are given in the following table.

Item	1	2
λ_i	1000	500
C_i	20	100
K_i	50	75

The optimal solution is

$$Q_1^* = 238.38$$
$$Q_2^* = 92.324$$

with

$$\theta^* = .0879895.$$

Thus as the budget level is increased the average annual cost of placing orders decreases by the value of θ^*.

Many other constrained multi-item lot sizing problems can be formulated and solved. Some examples are given in the exercises.

2.5 Exercises

2.1. Derive an expression for the reorder point in the EOQ model when the lead time is greater than the reorder interval.

2.2. Show that the optimal order quantity for the incremental quantity discount model cannot be equal to any one of $\{q_1, q_2, \ldots, q_m\}$.

2.3. The demand for toilet paper in a convenience store occurs at the rate of 100 packets per month. Each packet contains a dozen rolls and costs $8. Assuming the order placement cost to be $20 per order and the holding cost rate to be .25, determine the optimal order quantity, reorder interval, and optimal cost using the EOQ model. Take one month to be the equivalent of four weeks.

2.4. Suppose now that backordering of demand is allowed in the situation described in the previous question. The cost of backordering is $10/packet/year. Use the EOQ model with backordering to determine the optimal order quantity, maximum backlog allowed, and optimal average annual cost. Also, compute the lengths of the portions of a cycle with positive on-hand inventory and backorders, respectively.

2.5. Compare the optimal average annual costs obtained in the last two exercises. Assume that the relationship between the optimal costs of the EOQ model and the EOQ model with backordering holds true in general with the same holding cost, fixed ordering cost, and demand rate (for example, you may find that the optimal cost of the EOQ model with backordering is lower than the optimal cost of the EOQ model.) Can you think of an intuitive explanation for your conclusion?

2.6. If the lead time in Exercise 2.3 is 1 week, determine the reorder point.

2.7. The demand for leather laptop cases in an electronics store is fairly regular throughout the year at 200 cases per year. Each leather case costs $50. Assuming the fixed order placement cost to be $5/order and holding cost rate to be .30, determine the optimal order quantity using the EOQ model. Compute the reorder point assuming the lead time to be 4 weeks.

2.8. Suppose in Exercise 2.3 the convenience store receives a discount of 5% per packet on all units purchased if the order quantity is at least 200 and 7.5% per packet if the order quantity is at least 250. Would your answer change? If so, what is the new recommended purchase quantity?

2.9. Solve the last exercise assuming the discount type to be an incremental quantity discount.

2.10. We illustrated that in the all units discount case, the cost curves for different discount levels are in increasing order of per-unit cost. That is, a curve with a higher per-unit cost always lies above another curve with lower per-unit cost for *any* order quantity. Establish this observation mathematically by comparing the total average annual costs for the two levels of discounts with purchasing costs C_1 and C_2 such that $C_1 < C_2$.

2.11. The purchasing agent of Doodaldoo Dog Food company can buy horsemeat from one source for $0.06 per pound for the first 1000 pounds and $0.058 for each additional pound. The company requires 50,000 pounds per year. The cost of placing an order is $1.00. The inventory holding cost rate is $I = .25$ per year. Compute the optimal purchase quantity and the total average annual cost.

2.12. Derive an expression for the reorder point in the EOQ model with backordering when the lead time is either greater than or less than the length of the reorder interval.

2.13. Show in the EOQ model with no backordering that it is optimal to place an order such that it arrives when the on-hand inventory is zero. Specifically, show that the cost of any other policy such that the order arrives when the on-hand inventory is positive is greater than the cost of the policy in which the order arrives when the on-hand inventory is zero.

2.14. For the all units discount case, let j^* be the largest discount index such that $Q_j^* = \sqrt{\frac{2\lambda K}{IC_j}} \in [q_j, q_{j+1})$. Prove that the optimal order quantity cannot be less than Q_j^*.

2.15. Consider the basic EOQ model. Suppose the order is delivered in boxes each of which holds A units. Thus, the order quantity can only be a multiple of A. Using the convexity of the cost function in Q, compute the optimal order quantity.

2.16. Recall that in the algorithm for the all units discount model we stop as soon as we identify a discount for which the optimal EOQ, Q_j^*, is feasible. An alternative stopping criterion is as follows. In Step 3, compute $Z(Q_j^*)$ and stop if $Z(Q_j^*) \geq Z_{min}$. While this stopping criterion involves additional computation, it may eliminate some iterations and save computational effort overall.

Prove that an algorithm that utilizes the above stopping criterion generates an optimal solution. That is, show that if j is the largest discount index for which $Z(Q_j^*) \geq Z_{min}$, then the optimal order quantity cannot be less than Q_j^*.

2.17. The basic EOQ model permitted the value of Q^* to be any positive real number. Suppose we restrict Q to be integer-valued. Develop an approach for finding Q^* in this case.

2.18. Llenroc Automotive produces power steering units in production lots. These units go into Jaguar's 4.2 liter engine models and have a demand rate estimated to be 20,000 units per year. Llenroc makes a variety of power steering units in addition to the one used in this model of Jaguars. Switching from producing one power steering model type to another incurs a cost of $400. The cost of carrying a power steering unit in stock for a year is estimated to be $35. While the goal is to minimize the average annual cost of changeovers and carrying inventory, management has decided that no more than 10 setups per year are permissible for the pump in question.

Develop a mathematical model and a solution approach for determining the optimal lot size given the constraint on production. What is the solution if 20 setups were possible?

2.19. Llenroc Electronics stocks n items. These items are ordered from a group of suppliers. The cost of placing an order for items of type i is K_i. The unit purchase cost for an item of type i is C_i. The holding cost factor I is the same for all item types. Suppose a maximum of b orders can be processed in total in a year. Derive formulas that can be used to determine the optimal procurement quantities for each item so that the average annual cost of holding inventories is minimized subject to a constraint on the total number of orders that can be processed per year. Also, construct formulas for each item when the objective is to minimize the sum of the average annual fixed ordering costs and the average annual holding costs subject to the same constraint.

2.20. The procurement department of Llenroc Electronics purchases the component types from a particular supplier. The fixed cost of placing an order for any of three types of components is $100. The inventory holding cost factor, I, is .20. Other relevant data are presented in the following table.

Item	1	2	3
Annual Demand Rate	10,000	30,000	20,000
Unit Cost	350	125	210
Space Reqm't $\left(\frac{ft^2}{unit}\right)$	10	15	12

The CFO has placed a $75,000 limit on the average value of inventory carried in stock. Furthermore, there is a limit on the amount of space available to store inventory of 10,000 ft^2.

Find the optimal order quantity for each item type so as minimize the total average annual fixed ordering and holding costs subject to the constraints.

3

Power-of-Two Policies

We will extend the results of the last chapter in three ways. First, we will consider multi-stage systems. These systems are characterized by multiple locations where inventory management decisions have to be made. Recall that in the last chapter, there was only a single stage, that is, the ordering and inventory management decisions involved only one location. Second, we will work with the reorder interval as the decision variable over which cost is optimized instead of the order quantity. We continue to assume that demand occurs at a deterministic and stationary rate. Thus, once we know the reorder interval, we can easily determine the corresponding order quantity. Therefore, the two decision variables are equivalent; our preference for the reorder interval is due to practical reasons, as we explain below. Third, instead of determining the optimal solution, we will develop algorithms that determine reorder intervals that are easier to use in practice but are not necessarily optimal. However, we will establish that the worst possible cost will not be higher than the optimal cost by more than 6%. These inventory management policies that we will develop are referred to as power-of-two (PO2) policies.

As we explained in Chapter 1, many of the inventory management models are applicable in production systems as well. In inventory systems, these models are used to determine the optimal order quantity and reorder interval; in production systems the objective is to find the lot size and the reorder interval between consecutive production runs. The PO2 policies that we will study in this chapter can be applied both in production and inventory systems. We will discuss the specific applications while studying various types of multi-stage systems.

The first advantage of the use of the reorder interval is motivated by a production system. Clearly, while planning production, it is natural to think in terms of producing items once every planning period (for example, shift, week, or month). When working with the order quantity as the decision variable, the corresponding reorder interval may be any real number; remember the relationship from last chapter, $T = Q/\lambda$. It could be

J.A. Muckstadt and A. Sapra, *Principles of Inventory Management: When You Are Down to Four, Order More*, Springer Series in Operations Research and Financial Engineering, DOI 10.1007/978-0-387-68948-7_3, © Springer Science+Business Media, LLC 2010

1.85 weeks, 2.3 weeks, or an irrational number, for example, $\sqrt{\frac{9}{7}}$ weeks. Such reorder intervals make resource planning over time a very difficult task.

Another advantage is that the mathematical formulation is simpler if we work with the reorder interval rather than with the order quantity as the decision variable. In a multi-stage production system, at each stage it is necessary to ensure that all the components required to produce the batch are available at the time of the production. This constraint is more easily formulated with the reorder interval as the decision variable.

There are practical reasons for using reorder intervals in a non-production setting as well. It is fairly common for retailers to place orders such that the order quantity is equal to n periods of forecast demand where n is chosen by the inventory manager. The length of a period can be measured in hours, days, weeks, etc., depending on the applications. Thus orders are placed only once every n periods.

3.1 Basic Framework

As in the previous chapter, we will assume that the demand occurs continuously over time at a constant and known rate λ. Whenever an order is placed, a fixed cost K is incurred, which is independent of the size of the order quantity. Holding inventory also incurs a cost, which is proportional to the average inventory. Throughout this chapter, we use $h = IC$. The holding cost h is the cost (in $) of holding one unit for one year. Backordering is not allowed.

Recall from the previous chapter that the EOQ model is given as follows:

$$\min_{Q \geq 0} Z(Q) = C\lambda + K\frac{\lambda}{Q} + \frac{1}{2}hQ, \tag{3.1}$$

where $Z(Q)$ is the average annual cost of managing the inventory and Q is the quantity purchased per order. In the subsequent analysis we will ignore the annual purchasing cost term $C\lambda$ because it is constant and therefore independent of Q.

Let us assume that the lead time is zero to simplify the discussion. Hence an order is placed and received at the same instant. Once Q units arrive, they are consumed at the rate λ. Recall that the reorder interval $T = Q/\lambda$. By substituting for Q by λT in (3.1) and eliminating the purchasing cost term $C\lambda$, we obtain the following formulation of the decision problem:

$$\min_{T \geq 0} Z(T) = \frac{K}{T} + \frac{1}{2}h\lambda T. \tag{3.2}$$

We abuse the notation here and continue to use Z to denote the cost function when the argument is T. To simplify notation further, let $g = \frac{1}{2}h\lambda$. Note that $g = \frac{1}{2}h\lambda$ is the annual holding cost if only one order of size λ is placed every year. Substituting g into (3.2), we get

$$\min_{T \geq 0} Z(T) = \frac{K}{T} + gT, \tag{3.3}$$

where T is measured in years. We next obtain the optimal solution by setting $\frac{dZ(T)}{dT} = 0$:

$$\frac{dZ(T)}{dT} = -\frac{K}{T^2} + g = 0,$$

$$\Rightarrow T^* = \sqrt{\frac{K}{g}}. \tag{3.4}$$

The optimal average annual cost $Z(T^*)$ is equal to

$$Z(T^*) = \frac{K}{\sqrt{\frac{K}{g}}} + g\sqrt{\frac{K}{g}} = 2\sqrt{Kg}.$$

Check to see that this optimal cost is same as found in Chapter 2.

3.1.1 Power-of-Two Policies

So far we have imposed only one constraint on the reorder interval T, the non-negativity constraint. Often the solution in (3.4) is impractical to implement; for example, it could be 1.09 weeks. For practical reasons, we assume that there exists a base planning period T_L and that the reorder interval must be an integer multiple of T_L. The base planning period could be equal to a shift, a day, a week, or even a year. Thus the minimum reorder interval is equal to T_L. This constraint can be expressed as follows:

$$T = nT_L, \quad n \in \{1, 2, 3, \dots\}.$$

We will further constrain n to be a power of two, that is, n can only be $1, 2, 4, 8, 16, \dots$. Therefore, the reorder interval T assumes only one of the following values:

$$\{T_L, 2T_L, 4T_L, 8T_L, \dots\}.$$

This type of inventory management policy in which the reorder interval is a power-of-two multiple of the base planning period is known as a *power-of-two policy* (or PO2 policy). Two questions arise at this point: What are the advantages of using PO2 policies, and why a power of two and not a power of three, four or five? The second question is easier to answer, and we discuss it first. The worst-case cost of using a PO2 policy is lower than the worst-case cost of using a PO3, PO4, or POn policy where n is any non-negative integer. In other words, if we compare the worst possible costs among the POn type policies, the PO2 policy's worst-case cost is closest to the optimal cost.

To discuss the advantages of PO2 policies, we reproduce here an example presented by Muckstadt and Roundy [240]. In their example, they cite the production control system used by a major automotive manufacturer in its stamping plants. The demand for products (doors, body panels, etc.) remained relatively constant over time. The decision regarding when to produce an item was divided into two stages. In the first stage, the week of the production was determined. This was done in such a way that each component was produced exactly once, twice, or four times during a four-week cycle. This is equivalent to the reorder interval being equal to one week, two weeks, or four weeks. The production quantity during each run was approximately the same. In the second stage, given the week of the production, the exact time of the production (day, shift, etc.) was determined. The job of the production planner was considerably easier because of this approach to production scheduling.

To see another advantage of the PO2 policies, let us consider a simple production system in which three products, A, B, and C, share a machine. Suppose that A, B, and C are produced every week, every three weeks, and every five weeks, respectively. In this case, the production cycle repeats every fifteen weeks; see the table below. In other words, all the three products will be produced in the same week every fifteen weeks. This causes a problem in balancing the load on the machine over time since the fifteenth week is likely to be overloaded.

	1	2	3	4	5	6	7	8	9	10	11	12	13	14	15	16
A	x	x	x	x	x	x	x	x	x	x	x	x	x	x	x	x
B	x			x			x			x			x			x
C	x					x					x					x

If, on the other hand, the reorder intervals of products B and C are reduced by 1 week to 2 and 4 weeks, respectively, then a production schedule can be determined such that B and C are never produced in the same week; see the table below.

	1	2	3	4	5	6	7	8	9	10	11	12	13	14	15	16
A	x	x	x	x	x	x	x	x	x	x	x	x	x	x	x	x
B		x		x		x		x		x		x		x		x
C	x				x				x				x			

To summarize, the PO2 policies make production scheduling easier. They are also useful in ensuring a more balanced load on machines in a production system. This is the case when many items are produced on a particular machine. This was the situation in the real automotive example discussed earlier in this subsection.

3.1.2 PO2 Policy for a Single-Stage System

The inventory management problem for a single-stage system in which the reorder interval is restricted to be a power-of-two multiple of the base planning period is as follows:

$$\min_{T \geq 0} Z(T) = \frac{K}{T} + gT,$$ (3.5)

subject to

$$T = 2^\ell T_L, \quad \ell = \{0, 1, 2, 3, \ldots\},$$

where T_L is known. To solve this problem, we cannot simply take the derivative of the objective function, since T can take only discrete values. Therefore, we will use a different approach to obtain an optimal solution.

Figure 3.1 illustrates that $Z(T)$ is a convex function of T. This causes $Z(T)$ to first decrease and then increase among the feasible values for T, $T_L, 2T_L, \ldots$. Let an optimal value of T be equal to $2^{\ell^*} T_L$. Then ℓ^* should satisfy the following inequalities:

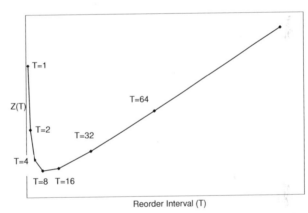

Fig. 3.1. Total average annual cost as a function of the reorder interval.

$$Z(2^{\ell^*-1}T_L) > Z(2^{\ell^*}T_L) \leq Z(2^{\ell^*+1}T_L), \quad \ell^* \geq 1,$$
$$Z(2^{\ell^*}T_L) \leq Z(2^{\ell^*+1}T_L), \quad \ell^* = 0.$$

For simplicity, we will assume that $\ell^* \geq 1$. The above inequalities can be interpreted in the following two equivalent ways:

1. Consider all the integers ℓ such that the cost corresponding to the reorder interval $2^{\ell}T_L$ is no greater than the cost corresponding to a reorder interval of length $2^{\ell+1}T_L$. That is, $Z(2^{\ell}T_L) \leq Z(2^{\ell+1}T_L)$. Then ℓ^* is the smallest among these values.

   For example, in Figure 3.1, possible values of ℓ such that $Z(2^{\ell}T_L) \leq Z(2^{\ell+1}T_L)$ are $3, 4, 5, \ldots$, and $\ell^* = 3$.

2. Consider all the integers $\ell \geq 1$ such that the cost corresponding to the reorder interval $2^{\ell}T_L$ is strictly lower than the cost corresponding to a reorder interval of length $2^{\ell-1}T_L$. Then ℓ^* is the largest among these values. Note that if $Z(2^{\ell-1}T_L) \leq Z(2^{\ell}T_L)$ for all $\ell \geq 1$, then the optimal PO2 reorder interval is equal to T_L. In this case, the cost function is increasing among the feasible values for T. Thus, the optimal solution occurs at the lowest feasible value, T_L.

   For example, in Figure 3.1, possible values of ℓ such that $Z(2^{\ell}T_L) \leq Z(2^{\ell-1}T_L)$ are $1, 2$, and 3. The largest among them is 3, so $\ell^* = 3$.

We will use the first statement to compute ℓ^*. Thus, ℓ^* is the smallest non-negative integer (including zero) such that

$$Z(2^{\ell^*}T_L) \leq Z(2^{\ell^*+1}T_L).$$

This relationship implies the following:

$$\frac{K}{2^{\ell^*}T_L} + g(2^{\ell^*}T_L) \leq \frac{K}{2^{\ell^*+1}T_L} + g(2^{\ell^*+1}T_L),$$

$$\frac{K}{2^{\ell^*}T_L}\left(1 - \frac{1}{2}\right) \leq g(2^{\ell^*}T_L)(2-1),$$

$$\frac{K}{g} \leq 2(2^{\ell^*}T_L)^2,$$

$$\sqrt{\frac{K}{2g}} \leq 2^{\ell^*}T_L,$$

$$\Rightarrow \frac{1}{T_L}\sqrt{\frac{K}{2g}} \leq 2^{\ell^*}.$$

Consequently ℓ^* is the smallest non-negative integer such that

$$2^{\ell^*} \geq \frac{1}{T_L}\sqrt{\frac{K}{2g}} = \frac{1}{\sqrt{2}T_L}T_E, \tag{3.6}$$

where $T_E = \sqrt{\frac{K}{g}}$ is the optimal reorder interval for the EOQ model.

Let us consider an example. For a base planning period equal to one week, let us obtain the optimal PO2 policy for pencil boxes for the office supplies retailer example in Chapter 2.

The solution to this problem can be obtained as follows. Given $T_L = 1$ week, T_L is $1/52$ of a year. From the solution to the example in Chapter 2, $T_E = 0.196$ years. Therefore, the optimal power-of-two multiple 2^{ℓ^*} is the smallest non-negative integer such that

$$2^{\ell^*} \geq \frac{1}{\sqrt{2} T_L} T_E = \frac{1}{\sqrt{2}(1/52)}(0.196) = 7.21,$$

and hence $2^{\ell^*} = 8$. The optimal PO2 reorder interval is equal to $2^{\ell^*} T_L = (8)(1/52) = 0.154$ years $= 8$ weeks.

3.1.2.1 Cost for the Optimal PO2 Policy

In the previous subsection we stated that the optimal PO2 reorder interval is the largest positive integer $\ell \geq 1$ such that the cost for a reorder interval $T = 2^{\ell} T_L$ is lower than the cost when $T = 2^{\ell-1} T_L$, or $T = T_L$. Thus, ℓ^* is also the largest non-negative integer such that

$$2^{\ell^*} \leq \frac{1}{T_L}\sqrt{\frac{2K}{g}} = \frac{\sqrt{2}}{T_L} T_E. \tag{3.7}$$

Equations (3.6) and (3.7) imply that

$$\frac{1}{\sqrt{2} T_L} T_E \leq 2^{\ell^*} \leq \frac{\sqrt{2}}{T_L} T_E,$$

or

$$\frac{1}{\sqrt{2}} T_E \leq 2^{\ell^*} T_L \leq \sqrt{2} T_E.$$

Therefore the optimal PO2 reorder interval is at least $\frac{1}{\sqrt{2}} T_E = 0.707 T_E$ but no higher than $\sqrt{2} T_E = 1.414 T_E$.

Since $Z(T)$ is a convex function of T, for any $T_1 \leq T_2 \leq T_3$, $Z(T_2) \leq \max\{Z(T_1), Z(T_3)\}$. As a result,

$$Z\left(2^{\ell^*} T_L\right) \leq \max\left\{Z\left(\sqrt{2} T_E\right), Z\left(\frac{1}{\sqrt{2}} T_E\right)\right\}.$$

Next we show that the costs corresponding to reorder intervals αT_E and $\frac{1}{\alpha} T_E$ are equal. Furthermore, this cost is equal to $\frac{1}{2}(\alpha + \frac{1}{\alpha}) Z(T_E)$ when either of these reorder intervals is put into effect. Note that

$$Z(\alpha T_E) = \frac{K}{\alpha T_E} + g(\alpha T_E).$$

For $T_E = \sqrt{\frac{K}{g}}$,

$$Z(\alpha T_E) = \frac{K}{\alpha\sqrt{\frac{K}{g}}} + g\left(\alpha\sqrt{\frac{K}{g}}\right)$$

$$= \frac{1}{\alpha}\sqrt{Kg} + \alpha\sqrt{Kg}$$

$$= \left(\alpha + \frac{1}{\alpha}\right)\sqrt{Kg} = \frac{1}{2}\left(\alpha + \frac{1}{\alpha}\right)Z(T_E).$$

Similarly $Z\left(\frac{1}{\alpha}T_E\right) = \frac{1}{2}\left(\alpha + \frac{1}{\alpha}\right)Z(T_E)$. Therefore, the cost of using a reorder interval which is $\sqrt{2}T_E$ or $\frac{1}{\sqrt{2}}T_E$ is the same and is equal to

$$\frac{1}{2}\left(\sqrt{2} + \frac{1}{\sqrt{2}}\right)Z(T^*) \approx (1.06)Z(T^*)$$

since $\frac{1}{2}(\sqrt{2} + \frac{1}{\sqrt{2}}) \approx 1.06$. Therefore, the worst possible cost for the PO2 policy is at most 6% greater than the optimal cost.

Let us now compare the cost of the optimal PO2 policy (excluding the purchasing cost) computed in our earlier example with that of the optimal policy computed in Chapter 2.

The sum of the average annual holding and order placement costs by following the optimal policy is equal to

$$Z(T_E) = \frac{K}{T_E} + \frac{1}{2}IC\lambda T_E = \frac{10}{0.196} + \frac{1}{2}(0.2)(5)(520)(0.196) = \$101.98.$$

Similarly, the sum of the annual holding and order placement costs incurred by following the optimal PO2 policy is equal to

$$Z(0.154) = \frac{10}{0.154} + \frac{1}{2}(0.2)(5)(520)(0.154) = \$104.98.$$

Therefore, the PO2 policy is more expensive than the optimal policy by

$$\frac{104.98 - 101.98}{101.98} = 2.94\%.$$

3.2 Serial Systems

The simplest multi-stage extension to the EOQ problem is a serial system consisting of n stages arranged linearly. Figure 3.2 shows an example of a serial system. Observe that each stage except stage n has exactly one predecessor stage and that each stage except stage 1 has exactly one successor stage. Our goal is to determine how to determine the reorder intervals for each of the n stages.

Fig. 3.2. Serial system.

In the following subsection we introduce some nomenclature that will facilitate our discussion.

3.2.1 Assumptions and Nomenclature

Conceptually we can represent a serial system as a directed graph. The stages form the nodes and the flow of the product is along the arcs of this graph. Let $N(G)$ and $A(G)$ denote the node set and the arc set, respectively. That is,

$$N(G) = \{1, \ldots, n\}$$

and

$$A(G) = \{(n, n-1), \ldots, (2, 1)\}.$$

For any stage i, λ_i represents the demand rate, measured in units per year. Since stage i could use several components produced at stage $i+1$, λ_i could be different for each stage. There could be demand for components completing a subset of the stages, too. However, for simplicity, we will assume λ_i is the same for all the stages. In other words, $\lambda_i = \lambda$ for all $i \in N(G)$. We assume there exists a base planning period, T_L, measured in years. All the reorder intervals must be power-of-two multiples of T_L. That is, if T_i represents the reorder interval for stage i, then

$$T_i = 2^{\ell_i} T_L, \quad \ell_i \in \{0, 1, 2, \ldots\}, \quad i = 1, \ldots, n.$$

There are many policies within the class of power-of-two policies that we could consider. We will consider only policies that are both *nested* and stationary. By a nested policy we mean that if stage i places an order at time t, then stages 1 through $i-1$ also place an order at the same time. Thus when stage n orders, all succeeding stages order and the system restarts its inventory planning cycle. By stationary we mean the reorder intervals do not change over time. The reorder interval for stage i is always T_i.

Nested policies have two advantages. First, for the same order frequency and order sizes per year, a nested policy generates lower annual holding cost than a non-nested policy for a serial system; we discuss this fact in more detail later. Second, since the order placement at a stage in a nested policy leads to the order placement at all the successor stages, a nested policy prevents buildup of work-in-process inventory.

We will consider two types of costs, the fixed ordering cost and the holding cost. Let K_i be the fixed cost at stage i. The average annual fixed cost of placing orders is computed in the same manner as for the EOQ model and is equal to K_i/T_i for stage i. The computation of the annual holding cost at each stage could also be done in a manner similar to the one used in the development of the EOQ model. Specifically, the average annual stage-i holding cost could be obtained by multiplying the average on-hand stock by the on-hand holding cost rate h_i. However, as we will see, this approach is difficult to use for complex networks including serial systems. Therefore, we will use a different but equivalent method to compute the average annual holding cost.

In the new approach, the average annual holding cost at stage i is computed as the product of the average *echelon holding cost* and *echelon stock*. The echelon holding cost, denoted by h'_i, is a measure of the value added *only* due to the activity at stage i. On the other hand, the conventional holding cost h_i is a measure of the *total* value added up to and including stage i. Therefore, the echelon holding cost for stage i, h'_i, is equal to the difference of the conventional holding cost for stage i, h_i, and conventional holding cost for the predecessor stage $i+1$, h_{i+1}. That is, $h'_i = h_i - h_{i+1}$. For stage n, the echelon and conventional holding costs are equal, that is, $h_n = h'_n$. If we add the echelon holding costs for stages $j = i, i+1, \ldots, n$, we get

$$h'_i + h'_{i+1} + \cdots + h'_n = (h_i - h_{i+1}) + (h_{i+1} - h_{i+2}) + \cdots + (h_{n-1} - h_n) + (h_n),$$

where, for stage n, we substitute $h'_n = h_n$. The right-hand side telescopes to h_i. Thus, $h_i = \sum_{j=i}^{n} h'_j$. This expression is intuitive; it says that the total value added up through stage i (of which h_i is the measure) is equal to the sum of the incremental values added in stages n through i (of which h'_i is the measure). Note that both the on-hand and echelon holding costs are measured in $ per unit per year.

Let us examine a three-stage serial system. Let the on-hand (conventional) holding costs be $h_1 = 5$, $h_2 = 4$, $h_3 = 1$. Let us compute the echelon holding costs. For stage 3,

$h'_3 = h_3 = 1$. For stages 1 and 2, we will use the relation $h'_i = h_i - h_{i+1}$. Therefore, for stage 2, $h'_2 = h_2 - h_3 = 3$ and for stage 1, $h'_1 = h_1 - h_2 = 1$.

We define the *echelon stock* for stage i to be the sum of the on-hand inventories at stages 1 through i. For example, if stage 1 has 2 units, stage 2 has 3 units, and stage 3 has 1 unit, then the echelon stocks for stage 3 and stage 2 are equal to 6 units and 5 units, respectively. The echelon stock for stage 1 is equal to its on-hand stock, 2 units.

Figure 3.3 shows both the on-hand and echelon stocks for a two-stage system. Observe that the echelon stock graphs for both the stages exhibit the familiar saw-tooth pattern. On the other hand, the graphs for the on-hand stock are not both saw-tooth in nature. Consequently, for more complex systems it is not always easy to calculate the average on-hand stock. However, the graph of echelon stock is always of the saw-tooth form no matter how complex the system. Hence it is easier to compute the average echelon stock compared to the average on-hand stock. But do the two approaches yield the same holding costs? The answer is yes. We show the equivalence between the two approaches for a two-stage system where $T_1 = \frac{1}{2}T_2$.

First, we compute the average annual holding cost using the average on-hand inventory and on-hand inventory holding cost. For stage 2, consider the cy-

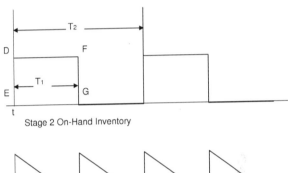

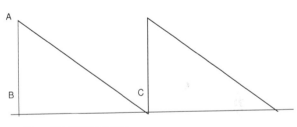

Fig. 3.3. Evolution of on-hand and echelon stocks over time for a two-stage system.

cle that starts at time t, as shown in Figure 3.3. At time t, an order of size λT_2 is placed. Since nested policies are followed, stage 1 also places an order of size $\lambda T_1 = \frac{1}{2}\lambda T_2$ with stage 2 at time t. Assuming the lead time is zero, the order sent from stage 2 arrives immediately at stage 1. Consequently, the leftover inventory at stage 2 at time t is $\lambda T_2 - \frac{1}{2}\lambda T_2 = \frac{1}{2}\lambda T_2$. Now at time $t + T_1$, stage 1 places another order of size $\lambda T_1 =$

$\frac{1}{2}\lambda T_2$ with stage 2, which again is sent and received immediately. This depletes all the remaining inventory at stage 2 and so the on-hand inventory at stage 2 is zero in the second half of the cycle. Therefore, the average annual inventory holding cost at stage 2 is

$$(h_2)\frac{\text{Area Under the Inventory Curve DEFG}}{\text{Length of the Cycle}} = (h_2)\frac{(\frac{1}{2}\lambda T_2)(T_1)}{T_2} = \frac{1}{2}h_2\lambda T_1.$$

For stage 1, the computation of the average annual inventory cost is the same as for the EOQ model and is equal to $\frac{1}{2}h_1\lambda T_1$. Therefore, the total average annual inventory cost for the two stages is

$$\frac{1}{2}h_2\lambda T_1 + \frac{1}{2}h_1\lambda T_1 = \frac{1}{2}(h_1 + h_2)\lambda T_1. \tag{3.8}$$

Next we compute the average annual cost of holding inventory using the echelon stocks and echelon holding costs. Figure 3.3 also shows how the echelon stock at stage 2 changes over time. Once again, let us consider the cycle that starts at time t. Even though half of the inventory that stage 2 received at time t is sent to stage 1 immediately, it still remains part of the echelon stock for stage 2. Thus, the echelon stock for stage 2 at time t is equal to λT_2. Similarly, at time $t + T_1$ the remaining on-hand inventory at stage 2 is sent to stage 1. But the on-hand stock at stage 1 is part of the echelon stock for stage 2 and therefore the echelon stock at $t + T_1$ for stage 2 is equal to λT_1. The echelon stock declines at the rate λ. The annual holding cost for stage 2 is thus equal to

$$(h'_2)\frac{\text{Area of triangle ABC}}{\text{Length of the Cycle}} = (h'_2)\frac{\frac{1}{2}(\lambda T_2)(T_2)}{T_2} = \frac{1}{2}h_2\lambda T_2$$

since $h'_2 = h_2$. For stage 1, the annual holding cost is equal to $\frac{1}{2}h'_1\lambda T_1 = \frac{1}{2}(h_1 - h_2)\lambda T_1$ since $h'_1 = h_1 - h_2$. The sum of the average annual holding costs for stages 1 and 2 is equal to

$$
\begin{aligned}
&= \frac{1}{2}h_2\lambda T_2 + \frac{1}{2}(h_1 - h_2)\lambda T_1 \\
&= \frac{1}{2}h_2\lambda(2T_1) + \frac{1}{2}(h_1 - h_2)\lambda T_1 \\
&= \frac{1}{2}(h_1 + h_2)\lambda T_1,
\end{aligned}
$$

which is the same as (3.8).

Theorem 3.1. *Nested policies are optimal for n-stage serial systems.*

Let us see why this is the case for a two-stage system. In Figure 3.4, suppose an order is placed at stage 2 at time t'. Furthermore, let t be the first time following t' at which stage 1 places an order. Consider another plan that differs from the first one only in that the order placed at stage 2 is moved from t' to t. All other ordering decisions remain the same. The number of setups is the same in both plans; however, the holding cost at the second stage is lower in the second plan by an amount equal to $h_2 \lambda (t' - t)$.

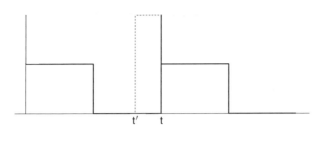

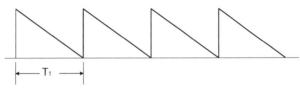

Fig. 3.4. Nested versus non-nested policies for a two-stage system.

Having laid the background, we will develop the model in the following subsection.

3.2.2   A Mathematical Model for Serial Systems

For a serial system, the mathematical model is as follows.

$$(P) \quad \min \sum_{i=1}^{n} \left\{ \frac{K_i}{T_i} + g_i T_i \right\}$$

subject to

$$T_i = 2^{\ell_i} T_L, \quad \ell_i \in \{0, 1, 2, \dots\},$$

$$T_i \geq T_{i-1} \geq 0,$$

where $g_i = \frac{1}{2} \lambda h_i'$. The above formulation has two types of constraints. The first type of constraint $(T_i = 2^{\ell_i} T_L)$ arises from the restriction of a PO2 policy. The second type of constraint $(T_i \geq T_{i-1})$ is necessary to ensure nestedness. Given nestedness, an order placed by stage i always requires $i - 1$ to place an order. As a result, the reorder interval at stage i cannot be less than the reorder interval at stage $i - 1$ $(T_i \geq T_{i-1})$.

To solve problem (P), we consider the following *relaxation*, that is, a problem with fewer constraints:

$$(RP) \quad \min \sum_{i=1}^{n} \left\{ \frac{K_i}{T_i} + g_i T_i \right\}$$

$$T_i \geq T_{i-1} \geq 0.$$

In problem (RP), the reorder intervals are not restricted to be power-of-two multiples of the base planning interval. As a result, (RP) is easier to solve. Our approach will be to first obtain the solution to (RP) and then use this solution to solve (P).

Let us introduce some additional definitions. Define a subgraph G of G' to consist of a node set $N(G)$ and an arc set $A(G)$ such that $N(G)$ and $A(G)$ are subsets of the sets of nodes and arcs of G', that is, $N(G) \subset N(G')$ and $A(G) \subset A(G')$, respectively. Further, an arc (i, j) is an element of $A(G)$ if and only if the nodes i and j are in $N(G)$. This condition is a *sensibility* condition that ensures that there are no arcs in G whose beginning and ending nodes are not contained in $N(G)$. In Figure 3.5, G is a subgraph of graph G'. Also, $N(G) = \{1, 2, 3\}$ and $A(G) = \{(3, 2), (2, 1)\}$.

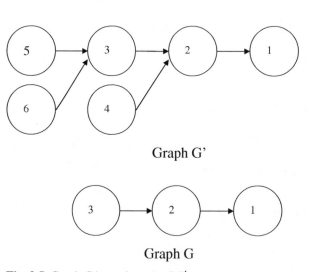

Graph G'

Graph G

Fig. 3.5. Graph G is a subgraph of G'.

Now, suppose G_1 and G_2 are subgraphs of G'. We say that G_1 and G_2 are *ordered by precedence* if there does not exist an arc $(j, i) \in A(G')$ such that $j \in N(G_1)$ and $i \in N(G_2)$. In other words, if G_1 and G_2 are ordered by precedence, then no arc can start from a node in $N(G_1)$ and end at a node in $N(G_2)$; any arcs that connect G_1 and G_2 must originate in G_2. The upper portion of Figure 3.6 shows two subgraphs G_1 and G_2 for graph G' shown in Figure 3.5. Subgraphs G_1 and G_2 are ordered by precedence since the only arc that connects the two subgraphs goes from node 3, belonging to higher-indexed G_2, to node 2, belonging to lower-indexed G_1.

The lower portion of Figure 3.6 also shows two subgraphs G_1 and G_2 for graph G' shown in Figure 3.5. The only connecting arc between the two subgraphs starts from

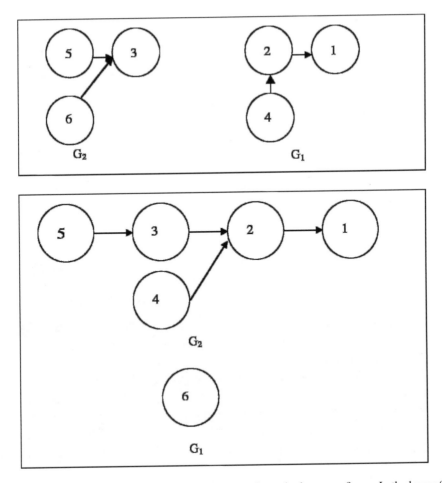

Fig. 3.6. The subgraph pair (G_1, G_2) is ordered by precedence in the upper figure. In the lower figure, the subgraph pair (G_1, G_2) is not ordered by precedence.

node 6 and ends at node 3. Since node 6 belongs to lower-indexed G_1 and node 3 belongs to higher-indexed G_2, subgraphs G_1 and G_2 are *not* ordered by precedence.

This definition can be extended to include multiple subgraphs. A collection of subgraphs $G_1, \ldots, G_M$ are *ordered by precedence* if for $\ell, k \in \{1, 2, \ldots, M\}$ such that $\ell < k$, there does not exist a node $j \in N(G_\ell)$ and $i \in N(G_k)$ such that $(j, i) \in A(G')$. Once again, arcs that connect two subgraphs cannot originate in a subgraph with a lower index; they must always originate in a subgraph with a higher index when the subgraphs are ordered by precedence.

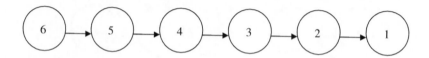

Fig. 3.7. Example of a serial system with 6 stages.

Next, we define the term *ordered partition*. Subgraphs $G_1, \ldots, G_M$ form an *ordered partition* of G' if

1. The subgraphs $G_1, G_2, \ldots, G_M$ are ordered by precedence.
2. Node subsets $N(G_i)$, $i = \{1, 2, \ldots, M\}$, form a partition of $N(G')$, that is, $\bigcup_{i=1}^{M} N(G_i) = N(G')$, and for any $i, j \in \{1, 2, \ldots, M\}, N(G_i) \cap N(G_j) = \emptyset$ for $i \neq j$. This means that all the nodes in $N(G')$ are split into M different sets $N(G_1)$, $N(G_2)$, $\ldots, N(G_M)$ in such a way that each node is in exactly one of these sets.

In general, for a serial system, if $G_1, \ldots, G_M$ is an ordered partition, then $N(G_M) = \{n_{M-1} + 1, \ldots, n\}$, $N(G_{M-1}) = \{n_{M-2} + 1, \ldots, n_{M-1}\}$, $\ldots$, $N(G_1) = \{1, \ldots, n_1\}$. Each subgraph contains a certain number of consecutive nodes.

Finally, a *directed cut* of a graph G_1 is an ordered partition of G_1, (G_1^-, G_1^+). As an example, consider the serial system in Figure 3.7. Suppose G_1 with node set $N(G_1) = \{3, 4, 5, 6\}$ is a subgraph. An ordered partition of $G_1 = G_1^- \cup G_1^+$, where the node set of G_1^- is $N(G_1^-) = \{3, 4\}$ and the node set of G_1^+ is $N(G_1^+) = \{5, 6\}$. A directed cut is just like any other ordered partition with one major distinguishing feature: the ordered partition has only two subgraphs.

We will now proceed to solving our problem. Consider a smaller version of the optimization problem (RP) that is defined on a given subgraph G_k of the serial system:

$$(RP_k) \quad \min_{\{T_i : i \in N(G_k)\}} \sum_{i \in N(G_k)} \left\{ \frac{K_i}{T_i} + g_i T_i \right\}$$

subject to

$$T_i \geq T_{i-1} \geq 0, \quad (i, i-1) \in A(G_k).$$

We further assume that the reorder interval for each stage in G_k is the same, that is, $T_i = T(k)$ for all $i \in N(G_k)$. Then (RP_k) becomes

$$(RP_k) \quad \min_{T(k)} \sum_{i \in N(G_k)} \left\{ \frac{K_i}{T(k)} + g_i T(k) \right\}$$

subject to

$$T(k) \geq 0,$$

or

$$(RP_k) \quad \min_{T(k)} \left\{ \frac{\sum_{i \in N(G_k)} K_i}{T(k)} + \left(\sum_{i \in N(G_k)} g_i \right) T(k) \right\}$$

subject to

$$T(k) \geq 0.$$

The above formulation is similar to that of the EOQ problem and its solution can be found by taking the derivative of the objective function and setting it equal to zero. The solution is

$$T(k) = \sqrt{\frac{\sum_{i \in N(G_k)} K_i}{\sum_{i \in N(G_k)} g_i}}.$$

To simplify, let

$$K(G_k) = \sum_{i \in N(G_k)} K_i \quad \text{and} \quad g(G_k) = \sum_{i \in N(G_k)} g_i.$$

Then

$$T(k) = \sqrt{\frac{K(G_k)}{g(G_k)}}.$$

How can we use the above observations? Suppose we generate an ordered partition of the serial system into M subgraphs such that the stages in a subgraph will have the same reorder interval. Then we can use the solution to (RP_k) to compute these reorder intervals. Further, suppose these reorder intervals are feasible in the sense that the reorder intervals are non-decreasing as we go upstream in the serial system. Will this solution be optimal to (RP)? If yes, then how do we generate this type of ordered partition? The answer to the first question is a qualified yes, and we discuss it in the following theorem. The answer to the second question, that is, the computational approach to generate the optimal partition, is discussed after the theorem.

Theorem 3.2. *Suppose there are M reorder intervals $T(1), \ldots, T(M)$. The necessary and sufficient conditions for the reorder intervals to be optimal for (RP) are*

(i) There exists an ordered partition $(G_1, \ldots, G_M)$ of G such that

$$T(k) = \sqrt{K(G_k)/g(G_k)},$$

(ii) $T(1) \le T(2) \le \cdots \le T(M)$ *(feasibility), and*

(iii) For any subgraph G_k, $k = 1, \ldots, M$, there does not exist a directed cut (G_k^-, G_k^+) of G_k for which

$$\frac{K(G_k^-)}{g(G_k^-)} < \frac{K(G_k^+)}{g(G_k^+)}.$$

Condition (i) states that every reorder interval in the optimal solution must be equal to the common reorder interval $\sqrt{K(G_k)/g(G_k)}$ of some subgraph G_k in an ordered partition. Condition (ii) is a feasibility condition that requires that the reorder intervals satisfy the constraints in problem (RP). The most important requirement in the above theorem is conveyed in Condition (iii) which characterizes the ordered partition that gives us the optimal solution to (RP). According to it, if we are able to partition a subgraph G_k further into two subgraphs (G_k^-, G_k^+) in such a way that the reorder interval for G_k^- is less than that for G_k^+, then the solution will not be optimal. The solution will be optimal if and only if we are unable to generate such a directed cut for any subgraph in the optimal partition.

We now use this condition to design an algorithm that will give us an optimal solution to (RP).

3.2.3 Algorithm to Obtain an Optimal Solution to (RP)

Step 1: Set $k = 1, i = 1$. Set $N(G_k) = \{1\}$.

Step 2: Set $i \leftarrow i + 1$. If $i \le n$, go to Step 3. Otherwise, we are done.

Step 3: If $\frac{K(G_k)}{g(G_k)} \le \frac{K_i}{g_i}$, then $k \leftarrow k + 1$ and $N(G_k) = \{i\}$. Go to Step 2. Otherwise, $N(G_k) \leftarrow N(G_k) \cup \{i\}$. If $k > 1$, go to Step 4; otherwise, return to Step 2.

Step 4: Set $l = k$.

Step 5: If $\frac{K(G_{l-1})}{g(G_{l-1})} \le \frac{K(G_l)}{g(G_l)}$, then set $k = l$ and go to Step 2. Otherwise, go to Step 6.

Step 6: $N(G_{l-1}) \leftarrow N(G_{l-1}) \cup N(G_l)$ and $l \leftarrow l - 1$. If $l > 1$, go to Step 5. Otherwise, $k = l$ and go to Step 2.

We will refer to this algorithm as the string algorithm.

The crucial step in the above algorithm is Step 3. In this step, we consider adding node i to subgraph G_k. If the square of the reorder interval for stand-alone node i, $\frac{K_i}{g_i}$, is greater than or equal to that of the subgraph G_k, $\frac{K(G_k)}{g(G_k)}$, then node i is not included in the subgraph G_k, and it becomes the first node in a new subgraph. This step ensures feasibility of the reorder intervals. Since node $i - 1$ belongs to subgraph G_k, this step ensures that $T_i \ge T_{i-1}$. On the other hand, when $\frac{K_i}{g_i} < \frac{K(G_k)}{g(G_k)}$, not adding node i to G_k

would violate the feasibility condition. Therefore, we add node i to G_k. However, this changes the reorder interval for G_k, which now is lower than when i was not part of it. As a result, the reorder interval for G_k may now become lower than that for G_{k-1}. Thus, we may have to merge some of the subgraphs $G_1, G_2, \ldots, G_k$, which we do in Steps 4–6, to ensure that the reorder intervals remain feasible.

The following example shows an implementation of the above algorithm. The production of an item requires it to go through five processes sequentially. The fixed cost of production at each stage is obtained by multiplying the setup time with the labor cost per unit time. The following table contains the estimates of the fixed cost of production along with the holding cost per year.

Process No.	K_i	h_i
1	10	0.6
2	12	0.5
3	25	0.35
4	7	0.1
5	2	0.05

If the annual demand is 500 units, let us obtain an optimal partition and use it to find a solution to problem (RP). To solve this problem, we first obtain the echelon holding costs and use them to calculate g_i for each process.

Process No.	h_i'	g_i
1	0.1	25
2	0.15	37.5
3	0.25	62.5
4	0.05	12.5
5	0.05	12.5

Next, we apply the string algorithm.

Iteration 1:

Step 1: Set $k = 1$, $i = 1$, $N(G_1) = \{1\}$.
Step 2: Set $i = 2$.
Step 3: We calculate $\frac{K(G_1)}{g(G_1)} = \frac{10}{25} = 0.4$ and compare it with $\frac{K_2}{g_2} = \frac{12}{37.5} = 0.32$. Since $\frac{K(G_1)}{g(G_1)} > \frac{K_2}{g_2}$, we add process 2 to $N(G_1)$. Since $k = 1$, we do not need to go to Step 4.

Iteration 2:

Step 2: Set $i = 3$.

Step 3: We calculate $\frac{K(G_1)}{g(G_1)} = \frac{10+12}{25+37.5} = 0.352$ and compare it with $\frac{K_3}{g_3} = \frac{25}{62.5} = 0.4$. Since $\frac{K(G_1)}{g(G_1)} < \frac{K_3}{g_3}$, we do not add process 3 to $N(G_1)$. Instead, we create a new subgraph $N(G_2) = \{3\}$, and set $k = 2$.

Iteration 3:

Step 2: Set $i = 4$.

Step 3: We compute $\frac{K(G_2)}{g(G_2)} = \frac{25}{62.5} = 0.4$ and compare it with $\frac{K_4}{g_4} = \frac{7}{12.5} = 0.56$. Since $\frac{K(G_2)}{g(G_2)} < \frac{K_4}{g_4}$, we do not add process 4 to $N(G_2)$. Instead, we create a new subgraph $N(G_3) = \{4\}$, and set $k = 3$.

Iteration 4:

Step 2: Set $i = 5$.

Step 3: We compute $\frac{K(G_3)}{g(G_3)} = 0.56$ and compare it with $\frac{K_5}{g_5} = \frac{2}{12.5} = 0.16$. Since $\frac{K(G_3)}{g(G_3)} > \frac{K_5}{g_5}$, we add process 5 to $N(G_3)$ and go to Step 4.

Step 4: Set $l = 3$.

Step 5: Recall that $\frac{K(G_2)}{g(G_2)} = 0.4$. Next compute $\frac{K(G_3)}{g(G_3)} = \frac{K_4+K_5}{g_4+g_5} = \frac{9}{25} = 0.36$. Since $\frac{K(G_2)}{g(G_2)} > \frac{K(G_3)}{g(G_3)}$, we go to Step 6.

Step 6: Set $N(G_2) = N(G_2) \cup N(G_3) = \{3,4,5\}$. Also, set $l = 2$ and go to Step 5 in the next iteration.

Iteration 5:

Step 5: We compute $\frac{K(G_2)}{g(G_2)} = \frac{K_3+K_4+K_5}{g_3+g_4+g_5} = \frac{34}{87.5} = 0.39$ and compare it with $\frac{K(G_1)}{g(G_1)} = 0.352$ as computed in Iteration 2. Since $\frac{K(G_2)}{g(G_2)} > \frac{K(G_1)}{g(G_1)}$, we set $k = 2$ and go to Step 2. The algorithm terminates since $i+1 = 6 > n$.

Therefore, the ordered partition is $N(G_1) = \{1,2\}, N(G_2) = \{3,4,5\}$. The reorder intervals that solve the relaxed problem are as follows,

$$T(1) = \sqrt{\frac{K(G_1)}{g(G_1)}} = \sqrt{(0.35)} = 0.59 \text{ years, and}$$

$$T(2) = \sqrt{\frac{K(G_2)}{g(G_2)}} = \sqrt{(0.39)} = 0.62 \text{ years.}$$

Suppose now that we have an optimal solution to problem (RP). That is, we have an optimal ordered partition $G_1, \ldots, G_M$ of the serial system such that the reorder intervals for all stages in G_k are equal to $T(k)$, $k = 1, \ldots, M$. To find the optimal PO2 solution for the serial system, we solve the following problem for each of the subgraphs G_k, $k = 1, \ldots, M$:

$$\min_{\ell \in \{0,1,\ldots\}} \sum_{i \in N(G_k)} \left\{ \frac{K_i}{T_P(k)} + g_i T_P(k) \right\}$$

subject to

$$T_P(k) = 2^\ell T_L, \; \ell = 0, 1, 2, \ldots.$$

Observe that the above formulation is equivalent to the single-stage problem we examined in Section 3.1.2. Since its solution can be found using a first differencing approach as before, the optimal value for ℓ is the smallest non-negative integer for which

$$2^\ell \geq \frac{1}{\sqrt{2}T_L} \sqrt{\frac{K(G_k)}{g(G_k)}} = \frac{1}{\sqrt{2}T_L} T(k). \tag{3.9}$$

It can be shown that this solution is optimal for problem (P), which can be found in Muckstadt and Roundy [240]. To summarize, the optimal solution to (P) can be found using a 3-step procedure.

1. Find the optimal partition using the string algorithm.
2. Compute $T(k) = \sqrt{K(G_k)/g(G_k)}$, the solution to (RP).
3. Find the optimal PO2 reorder interval using (3.9).

Assuming a base planning period of 1 week, let us now compute the optimal PO2 production policy for the earlier example. For subgraph G_1, the optimal power-of-two multiple 2^ℓ corresponds to the smallest ℓ such that $2^\ell \geq \frac{1}{\sqrt{2}T_L} T(1) = \frac{1}{\sqrt{2}(1/52)}(0.59) = 21.7$. Thus, $\ell^* = 5$ and the optimal PO2 reorder interval $T_P^* = (2^5)(1/52) = 0.62$ years.

Similarly, for subgraph G_2, the optimal power-of-two multiple 2^ℓ corresponds to the smallest ℓ such that $2^\ell \geq \frac{1}{\sqrt{2}T_L} T(2) = \frac{1}{\sqrt{2}(1/52)}(0.62) = 22.8$. Thus, $\ell^* = 5$ again and the optimal PO2 reorder interval $T_P^* = (2^5)(1/52) = 0.62$ years. Thus all stages will follow the same plan in the power-of-two solution even though not all the reorder intervals assumed the same value in the solution to the relaxed problem.

3.3 Multi-Echelon Distribution Systems

In this section, we consider a simple distribution system consisting of one central warehouse and multiple regional warehouses. The central warehouse routinely replenishes inventory to the regional warehouses. Assume that shipments made from the central warehouse arrive instantaneously at the regional warehouses. The inventory at the central warehouse is replenished by an outside supplier. The question we are interested in answering is how often should the inventory at the regional warehouses and central warehouse be replenished.

Suppose there are a total of n regional warehouses. Suppose further that the demand rate at warehouse i is λ_i units/year, and that holding costs are charged in proportion to echelon stock. At the central warehouse, the echelon stock is equal to the total of on-hand stock at the central warehouse as well as at the n regional warehouses, that is, all the inventory in the system. At a regional warehouse, the echelon stock is equal to its on-hand stock. Let h_i' be the echelon holding cost rate for warehouse i, $i = 0, 1, \ldots, n$, where 0 denotes the central warehouse. For the central warehouse, $h_0 = h_0'$, where h_0 is the conventional on-hand inventory holding cost rate. If h_i is the conventional on-hand inventory cost rate at regional warehouse i, then $h_i' = h_i - h_0$. The reason we charge costs proportional to the echelon stock is the same as for a serial system; the graph of echelon stock follows a saw-tooth pattern which makes it easy to compute average inventory levels over time. A fixed cost of K_i is incurred for each order placed by warehouse i.

As we did for the serial system, we assume that the inventory management policy is both nested and stationary. We assume that a base planning period exists, which we denote by T_L. The reorder interval at each location is a power-of-two multiple of T_L (measured in years). That is, the reorder interval at warehouse i is

$$T_i = 2^{\ell_i} T_L, \quad \ell_i \in \{0, 1, 2, \ldots\}, \quad i = 0, \ldots, n.$$

Let us now state the mathematical model and the solution algorithm.

3.3.1 A Mathematical Model for Distribution Systems

The mathematical model for a distribution system is

$$(P) \quad \min \sum_{i=0}^{n} \left\{ \frac{K_i}{T_i} + g_i T_i \right\}$$

subject to

$$T_i = 2^{\ell_i} T_L, \; \ell_i \in \{0, 1, \dots\},$$
$$T_0 \geq T_i \geq 0, \quad i = 1, \dots, n,$$

where $g_i = \frac{1}{2}\lambda_i h_i'$. The constraint $T_0 \geq T_i$ is necessary to ensure nestedness of the policy.

As was the case for the serial system problem, this problem is also solved in a two-step process. We first solve a relaxation to the above problem in which we ignore the power-of-two restriction of the base planning period. Once we find the solution to the relaxed problem, we use it to find the optimal solution to (P). We will now examine the relaxation and then show how the optimal solution to (P) can be found.

3.3.1.1   Relaxed Problem

The relaxation to (P) is

$$(RP) \quad \min \sum_{i=0}^{n} \left\{ \frac{K_i}{T_i} + g_i T_i \right\}$$

subject to

$$0 \leq T_i \leq T_0, \qquad i = 1, \dots, n.$$

Since (RP) has fewer constraints than (P), the optimal cost to (RP) provides a lower bound estimate on the objective function of (P).

Our approach to finding the solution to (RP) remains the same as for a serial system. We want to find an *ordered partition* of the graph G with node set $N(G) = \{0, 1, 2, \dots, n\}$ and arc set $A(G) = \{(0, 1), (0, 2), \dots, (0, n)\}$. The optimal solution to (RP) has the following form. All the regional warehouses can be divided into two categories. The regional warehouses in the first category share a common reorder interval with the central warehouse. The remaining regional warehouses lie in the second category. These warehouses follow their natural reorder intervals, that is, unconstrained reorder intervals $\sqrt{\frac{K_i}{g_i}}$.

Without any loss of generality, suppose the regional warehouses are numbered such that $\frac{K_1}{g_1} \leq \frac{K_2}{g_2} \leq \dots \leq \frac{K_n}{g_n}$. Let $\mathcal{C}$ be the set of nodes corresponding to the regional warehouses sharing a common reorder interval with the central warehouse. If $\frac{K_n}{g_n} \leq \frac{K_0}{g_0}$, then $\mathcal{C} = \emptyset$ and $N(G)$ is partitioned such that $N(G_i) = \{i\}$ and $T_i = \sqrt{\frac{K_i}{g_i}}$, $i = 0, 1, \dots, n$. Therefore, all the warehouses follow their own reorder interval.

If $\mathcal{C}$ is not an empty set, let $\bar{i}$ be the smallest value of i for which

$$\frac{\sum_{j=\bar{i}+1}^{n} K_j + K_0}{\sum_{j=\bar{i}+1}^{n} g_j + g_0} < \frac{K_{\bar{i}}}{g_{\bar{i}}}, \tag{3.10}$$

and $\mathcal{C} = \{\bar{i}, \bar{i}+1, \ldots, n\}$. Then the ordered partition is $N(G_1) = \{1\}, \ldots, N(G_{\bar{i}-1}) = \{\bar{i} - 1\}$, $N(G_{\bar{i}}) = \{\bar{i}, \ldots, n, 0\}$. Note that $\mathcal{C}$ is always a subset of $N(G_{\bar{i}})$ and $N(G_{\bar{i}}) = \mathcal{C} \cup \{0\}$.

The formation of $\mathcal{C}$ is based on the following simple rule: add a node to $\mathcal{C}$ only when leaving it alone would result in infeasibility. Suppose $\mathcal{C} = \{j+1, \ldots, n\}$ and we are considering whether or not to add node j to $\mathcal{C}$. If node j is not added to $\mathcal{C}$, then the reorder interval of the subgraph containing node 0 is equal to

$$T_0 = \sqrt{\frac{K_{j+1} + K_{j+2} + \cdots + K_n + K_0}{g_{j+1} + g_{j+2} + \cdots + g_n + g_0}}$$

and the reorder interval of node j would equal $T_j = \sqrt{\frac{K_j}{g_j}}$. If $\frac{K_j}{g_j}$ were greater than $\frac{K_{j+1} + K_{j+2} + \cdots + K_n + K_0}{g_{j+1} + g_{j+2} + \cdots + g_n + g_0}$, then T_j would be greater than T_0 which would not be feasible. As a result, we add node j also to $\mathcal{C}$. This ensures that nodes j and 0 have the same reorder interval thus preserving feasibility. We keep adding such nodes until we hit upon the node $\bar{i}$ which is the last node that if left alone would have a higher reorder interval than node 0. All nodes having an index less than $\bar{i}$, that is, nodes $1, 2, \ldots, \bar{i}-1$, have reorder intervals that are equal to their natural reorder intervals.

We next show that the above approach results in an optimal solution by demonstrating that the resulting solution satisfies Theorem 3.2.

1. The subgraphs $G_1, G_2, \ldots$ form an ordered partition of G; these subgraphs must satisfy two conditions to form an ordered partition of G. First, arcs connecting any two subgraphs must start from a higher-indexed subgraph (ordering by precedence) and the set of nodes $N(G_1), N(G_2), \ldots, N(G_{\bar{i}})$ must form a partition to $N(G)$. Regarding the first requirement, since arcs go only from node 0 (which is in $G_{\bar{i}}$, the highest-indexed subgraph) to other nodes, the subgraphs are ordered by precedence. The second requirement is fulfilled because $\bigcup_{j=1}^{\bar{i}} N(G_j) = N(G)$ and each node belongs to exactly one subgraphs's node set.

2. The feasibility condition is satisfied by construction. Since the regional warehouses are indexed such that $\frac{K_1}{g_1} \le \frac{K_2}{g_2} \le \cdots \le \frac{K_n}{g_n}$, $T(1) = \sqrt{\frac{K_1}{g_1}} \le \cdots \le T(\bar{i}-1) = \sqrt{\frac{K_{\bar{i}-1}}{g_{\bar{i}-1}}}$. Using (3.10), $\bar{i}-1$ is the largest-indexed warehouse that does not satisfy the inequality in (3.10). Thus,

$$\frac{\sum_{j=\bar{i}}^{n} K_j + K_0}{\sum_{j=\bar{i}}^{n} g_j + g_0} = T(\bar{i}) \ge \frac{K_{\bar{i}-1}}{g_{\bar{i}-1}} = T(\bar{i}-1).$$

3. There is no directed cut of $G_{\bar{i}}$ for which $\frac{K(G_{\bar{i}}^-)}{g(G_{\bar{i}}^-)} < \frac{K(G_{\bar{i}}^+)}{g(G_{\bar{i}}^+)}$. This condition states that $G_{\bar{i}}$ cannot be split into two subgraphs that are ordered by precedence such that the subgraph that does not contain 0 has a $\frac{K}{g}$-ratio (and hence the reorder interval) that is lower than the $\frac{K}{g}$-ratio for the other subgraph. Our method of creating $G_{\bar{i}}$ automatically precludes such a possibility.

To summarize, the optimal solution is

$$T_0 = \sqrt{(\sum_{j=\bar{i}}^{n} K_j + K_0)/(\sum_{j=\bar{i}}^{n} g_j + g_0)}$$

$$= T_i, \quad i = \bar{i}, \bar{i}+1, \ldots, n,$$

$$T_i = \sqrt{\frac{K_i}{g_i}}, \quad i = 1, \ldots, \bar{i}-1,$$

when $\mathcal{C} \neq \emptyset$. If $\mathcal{C} = \emptyset$, then $T_i = \sqrt{\frac{K_i}{g_i}}, i = 0, 1, \ldots, n$.

Let us now illustrate this solution approach. Suppose the supply chain for spare parts of an automobile manufacturer consists of two echelons, 2 large national warehouses and 6 regional warehouses spread all over the US. Each part is stocked in exactly one national warehouse and in each of the 6 regional warehouses. The regional warehouses fulfill the orders placed by the dealerships in their region. The regional warehouses get their supplies for each part from the national warehouse that stocks them; the national warehouse places orders with an external supplier who is the manufacturer.

Brake pads see steady demand throughout the year (200 per year) at each of the regional warehouses. Given a base planning period of 1 week, let us determine the reorder intervals for brake pads at the appropriate national warehouse and the regional warehouses without imposing the power-of-two restriction on the reorder intervals. The remaining data are as follows:

Warehouse	K_i	h_i
0	500	10
1	100	15
2	125	12
3	110	20
4	150	18
5	75	20
6	90	17

We begin by computing the echelon holding cost at each warehouse h_i', the value of $g_i = \frac{1}{2}\lambda_i h_i'$, and $\frac{K_i}{g_i}$. Note that $\lambda_0 = \lambda_1 + \lambda_2 + \cdots + \lambda_6 = 6(200) = 1,200$ per year.

Warehouse	h'_i	g_i	$\frac{K_i}{g_i}$	Rank of $\frac{K_i}{g_i}$
0	10	6,000	0.083	
1	5	500	0.2	5
2	2	200	0.625	6
3	10	1000	0.11	2
4	8	800	0.19	4
5	10	1000	0.075	1
6	7	700	0.13	3

The new indexes of the regional warehouses are as follows: $1' \leftarrow 5$, $2' \leftarrow 3$, $3' \leftarrow 6$, $4' \leftarrow 4$, $5' \leftarrow 1$, $6' \leftarrow 2$.

Iteration 1:

Since $\frac{K_{6'}}{g_{6'}} > \frac{K_0}{g_0}$, $\mathcal{C}$ is not an empty set. Add node $6'$ to $\mathcal{C}$.

Iteration 2:

Next, we check if node $5'$ should be added to $\mathcal{C}$. Since $\frac{K_{5'}}{g_{5'}} = 0.2 > \frac{K_0 + K_{6'}}{g_0 + g_{6'}} = \frac{625}{6200} = 0.1$, we add node $5'$ to $\mathcal{C}$.

Iteration 3:

To check if node $4'$ should be added to $\mathcal{C}$, we compare $\frac{K_{4'}}{g_{4'}} = 0.19$ with $\frac{K_0 + K_{6'} + K_{5'}}{g_0 + g_{6'} + g_{5'}} = \frac{725}{6700} = 0.108$. Since $.19 > .108$, we add node $4'$ to $\mathcal{C}$.

Iteration 4:

To check if node $3'$ should be added to $\mathcal{C}$, we compare

$$\frac{K_{3'}}{g_{3'}} = 0.13 \quad \text{with} \quad \frac{K_0 + K_{6'} + K_{5'} + K_{4'}}{g_0 + g_{6'} + g_{5'} + g_{4'}} = \frac{875}{7500} = 0.117.$$

Since $.13 > .117$, we add node $3'$ to $\mathcal{C}$.

Iteration 5:

To check if node $2'$ should be added to $\mathcal{C}$, we compare

$$\frac{K_{2'}}{g_{2'}} = 0.11 \quad \text{with} \quad \frac{K_0 + K_{6'} + K_{5'} + K_{4'} + K_{3'}}{g_0 + g_{6'} + g_{5'} + g_{4'} + g_{3'}} = \frac{965}{8200} = 0.118.$$

Therefore, node $2'$ is not added to $\mathcal{C}$ and $\bar{i} = 3'$. Thus, $\mathcal{C} = \{3', 4', 5', 6'\}$. At this point, we revert back to the original indexes. Thus, with original indexes, $\mathcal{C} = \{6, 4, 1, 2\}$.

The reorder intervals are

$$T_0 = T_6 = T_4 = T_1 = T_2 = \sqrt{0.118} = 0.343 \text{ years}$$
$$T_3 = T_{2'} = \sqrt{0.11} = 0.33 \text{ years}$$
$$T_5 = T_{1'} = \sqrt{0.075} = 0.274 \text{ years}.$$

3.3.2 Powers-of-Two Solution

The solution to (RP) provides the basis for establishing the optimal solution to (P). Suppose we have found $\mathcal{C}$ using the rules discussed above. Define

$$K(G_{\bar{i}}) = \sum_{j=\bar{i}}^{n} K_j + K_0$$

and

$$g(G_{\bar{i}}) = \sum_{j=\bar{i}}^{n} g_j + g_0.$$

Then ℓ_0^*, the optimal power of two for node 0, is the smallest non-negative integer such that

$$2^{\ell_0^*} \geq \frac{1}{T_L} \left[\frac{K(G_{\bar{i}})}{2g(G_{\bar{i}})} \right]^{1/2}.$$

Since the reorder interval for every node in $G_{\bar{i}}$ is the same in the solution to (RP), their optimal powers-of-two solutions are equal, that is,

$$\ell_0^* = \ell_{\bar{i}}^* = \ell_{\bar{i}+1}^* = \cdots = \ell_n^*.$$

For a node i that is not in $\mathcal{C}$, ℓ_i^* is the smallest non-negative integer satisfying

$$2^{\ell_i^*} \geq \frac{1}{T_L} \left[\frac{K_i}{2g_i} \right]^{1/2}, \quad i = 1, \ldots, \bar{i} - 1.$$

To summarize, the steps for computing the optimal solution to (P) are stated in the following algorithm:

Step 1: Sort the $\frac{K_i}{g_i}$, $i = 1, \ldots, n$, from smallest to largest and renumber the regional warehouses accordingly such that warehouse number 1 is the one with the smallest value of $\frac{K_i}{g_i}$, warehouse number 2 is the one with the second-smallest value of $\frac{K_i}{g_i}$, and so on.

Step 2: Find $\bar{i}$ using condition (3.10). Define $\mathcal{C} = \{\bar{i}, \bar{i}+1, \ldots, n\}$.

Step 3: Calculate $K(G_{\bar{i}})$ and $g(G_{\bar{i}})$. Compute ℓ_0^* as the smallest non-negative integer that satisfies

$$2^{\ell_0^*} \geq \frac{1}{T_L} \left[\frac{K(G_{\bar{i}})}{2g(G_{\bar{i}})} \right]^{1/2}.$$

Then the optimal PO2 reorder interval for the central warehouse and regional warehouses with redefined indexes (in Step 1) $\bar{i}, \bar{i}+1, \ldots, n$ is $2^{\ell_0^*} T_L$.

Step 4: Set the PO2 reorder interval for regional warehouses with indexes $i = 1, 2, \ldots, \bar{i}-1$ as $2^{\ell_i^*} T_L$, where ℓ_i^* is the smallest non-negative integer that satisfies $2^{\ell_i^*} \geq \frac{1}{T_L} \left[\frac{K_i}{2g_i} \right]^{1/2}$.

Earlier we demonstrated Steps 1–2. We now show how to compute the optimal PO2 policy for the same system. Assume the base planning period is one week.

Earlier we found that nodes $3', 4', 5', 6'$, and 0 have a common reorder interval, that is, $\ell_0^*, \ell_{3'}^*, \ell_{4'}^*, \ell_{5'}^*, \ell_{6'}^*$ are equal. Specifically, ℓ_0^* is the smallest non-negative integer such that

$$2^{\ell_0^*} \geq \frac{1}{\sqrt{2}T_L} \sqrt{\frac{K(G_{\bar{i}})}{g(G_{\bar{i}})}} = \frac{1}{\sqrt{2}(1/52)}(0.343) = 12.61,$$

where $\sqrt{\frac{K(G_{\bar{i}})}{g(G_{\bar{i}})}}$ is 0.343 years. Thus, $2^{\ell_0^*} = 16$ and $\ell_0^* = 4$. The optimal power-of-two reorder interval for nodes $0, 3', 4', 5'$, and $6'$ is equal to $2^4(1/52) = 0.31$ years.

Similarly, for node $2'$, $\ell_{2'}^*$ is the smallest non-negative integer such that

$$2^{\ell_{2'}^*} \geq \frac{1}{\sqrt{2}T_L} \sqrt{\frac{K_{2'}}{g_{2'}}} = \frac{1}{\sqrt{2}(1/52)}(0.33) = 12.19,$$

where $\sqrt{\frac{K_{2'}}{g_{2'}}} = 0.33$ years. Thus $2^{\ell_{2'}^*} = 16$ and $\ell_{2'}^* = 4$. The optimal power-of-two reorder interval for node $2'$ is also equal to $2^4(1/52) = 0.31$ years.

Finally, for node $1'$, $\ell_{1'}^*$ is the smallest non-negative integer such that

$$2^{\ell_{1'}^*} \geq \frac{1}{\sqrt{2}T_L} \sqrt{\frac{K_{1'}}{g_{1'}}} = \frac{1}{\sqrt{2}(1/52)}(0.274) = 10.07,$$

where $\sqrt{\frac{K_{1'}}{g_{1'}}} = 0.274$ years. Thus $2^{\ell_{1'}^*} = 16$ and $\ell_{1'}^* = 4$. The optimal power-of-two reorder interval for node $1'$ is also equal to $2^4(1/52) = 0.31$ years.

3.4 Joint Replenishment Problem (JRP)

Oftentimes a supplier provides multiple products to a customer. When the customer places an order, the supplier incurs two types of fixed costs. The first type of fixed cost is independent of the number of products ordered. This cost may arise from payment to the shipping company whose charges do not depend on the assortment of products inside the truck. The second type of fixed cost is incurred for each product shipped. This cost may arise from postage, invoice preparation, testing upon arrival, any special handling while unloading the product, etc. As before, both types of fixed costs are independent of the order sizes. Given that the first type of fixed cost is the same no matter how many products are ordered, it might be economical to combine orders of several products. Our goal in this section is to identify the ordering policy for each product.

The graphical structure of this problem is shown in Figure 3.8. Arcs originate from n nodes that form the first echelon to node 0, which is the single node in the lower echelon. Thus, we consider a two-echelon system here. Observe that the direction of flow is opposite to that of the distribution system we studied in the last section.

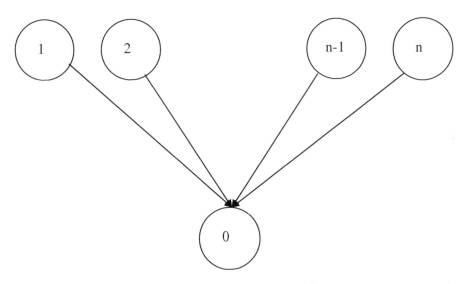

Fig. 3.8. Graphical representation of the joint replenishment problem.

3.4.1   A Mathematical Model for Joint Replenishment Systems

As Figure 3.8 indicates, the sets of nodes and arcs in the system are

$$N(G) = \{0, 1, \ldots, n\},$$
$$A(G) = \{(1,0), (2,0), \ldots, (n,0)\}.$$

The nodes $1, 2, \ldots, n$ represent different products. The purpose of the arcs is to represent precedence; a *real* flow of products does not occur. Whenever an order is placed by node i, $i = 1, \ldots, n$, that is, for product i, a fixed cost K_i is incurred. This cost is independent of the order quantities for any product including product i. By having arcs go from other nodes to node 0, we ensure that the reorder interval for node 0 will not be higher than the reorder intervals for the other nodes. Whenever an order is placed for any of the products, an order is placed for node 0, too, and the setup cost K_0 is incurred. Now, the value added for this setup is 0 and thus $h_0' = 0$. For other nodes, the annual holding costs are charged in proportion to the echelon stock; since inventory is not held at node 0, the echelon stock at node i, $i = 1, 2, \ldots, n$, is equal to the on-hand stock at node i. Finally, the demand rate for product i is λ_i units/year $i = 1, \ldots, n$.

We again employ a nested and stationary policy in which the reorder interval for each node is a power-of-two multiple of a base planning period. Any order for any product i, $i = 1, 2, \ldots, n$, requires an order to be placed by node 0. Thus nestedness of the ordering policy is a necessity for this system. The following is the model for the system problem.

$$(P) \quad \min \sum_{j=0}^{n} \left(\frac{K_i}{T_i} + g_i T_i \right)$$

subject to

$$T_i \geq T_0,$$
$$T_i = 2^{\ell_i} T_L, \quad \ell_i = 0, 1, 2, \ldots, \quad i = 1, 2, \ldots, n,$$

where $g_i = \frac{1}{2} \lambda_i h_i$. The constraints $T_i \geq T_0$ are necessary for nestedness, and the constraints $T_i = 2^{\ell_i} T_L$ ensure a PO2 policy is followed.

Similarly to the serial system and distribution system problems, this problem is solved in two steps. First, we construct a relaxation in which the power-of-two restriction is dropped. In the second step, we use a rounding rule to establish the power-of-two solution. The relaxed problem is

$$(RP) \quad \min \sum_{i=0}^{n} \left\{ \frac{K_i}{T_i} + g_i T_i \right\}$$

subject to

$$T_i \geq T_0 \geq 0.$$

Suppose the nodes are indexed such that

$$\frac{K_1}{g_1} \leq \frac{K_2}{g_2} \leq \cdots \leq \frac{K_n}{g_n}.$$

If the nodes are not indexed in this way, then we must re-index them such that node 1 is the one with the smallest $\frac{K}{g}$-ratio, node 2 is the one with the second-smallest $\frac{K}{g}$-ratio, and so on.

To obtain the optimal solution to the relaxed problem (RP), we will establish an ordered partition of the graph in Figure 3.8 to create two categories of subgraphs. The first category consists of only one subgraph, which contains the nodes 0 and 1 and possibly some other nodes from among $\{2,3,\ldots,n\}$. All the nodes in this subgraph have the same reorder interval. Each subgraph in the second category contains a single node.

To obtain this ordered partition, we identify the nodes that are in the same subgraph as node 0. Let $\bar{i} \geq 2$ be the largest (sorted) index for which

$$\frac{\sum_{j=0}^{\bar{i}-1} K_j}{\sum_{j=1}^{\bar{i}-1} g_j} > \frac{K_{\bar{i}}}{g_{\bar{i}}}. \tag{3.11}$$

Then define $\mathcal{C} = \{2,\ldots,\bar{i}\}$ and $N(G_1) = \mathcal{C} \cup \{0,1\} = \{0,1,2,\ldots,\bar{i}\}$. Note that node 1 will always be contained in G_1. Let $K(G_1) = \sum_{j=0}^{\bar{i}} K_j$ and $g(G_1) = \sum_{j=1}^{\bar{i}} g_j$. For $i > \bar{i}$, $N(G_{i-\bar{i}+1}) = \{i\}$. Then the optimal solution to the relaxed problem is

$$T_0 = T_1 = \cdots = T_{\bar{i}} = \sqrt{\frac{K(G_1)}{g(G_1)}}$$

and

$$T_j = \sqrt{\frac{K_j}{g_j}}, \quad j > \bar{i}.$$

The formation of $\mathcal{C}$ is based on the following simple rule: add a node to $\mathcal{C}$ only when not including it in $\mathcal{C}$ would result in infeasibility. Suppose $\mathcal{C} = \{2,3,\ldots,j-1\}$ and we are considering whether or not to add node j to $\mathcal{C}$. Let T_0^{j-1} be the reorder interval of node 0 when $\mathcal{C}$ contains nodes $2,3,\ldots,j-1$. If node j is not added to $\mathcal{C}$, then the reorder interval of the subgraph containing node 0 is equal to

$$T_0^{j-1} = \sqrt{\frac{K_0 + K_1 + \cdots + K_{j-1}}{g_1 + g_2 + \cdots + g_{j-1}}}$$

and the reorder interval of node j would be equal to $T_j = \sqrt{\frac{K_j}{g_j}}$. If $\frac{K_j}{g_j}$ were less than $\frac{K_0 + K_1 + \cdots + K_{j-1}}{g_1 + g_2 + \cdots + g_{j-1}}$, then T_j would be less than T_0^{j-1} which would violate the feasibility constraint. As a result, we add node j to $\mathcal{C}$ as well, which makes j and 0 share a common reorder interval. We keep adding nodes until we reach node $\bar{i}$. After we add node $\bar{i}$ to $\mathcal{C}$, $T_0^{\bar{i}}$ is less than or equal to $T_{\bar{i}+1}, T_{\bar{i}+2}, \ldots, T_n$.

The optimality of this solution can be checked in two ways. The first method is by applying Theorem 3.2 and the second method is by using the Karush–Kuhn–Tucker or KKT conditions. Let us consider the three conditions in Theorem 3.2:

1. The subgraphs as constructed form an ordered partition of G. Arcs go from nodes $1, 2, \ldots, \bar{i}$ to node 0, which is subgraph G_1. $G_2, G_3, \ldots$ consist of a single node and no arcs. Hence the subgraphs are ordered by precedence. Furthermore, the sets of nodes $N(G_1), N(G_2), \ldots$ form a partition to $N(G)$ since their union is equal to $N(G)$ and each node is contained in exactly one such node set.

2. The feasibility condition is satisfied by construction. Using (3.11), $\bar{i}+1$ is the smallest index that does not satisfy the inequality in (3.11). Thus,

$$\sqrt{\frac{\sum\limits_{j=0}^{\bar{i}} K_j + K_0}{\sum\limits_{j=1}^{\bar{i}} g_j}} = T(0) = T(1) = \cdots = T(\bar{i}) \leq \sqrt{\frac{K_{\bar{i}+1}}{g_{\bar{i}+1}}} = T(\bar{i}+1).$$

Further,

$$T(\bar{i}+1) = \sqrt{\frac{K_{\bar{i}+1}}{g_{\bar{i}+1}}} \leq \cdots \leq T(n) = \sqrt{\frac{K_n}{g_n}}.$$

3. There is no directed cut of G_1 for which $\frac{K(G_1^-)}{g(G_1^-)} < \frac{K(G_1^+)}{g(G_1^+)}$. This condition requires that G_1 cannot be split into two ordered subgraphs such that the subgraph that contains 0 has a $\frac{K}{g}$-ratio (and hence the reorder interval) lower than the other subgraph. Our method of creating G_1 automatically precludes such a possibility.

The second approach for verifying the optimality of the solution requires evaluating the KKT conditions at this calculated solution. Recall the formulation of the problem (RP):

$$\min \sum_{j=0}^{n} \left\{ \frac{K_j}{T_j} + g_j T_j \right\}$$

subject to

$$T_j \geq T_0 \geq 0, \quad j = 1, \ldots, n.$$

The constraints can be equivalently written as $T_0 - T_j \leq 0, -T_0 \leq 0$. Let us now write the Karush–Kuhn–Tucker conditions, where θ_j is the multiplier associated with the constraint $T_j - T_0 \geq 0$ and θ_0 is the multiplier associated with the constraint $-T_0 \leq 0$.

Condition 1 (Optimality):

$$\frac{-K_j}{T_j^2} + g_j - \theta_j = 0, \quad j = 1, \dots, n \tag{3.12}$$

$$\frac{-K_0}{T_0^2} - \theta_0 + \sum_{j=1}^{n} \theta_j = 0. \tag{3.13}$$

Condition 2 (Complementary Slackness):

$$-\theta_0 T_0 = 0, \tag{3.14}$$

$$\theta_j (T_0 - T_j) = 0, \quad j = 1, \dots, n, \tag{3.15}$$

$$T_j \left(\frac{-K_j}{T_j^2} + g_j - \theta_j \right) = 0, \quad j = 1, \dots, n, \tag{3.16}$$

$$T_0 \left(\frac{-K_0}{T_0^2} - \theta_0 + \sum_{j=1}^{n} \theta_j \right) = 0. \tag{3.17}$$

Condition 3 (Feasibility):

$$\theta_j \geq 0, \tag{3.18}$$

$$0 \leq T_0 \leq T_j, \quad j = 1, \dots, n. \tag{3.19}$$

If we can identify a non-negative value for each θ_j such that these conditions are satisfied, we will be done. Accordingly, set

$$\theta_j = g_j - \frac{K_j}{T_j^2}, \quad j = 1, 2, \dots, \bar{i},$$

$$= 0, \quad j = 0, \bar{i} + 1, \bar{i} + 2, \dots, n.$$

This selection is not mere speculation. We obtain $\theta_1, \theta_2, \dots, \theta_n$ from (3.15) and (3.12). Since $T_0 \neq 0$, using (3.14), $\theta_0 = 0$.

We chose θ_j to ensure that (3.16) is satisfied. For the same reason, observe that (3.12) is also satisfied. Next, for $j = 0, 1, 2, \dots, \bar{i}$, $T_j = T_0$ and so (3.15) holds for these values of j. The same equation is also satisfied for $j = \bar{i} + 1, \dots, n$, since $\theta_j = 0$. By substituting for θ_j in (3.17), we obtain

$$T_0 \left(-\frac{K_0}{T_0^2} + \sum_{j=1}^{\bar{i}} \left(g_j - \frac{K_j}{T_j^2} \right) \right),$$

which is equal to

$$T_0 \left(\sum_{j=1}^{\bar{i}} g_j - \sum_{j=0}^{\bar{i}} \frac{K_j}{T_0^2} \right)$$

since $T_0 = T_1 = \cdots = T_{\bar{i}}$. But the above expression is equal to 0 since $T_0^2 = \dfrac{\sum_{j=0}^{\bar{i}} K_j}{\sum_{j=1}^{\bar{i}} g_j}$ and

hence (3.17) is satisfied. Obviously, (3.13) is also satisfied.

Since T_j and T_0 are feasible, the inequalities in (3.19) are also satisfied. Our final objective is to show that $\theta_j \geq 0$ for $j = 1, \ldots, \bar{i}$. For $j = 1, \ldots, \bar{i}$,

$$\theta_j = g_j - \frac{K_j}{T_j^2} = \frac{g_j}{T_0^2} \left(T_0^2 - \frac{K_j}{g_j} \right),$$

where $T_0 = T_j$. Since $\frac{K_j}{g_j} \leq T_0^2$, $\theta_j \geq 0$. Since all the Karush–Kuhn–Tucker conditions hold, the conjectured solution is optimal. Having established the optimality of the solution to problem (RP), we next show how to use this solution to obtain the solution to (P).

3.4.2 Rounding the Solution to the Relaxed Problem

The common PO2 reorder interval for nodes $1, \ldots, \bar{i}$ is $2^{\ell_0} T_L$, where ℓ_0 is the smallest non-negative integer for which

$$2^{\ell_0} \geq \frac{1}{\sqrt{2} T_L} \sqrt{\frac{K(G_1)}{g(G_1)}}.$$

For $i > \bar{i}$, the optimal reorder interval is $2^{\ell_i} T_L$, where ℓ_i is the smallest non-negative integer for which

$$2^{\ell_i} \geq \frac{1}{\sqrt{2} T_L} \sqrt{\frac{K_i}{g_i}}.$$

In summary, we find the optimal PO2 solution using the following algorithm.

Step 1: Number the nodes such that $\frac{K_1}{g_1} \leq \frac{K_2}{g_2} \leq \cdots \leq \frac{K_n}{g_n}$.

Step 2: Find the optimal ordered partition, that is, find the largest non-negative integer, $\bar{i}$, for which

$$\frac{\sum_{j=0}^{\bar{i}-1} K_j}{\sum_{j=1}^{\bar{i}-1} g_j} > \frac{K_{\bar{i}}}{g_{\bar{i}}}.$$

Step 3: Find the smallest non-negative integer, ℓ_0^*, for which

$$2^{\ell_0^*} \geq \frac{1}{\sqrt{2T_L}} \left[\frac{K(G_1)}{g(G_1)}\right]^{1/2},$$

and set $T_0 = T_1 = \cdots = T_{\bar{i}} = 2^{\ell_0^*} T_L$.

Step 4: For $i > \bar{i}$, find the smallest non-negative integer, ℓ_i^*, for which

$$2^{\ell_i^*} \geq \frac{1}{\sqrt{2T_L}} \sqrt{\frac{K_i}{g_i}},$$

and set $T_i = 2^{\ell_i^*} T_L$.

Let us demonstrate how this algorithm works by solving the following problem. Suppose a retailer orders five different products from a supplier. The relevant data are provided in the following table. Taking the base planning period to be 1 week, let us compute the optimal PO2 policy.

Item	K_i	g_i
0	5	0
1	1	1
2	2	1
3	3	1
4	6	1
5	16	1

Step 1: We compute the ratio $\frac{K_i}{g_i}$ for each node $i \geq 1$ and rank them in increasing order.

Item	K_i	g_i	$\frac{K_i}{g_i}$	Rank
0	5	0	–	–
1	1	1	1	1
2	2	1	2	2
3	3	1	3	3
4	6	1	6	4
5	16	1	16	5

All the nodes are indexed in the proper order so we do not need to re-index them.

Step 2: This is an iterative step in which we determine the set $\mathcal{C}$. We know that $\mathcal{C}$ contains at least one node, node 1.

Iteration 1:

Begin with node 2. Since the ratio $\frac{K_2}{g_2} = 2$ is less than $\frac{K_0+K_1}{g_1} = \frac{6}{1} = 6$, add node 2 to $\mathcal{C}$.

Iteration 2:

Next, we consider node 3. Once again, the ratio $\frac{K_3}{g_3} = 3$ is less than $\frac{K_0+K_1+K_2}{g_1+g_2} = \frac{8}{2} = 4$, so we also add node 3 to $\mathcal{C}$.

Iteration 3:

Next, we consider node 4. The ratio $\frac{K_4}{g_4} = 6$ is greater than $\frac{K_0+K_1+K_2+K_3}{g_1+g_2+g_3} = \frac{11}{3} = 3.67$, so we do not add node 4 to $\mathcal{C}$. Therefore, $\bar{i} = 3$, $N(G_1) = \{0,1,2,3\}$, and $K(G_1) = 11$ and $g(G_1) = 3$.

Step 3: The optimal power-of-two value ℓ_0 for nodes $0, 1, 2$, and 3 is the smallest non-negative integer such that $2^{\ell_0} \geq \frac{1}{\sqrt{2}T_L} \sqrt{\frac{K(G_1)}{g(G_1)}} = \frac{1}{\sqrt{2}(1/52)} \sqrt{3.67} = 70.41$. Thus $\ell_0 = 7$, and the optimal PO2 reorder intervals $T_0 = T_1 = T_2 = T_3 = (2^7)\left(\frac{1}{52}\right) = 2.46$ years.

Step 4: The optimal power-of-two value for node 4 is the smallest non-negative integer such that $2^{\ell_4} \geq \frac{1}{\sqrt{2}T_L} \sqrt{\frac{K_4}{g_4}} = \frac{1}{\sqrt{2}(1/52)} \sqrt{6} = 90.07$ and $\ell_4 = 7$ as well. Therefore, the optimal power-of-two reorder interval for node 4 is equal to $T_4 = (2^7)\left(\frac{1}{52}\right) = 2.46$ years.

In a similar way, we find that the optimal PO2 reorder interval for node 5 is 4.92 years.

3.5 Exercises

3.1. Compute the optimal power-of-three reorder interval for a single-stage system. Show that the cost of this policy is at most 15.47% more than the cost of the optimal policy.

3.2. The Campus Store of a university notes that the bi-weekly demand for folders is almost steady throughout the year and is equal to 30 per 2 weeks; the number of such bi-weekly periods is 26 per year. The demand slumps when the university is closed for short breaks but the surge in demand right after the break almost always makes up for it. The store has a warehouse within the university town. The deliveries from the suppliers first arrive in this warehouse and then are repackaged appropriately for shipment to the store. The time to transfer material from the warehouse to the store is at most a few hours and thus can be ignored. The fixed cost of placing an order with the supplier is $20 per order and the fixed cost of transferring items from the warehouse to the store is $10 per order. Taking the holding cost at the warehouse to be $0.30/unit/year and at the store to be $0.50/unit/year, determine the optimal PO2 policy for placing orders with the supplier and for transferring items to the store. Assume a base planning period of a week.

3.3. Derive inequality (3.7).

3.4. Consider a five-stage serial system. Suppose the on-hand holding costs are as follows: $h_1 = 4.5$, $h_2 = 4.25$, $h_3 = 3$, $h_4 = 2$, and $h_5 = 0.5$ (in $ per unit per year). Compute the echelon holding cost for each stage.

3.5. For a four-stage serial system, compute the on-hand holding costs if the echelon holding costs are $h'_1 = 1$, $h'_2 = 1.25$, $h'_3 = 0.4$, and $h'_4 = 0.6$ (in $ per unit per year).

3.6. A regional retail chain receives shipments from manufacturers at its central warehouse and in turn sends the shipments to the retail stores. Consider a miniature version of this problem having three retail stores and a single item, ceiling lamps. The annual demand for ceiling lamps is 100 at store 1, 80 at store 2, and 110 at store 3. The holding cost rate at each of the stores is .20 and the cost of a ceiling lamp at the time of entering a store is $15 per unit. This cost includes the original purchasing cost and all the value added. The value can be added, for example, by transportation, testing, unpackaging, etc. The holding cost rate for the central warehouse is .15 and the cost of an item at the time it enters the warehouse is estimated to be $12. The fixed cost to place an order is $100 for the warehouse and $10 for the retail stores. Assuming a base planning period of 2 weeks and a year consisting of 52 weeks, determine the optimal PO2 policy for each of the retail stores and the central warehouse.

3.7. In a serial system, is it enough to constrain the reorder intervals such that $T_i \geq T_{i-1}$ to ensure nestedness?

3.8. A deli orders three different types of patties (vegan, tuna, and beef) from a local manufacturer for use in sandwiches. The demand for vegan, tuna, and beef patties is 20, 30, and 125 per week, respectively. The order placement is rather easy; the store manager just goes to the manufacturer and collects as many patties as he wants. Each trip takes 30 minutes; assume the worth of the manager's time to be $30 per hour. The store manager, being a conscientious person, ensures that the patties are safe for consumption. This takes 30 minutes per order of the vegan and tuna patties and 20 minutes per order of the beef patties. This task is performed by an hourly-wage employee who earns $7 per hour. Taking the holding cost of these patties to be $0.80/patty/year, determine the optimal PO2 replenishment policy. Assume a base planning period of a week.

3.9. Suppose the demands for each of the three products in the last problem were doubled. Compute the new optimal PO2 policy.

3.10. Consider the system in Section 3.3.1. Suppose the demands in the regional warehouses are not identical, but are equal to 100, 125, 200, 80, 150, and 225, respectively. Assume the fixed cost and holding costs are as in the example. Determine the optimal PO2 reorder intervals assuming the base planning period is 2 weeks.

3.11. (Source: http://bmrc.berkeley.edu/courseware/ICMfg92/text/fab-6) There are four major manufacturing stages in the production of an Integrated Chip (IC)

- Wafer Fabrication
- Wafer Probe
- Chip Assembly
- Chip Test

Consider an IC that is used in a mature product and whose demand for the next year is projected to be 100,000 units. The setup cost for the four processes is $500, $260, $425, and $700 per setup, respectively. The on-hand inventory holding cost at these stages is equal to $1, $1.50, $1.75, and $2, respectively. Assuming the production system to be a serial system, determine the optimal PO2 policy. Assume a base planning period of a week.

3.12. Let $f : \Re \to \Re$ be a convex function and let x_1, x_2, and x_3 be three points in its domain such that $x_1 \le x_2 \le x_3$. Show that $f(x_2) \le \max\{f(x_1), f(x_3)\}$.

3.13. Show that the joint replenishment problem can be represented as a serial system problem.

4

Dynamic Lot Sizing with Deterministic Demand

To this point we have examined environments in which demand was assumed to occur at a known, constant rate over an infinite horizon. We now turn our attention to developing a finite-horizon, discrete-time model with deterministic but non-stationary demand for a single product at a single stage. In a finite-horizon discrete-time model, as the name suggests, the length of the planning horizon is finite and the order placement decisions are made at discrete intervals of time. Inventory is reviewed only at the beginning of a period, hence we can call this model a periodic review model. Backorders are not permitted.

There are three types of costs considered in this environment, the fixed ordering cost, the variable procurement cost (or payment to the supplier) and the holding cost. If there were no fixed cost, it would be optimal to place an order in every period. The fixed cost provides an economic incentive to combine several periods' demands into a single order. The variable procurement cost is also incurred only when an order is placed. The magnitude of this cost is proportional to the order quantity. Unlike the fixed and variable costs of order placement, the holding cost is not incurred when an order is placed. The holding cost is charged every period in proportion to the amount of on-hand inventory at a period's end.

We will present three ways to solve this type of problem. The first method is called the Wagner–Whitin [360] algorithm. The second approach is called the Wagelmans–Hoesel–Kolen [357] algorithm. Both of these methods will find an optimal solution. While these two algorithms produce an optimal solution, they do differ in the computational complexity required to compute the optimal procurement plan. The Wagner–Whitin algorithm may require a number of calculations proportional to T^2, where T is the length of the planning horizon. For a general planning environment, the Wagelmans–Hoesel–Kolen algorithm requires calculations proportional to $T \log T$. For the assumptions on which the dynamic lot sizing problem is based, the latter algorithm

J.A. Muckstadt and A. Sapra, *Principles of Inventory Management: When You Are Down to Four, Order More*, Springer Series in Operations Research and Financial Engineering, DOI 10.1007/978-0-387-68948-7_4, © Springer Science+Business Media, LLC 2010

requires calculations that are only proportional to T. Hence the second method is more efficient in finding the optimal solution. It is important to understand that not all algorithms are equally efficient. Efficient implementation in ongoing operations is essential to effective management of inventory. The third way is to employ a heuristic method to create a procurement plan. We will describe two such heuristics. The first is called the Silver–Meal heuristic and the second is called the least unit cost heuristic. Both heuristics are order T methods for computing a procurement plan. Although these approaches are very simple to implement, they do not necessarily obtain an optimal solution. We note that the Wagner–Whitin algorithm was the first approach in the literature to solve this lot sizing problem to optimality.

We begin by discussing the Wagner–Whitin algorithm. We will then present the Wagelmans–Hoesel–Kolen algorithm and then discuss the two heuristics.

4.1 The Wagner–Whitin (WW) Algorithm

The notation and the remaining assumptions for the model are as follows. Denote the fixed ordering cost and the holding cost in period t by K_t and h_t, respectively. In general, the Wagner–Whitin analysis applies to more general cases than we will discuss. Specifically, if C_t is the per-unit purchasing cost, then the analysis requires that $C_t + h_t \geq C_{t+1}$ for all t. For simplicity we will assume that these parameters are stationary over time. Consequently, $C_t \equiv C$, $K_t \equiv K$, and $h_t \equiv h$ for any period t, $t = 1, \ldots, T$. It is not necessary for all the periods to be of equal length even though that is often the case in the real world. When periods vary in length, the values of h_t would likely vary, too. We denote the known deterministic demand in period t by d_t. We use x_t to represent the inventory at the beginning of period t before the order-placement decision is made. We also assume that the lead time is zero to simplify the discussion.

We use y_t to denote the on-hand inventory after the order-placement decision is made and the order is received. Equivalently, y_t is equal to the sum of x_t and the order quantity. Because the order size is nonnegative, y_t is always greater than or equal to x_t. If an order is placed in period t, y_t is strictly greater than x_t, that is, $y_t > x_t$. Otherwise, when no order is placed in period t, y_t is equal to x_t. Also, y_t must be large enough to satisfy the demand in period t since backorders are not allowed.

The events in any period are as follows. At the beginning of the period, it is decided whether or not to place an order. If an order is placed, fixed and variable purchase costs are incurred. Since the lead time is zero, the order is received immediately. The demand is realized through the rest of the period. At the end of the period, a holding cost is charged in proportion to the amount of remaining on-hand inventory.

We next develop an expression for the cost in a generic period t. The fixed cost is equal to K if $y_t > x_t$; otherwise it is 0. Equivalently, we can write the fixed cost as $K \cdot \delta_t$ where

$$\delta_t = \begin{cases} 1, & y_t > x_t, \\ 0, & \text{otherwise.} \end{cases}$$

The purchasing cost is equal to the product of the unit purchasing cost C and the order quantity $y_t - x_t$. Finally, the leftover inventory at the end of period t is $y_t - d_t$, and the corresponding holding cost is $h(y_t - d_t)$. Combining the three terms, we obtain the total cost in period t as

$$K\delta_t + C(y_t - x_t) + h(y_t - d_t).$$

The cumulative cost over the whole horizon is equal to

$$\sum_{t=1}^{T} \{ K\delta_t + C(y_t - x_t) + h(y_t - d_t) \}.$$

We also impose two types of constraints. Since the order quantity is non-negative, y_t must be greater than or equal to x_t in any period t. To ensure that there is no backlogging, $y_t \geq d_t$.

Let $Z_1(x_1)$ be the optimal cost for the whole horizon given that the inventory at the beginning of the horizon is x_1. The complete formulation of the dynamic lot sizing problem is as follows:

$$Z_1(x_1) = \min \sum_{t=1}^{T} \{ K\delta_t + C(y_t - x_t) + h(y_t - d_t) \}$$

subject to

$$y_1 \geq x_1, \; y_2 \geq x_2, \ldots, \; y_T \geq x_T,$$
$$y_1 \geq d_1, \; y_2 \geq d_2, \ldots, \; y_T \geq d_T,$$

where

$$x_{t+1} = y_t - d_t, \quad t = 1, 2, \ldots, T.$$

The cumulative variable purchasing cost $\sum_{t=1}^{T} C(y_t - x_t)$ is independent of the decision variables, $y_1, y_2, \ldots, y_T$. The sum total of the order quantities over the horizon is

$$\sum_{t=1}^{T} (y_t - x_t) = (y_1 - x_1) + (y_2 - x_2) + \cdots + (y_{T-1} - x_{T-1}) + (y_T - x_T).$$

Using the relationship, $x_{t+1} = y_t - d_t$, we can substitute for y_1 with $x_2 + d_1$, y_2 with $x_3 + d_2$, and, in general, y_t with $x_{t+1} + d_t$ in the above expression. After the substitution, the above expression becomes

$$\sum_{t=1}^{T} (y_t - x_t) = (x_2 + d_1 - x_1) + (x_3 + d_2 - x_2) + \cdots$$
$$+ (x_T + d_{T-1} - x_{T-1}) + (x_{T+1} + d_T - x_T),$$

which simplifies to

$$\sum_{t=1}^{T} (y_t - x_t) = d_1 + d_2 + \cdots + d_T - x_1 + x_{T+1}. \tag{4.1}$$

Observe that all the terms on the right-hand side except x_{T+1} are constant and thus independent of the decision variables. The term x_{T+1} is the inventory remaining at the end of the horizon. This inventory is unutilized and is wasted. Therefore, in any optimal policy, x_{T+1} must be equal to zero. This implies that the cumulative order quantity over the horizon is independent of the decision variables. Since the total purchasing cost over the horizon is equal to the product of this cumulative order quantity and a constant C, the variable cost of order placement is also independent of the decision variables. Therefore, we exclude it from the cost function. The revised formulation is as follows:

$$Z_1(x_1) = \min \sum_{t=1}^{T} \{K\delta_t + h(y_t - d_t)\}$$

subject to

$$y_1 \geq x_1, \ y_2 \geq x_2, \ldots, \ y_T \geq x_T,$$
$$y_1 \geq d_1, \ y_2 \geq d_2, \ldots, \ y_T \geq d_T,$$

where

$$x_{t+1} = y_t - d_t, \ x_{t+1} \geq 0, \quad t = 1, 2, \ldots, T-1, \ x_{T+1} = 0.$$

Having developed the model, we now proceed to solving it. We develop the foundation for the algorithm in the following subsection.

4.1.1 Solution Approach

Given the inventory level x_t at the beginning of period t, let the optimal cost for period t and onwards through the end of the horizon be denoted by $Z_t(x_t)$. Then

$$Z_t(x_t) = \min \sum_{k=t}^{T} (K\delta_k + h(y_k - d_k))$$

subject to

$$y_t \geq x_t, \ldots, y_T \geq x_T,$$
$$x_{t+1} = y_t - d_t, \ldots, x_T = y_{T-1} - d_{T-1}, x_{T+1} = 0,$$
$$x_j \geq 0, j = t+1, \ldots, T.$$

Equivalently, $Z_t(x_t)$ is equal to the sum of the optimal cost incurred in period t, $K\delta_t + h(y_t - d_t)$, and the cost incurred over periods $t+1$ through the end of the horizon. The inventory at the beginning of period $t+1$ is equal to $y_t - d_t$ and so the optimal cost for periods $t+1$ onwards is equal to $Z_{t+1}(y_t - d_t)$. Formally,

$$Z_t(x_t) = \min_{y_t \geq x_t, y_t \geq d_t} K\delta_t + h(y_t - d_t) + Z_{t+1}(y_t - d_t).$$

For ease of discussion, we assume that the inventory at the beginning of the horizon x_1 is equal to zero. It is easy to convert a problem with $x_1 \neq 0$ to a problem with $x_1 = 0$. Just change the demand in the first period to $d_1 - x_1$, that is, $d_1 \leftarrow d_1 - x_1$, assuming $d_1 \geq x_1$. Thus, the first period demand is reduced by x_1 units. The new first period demand is the *actual* demand that we have to fulfill through the order placement. If $d_1 < x_1$, then find the first period $j > 1$ for which $\sum_{t=1}^{j} d_t > x_1$. Set $d_t = 0$ for $t = 1, 2, \ldots, j-1$ and $d_j \leftarrow \sum_{t=1}^{j} d_t - x_1$.

We next obtain some insights regarding the optimal solution. To illustrate the first idea, consider an example with $T = 5$. The following table provides the demand data and an order policy.

Period (t)	1	2	3	4	5
Demand (d_t)	25	30	15	45	30
Order Quantity ($y_t - x_t$)	55	0	30	30	30
Beginning-of-Period Inventory (x_t)	0	30	0	15	0
End-of-Period Inventory (x_{t+1})	30	0	15	0	0
Cost in Period t	$K + 30h$	0	$K + 15h$	K	K

Observe that the order placed in period 3 is larger than the demand in period 3 but is not large enough to cover the demand of period 4. The costs incurred in periods 3 and 4 are

$K + 15h$ and K, respectively. Suppose we tweak the above policy such that everything remains the same except for the order quantities in periods 3 and 4. Suppose we reduce the order quantity in period 3 by 15 units and increase the order quantity in period 4 by 15 to 45 units. The cost of the tweaked policy in period 3 is K which is lower than before. The cost incurred in other periods including period 4 remains the same. As a result, the revised policy has a lower overall cost.

The major characteristic of the revised policy is that whenever an order is placed, the quantity ordered is equal to the total demand over some number of future time periods. In the example, the order quantity in period 1 is equal to the sum of demands of the first two periods. The lower cost of the tweaked policy is not fortuitous. In general if an order satisfies a future period's demand only partially, it can be *tweaked* to reduce the overall costs.

The above discussion implies that in any period at least one of the two variables, the order quantity and the beginning of the period inventory, is zero. Equivalently, the product of the order quantity and the inventory at the beginning of the period is zero in any period. We state this observation formally in the following theorem.

Theorem 4.1. *In any optimal policy, $x_t(y_t - x_t) = 0$ for $t = 1, 2, \ldots, T$.*

The example also illustrates the fact that the order quantity in any period is equal to the sum of the demands of a certain number of future periods.

Theorem 4.1 can also be used to show the following result.

Proposition 4.1. *Suppose period t's demand is satisfied by an order placed in period j, $j < t$, in an optimal policy. Then the demands of periods $j+1, j+2, \ldots, t-1$ are also satisfied by the same order.*

Proposition 4.1 builds on Theorem 4.1 and tells us that not only the order size must equal the total demands over some number of future periods, but also that the demands should occur in consecutive periods. For example, it rules out placing an order in period t to satisfy the demands of periods t and $t+2$ and placing another order in period $t+1$ to satisfy its demand in that period. Clearly, such a policy will be a violation of Proposition 4.1. Therefore, if an order is placed in period t, then $y_t = d_t + d_{t+1} + \cdots + d_{k-1} + d_k$ for some $k \geq t$.

Suppose now that by some magic we discover that the beginning inventory in period $t+1, x_{t+1}$, is equal to 0 in an optimal policy. In that case, do the ordering decisions made in periods $1, 2, \ldots, t$ affect the cost and ordering decisions in periods $t+1, t+2, \ldots, T$? The answer is no. The reason is that the ordering decisions made in one period influence future decisions through the inventory that is carried into these periods. For example, if a large order is placed in a period, then there will be no need to place an order over the several subsequent periods. Furthermore, the holding cost incurred over the next several periods will be large owing to the amount of inventory being carried. When no inventory is carried from period t to period $t + 1$, there is no effect of the decisions made in periods up through t on the cost and ordering decisions made in period $t + 1$ and beyond.

We make use of the above observation when designing a solution algorithm. For every period t, we will assume that the inventory at its end (or the beginning of the next period $t + 1$) is zero. Thus, we consider a t-period problem. This absolves us of all the future responsibility and we can focus on periods $1, 2, \ldots, t - 1, t$ to determine when to place an order to satisfy period t's demand. It gets even simpler than this. Suppose we find that the demand for period $t - 1$ is satisfied by an order placed in period v in the optimal solution to the $t - 1$-period problem. That is, $y_v = d_v + d_{v+1} + \cdots + d_{t-1}$ in the $t - 1$-period problem's solution. Then the order that optimally satisfies period t's demand in the t-period problem must be placed in period v or later. See Subsection 4.1.4 for the technical details.

Before we outline the steps of the algorithm, let us understand the cost computations. Define $F(t)$ to be the optimal cost from period 1 through period t when inventory at the end of period t is zero. Also, define $\mathcal{C}_t^s$ to be the minimum cost over periods 1 through t when the inventory level at the end of period t is zero *and* period t's demand is satisfied by an order placed in period s. $\mathcal{C}_t^s$ can be written as the sum of the minimum costs incurred over periods 1 through $s - 1$ and periods s through t when $x_s = 0$ and $y_s = \sum_{j=s}^{t} d_j$. The optimal cost over periods 1 through $s - 1$ is equal to $F(s-1)$. The cost incurred between periods s and t includes the fixed cost incurred in period s and the holding costs incurred in periods $s, s + 1, \ldots, t$. The holding cost in period s is proportional to the inventory at the end of period s, which is equal to the sum of the demands in periods $s + 1, s + 2, \ldots, t$. Similarly, the holding cost in period $s + 1$ is proportional to the inventory carried forward to period $s + 2$ which is equal to the sum of the demands in periods $s + 2, s + 3, \ldots, t$. Since inventory at the end of period t is assumed to be 0, no holding cost is incurred in period t. Therefore, the total cost incurred in periods s through t is equal to

$$K + h(d_{s+1} + d_{s+2} + \cdots + d_t) + h(d_{s+2} + d_{s+3} + \cdots + d_t) + \cdots + h(d_t). \qquad (4.2)$$

We can now write an expression for $\mathcal{C}_t^s$ by adding $F(s - 1)$ to (4.2):

$$\mathcal{C}_t^s = F(s-1) + K + h(d_{s+1} + d_{s+2} + \cdots + d_t) + h(d_{s+2} + d_{s+3} + \cdots + d_t)$$
$$+ \cdots + h(d_t). \qquad (4.3)$$

As noted, if period $t - 1$'s demand is optimally satisfied by an order placed in period v in the $t - 1$-period problem, then it is sufficient to consider periods $v, v + 1, \ldots, t$ as periods in which to optimally place an order to satisfy period t's demand in a t-period problem. For each of these choices, we can compute the minimum cost over periods 1 through t. Hence, when period $t-1$'s demand is satisfied in period v in a $t-1$-period problem, the optimal cost over periods 1 through t, $F(t)$, is

$$F(t) = \min\{\mathcal{C}_t^v, \mathcal{C}_t^{v+1}, \ldots, \mathcal{C}_t^{t-1}, \mathcal{C}_t^t\}.$$

We are now prepared to state the Wagner–Whitin algorithm.

4.1.2 Algorithm

The steps of the algorithm are as follows.

Step 1: Set $t = 2$, $v = 1$ and $F(1) = K$.

Step 2: Since an order is placed in period 1, we need to determine whether to place an order in period 1 or in period 2 to satisfy period 2's demand. When the order is placed in period 2, the total cost is $F(1) + K_2 = K_1 + K_2$ since no inventory is carried into period 2. When the order placed in period 1 is for $d_1 + d_2$ units, the total cost is $\mathcal{C}_2^1 = F(1) + hd_2 = K_1 + hd_2$. Choose the decision with the least total cost for periods 1 and 2. That is,

$$F(2) = \min\{K_1 + K_2, K_1 + hd_2.\}$$

Set $v = 2$ if $K_1 + K_2 < K_1 + hd_2$. Otherwise, v remains unchanged.

Step 3: Consider the t-period problem. Given v, the demand for period t is satisfied by placing the order in one of the periods $v, v+1, v+2, \ldots, t$. Compute $\mathcal{C}_t^v, \mathcal{C}_t^{v+1}, \ldots, \mathcal{C}_t^{t-1}, \mathcal{C}_t^t$ using (4.3) and find

$$F(t) = \min\{\mathcal{C}_t^v, \mathcal{C}_t^{v+1}, \ldots, \mathcal{C}_t^{t-1}, \mathcal{C}_t^t\}.$$

Choose to place the order in the period with the lowest cost. Set a new value of v accordingly.

Step 4: Set $t \leftarrow t+1$. Stop if $t = T+1$. Otherwise, go to Step 3.

We illustrate the algorithm using the following example.

Consider an inventory system for a product whose life-cycle permits placement of five orders, that is, the length of the planning horizon is five periods. Whenever an order is placed, a fixed cost $K = \$100$ is incurred. The holding cost for any period t is equal to $h = \$ 1$/unit/period. The projected demands are as follows: $d_1 = 10$, $d_2 = 60$, $d_3 = 15$, $d_4 = 150$, $d_5 = 110$.

The algorithm proceeds as follows.

Iteration 1:

Step 1: We set $t = 2$ and $v = 1$. Since an order is placed in period 1, $F(1) = K$.

Step 2: $C_2^1 = K + hd_2 = 100 + (1)(60) = 160$ and $C_2^2 = F(1) + K = 200$. Thus, $F(2) = \min\{C_2^1, C_2^2\} = 160$ and the order is tentatively placed in period 1 to meet the demand for both periods. We continue to have $v = 1$.

Step 3: We next compute $C_3^1, C_3^2,$ and C_3^3.

$$C_3^1 = K + h(d_2 + d_3) + hd_3 = 100 + 1(60 + 15) + 1(15) = 190,$$
$$C_3^2 = F(1) + K + hd_3 = 100 + 100 + 1(15) = 215,$$
$$C_3^3 = F(2) + K = 160 + 100 = 260.$$

Therefore, $F(3) = 190$ and the order is tentatively placed in period 1 to meet the demand for the first three periods. The value of v remains unchanged at 1.

Step 4: Since $t < 5$, we go to the next iteration.

Iteration 2:

Step 3: We compute $C_4^1, C_4^2, C_4^3,$ and C_4^4 as follows:

$$C_4^1 = K + h(d_2 + d_3 + d_4) + h(d_3 + d_4) + h(d_4) = 640,$$
$$C_4^2 = F(1) + K + h(d_3 + d_4) + h(d_4) = 515,$$
$$C_4^3 = F(2) + K + h(d_4) = 410,$$
$$C_4^4 = F(3) + K = 290.$$

Hence $F(4) = 290$, and we produce in period 4 to meet period 4's demand. Set $v = 4$.

Step 4: Since $t < 5$, we perform one more iteration.

Iteration 3:

Step 3: We need to compute only C_5^4 and C_5^5.

$$C_5^4 = F(3) + K + h(d_5) = 400$$
$$C_5^5 = F(4) + K = 390$$

Thus we order in period 5 to meet period 5's demand.

Step 4: We stop since all the periods have been covered.

The solution to this problem is to produce 85 units in period 1, 150 units in period 4, and 110 units in period 5.

Once you understand the basic recursions, the above computations may be systematically tabulated (starting with the leftmost column) as illustrated in the following table.

–	1	2	3	4	5
1	**100**	**160**	**190**	640	–
2		200	215	515	–
3			260	410	–
4				**290**	400
5					**390**
Column Minimum	100	160	190	290	390
Production Period	1	1	1	4	5

The dynamic lot sizing problem can also be represented as a shortest-path network flow problem. In the following subsection, we demonstrate the formulation for such a representation.

4.1.3 Shortest-Path Representation of the Dynamic Lot Sizing Problem

In a shortest-path network flow problem, the objective is to identify a shortest path from the source to the destination. There are several intermediate nodes between the source and destination nodes. If a route exists between two nodes, it is indicated by a directed arc; thus the network is a directed graph. The path between the source and destination is a combination of such arcs. There is a cost corresponding to each arc/route. The shortest path is the one with the minimum total cost.

A dynamic lot sizing problem is represented using a graph having $T + 1$ stages. Node 1 is the source and node $T + 1$ is the destination. From each node t, an arc goes to all the downstream nodes, that is, to nodes $t + 1, t + 2, \ldots, T + 1$. Thus, the number of arcs that leave node t is equal to $T + 1 - t$. Similarly, arcs come into node t from nodes $1, 2, \ldots, t - 1$. An arc (s, t) represents an order placed in period s to satisfy the demands of periods $s, s + 1, \ldots, t - 1$. For example, an arc that connects nodes 1 and 3 represents an order placed in period 1 to satisfy the demands of periods 1 and 2.

The cost $c_{s,t}$ of an arc (s, t) is equal to the fixed ordering and holding costs incurred in periods s through $t - 1$. The costs incurred in periods s through $t - 1$ are

$$c_{s,t} = K + h(d_{s+1} + d_{s+2} + \cdots + d_{t-1}) + h(d_{s+2} + d_{s+3} + \cdots + d_{t-1}) + \cdots + hd_{t-1},$$

which can be written equivalently as

$$c_{s,t} = K + h \sum_{k=s+1}^{t-1} \sum_{i=k}^{t-1} d_i.$$

Let $y_{s,t} \in \{0,1\}$ be a binary variable corresponding to arc (s,t). If $y_{s,t} = 1$ in a solution, the arc (s,t) is part of the shortest path in that solution. When $y_{s,t} = 0$ in a solution, no order is placed in period s.

The complete formulation is as follows:

$$\min \sum_{s=1}^{T+1} \sum_{t=s+1}^{T+1} c_{s,t} y_{s,t}$$

subject to

$$\sum_{t=2}^{T+1} y_{1,t} = 1,$$

$$\sum_{s=1}^{k-1} y_{s,k} = \sum_{t=k+1}^{T+1} y_{k,t} \quad k = 2, 3, \ldots, T,$$

$$\sum_{s=1}^{T} y_{s,T+1} = 1.$$

Observe that to solve this shortest path problem, all values of $c_{s,t}$ must be computed in advance. Note that all these calculations are not required when solving the lot sizing problem using the Wagner–Whitin approach; refer to the table in the example found in Subsection 4.1.2. For example, to solve the shortest path problem, we would have to compute the costs $c_{1,5}, c_{2,5}$, and $c_{3,5}$ in the example, which were not computed when using the Wagner–Whitin algorithm. Hence, the dynamic programming approach that underlies the Wagner–Whitin algorithm is a more efficient way to solve the lot sizing problem.

4.1.4 Technical Appendix for the Wagner–Whitin Algorithm

The goal of this subsection is to provide certain technical details for the Wagner–Whitin algorithm. Again assume that $C_t + h_t \geq C_{t+1}$ for all t.

Previously, we stated that if $x_{t+1} = 0$ in the optimal solution, then the ordering decisions made prior to period $t+1$ do not affect the ordering decisions made in periods $t+1$ through T. Recall that we defined $F(t)$ as the optimal cost of running the system from periods 1 through t such that $x_{t+1} = 0$. Suppose period t's demand is satisfied by an order placed in period $j^*(t)$ in the t-period problem. The following theorem states that $j^*(t+1)$ cannot be less than $j^*(t)$.

Theorem 4.2. (Planning Horizon Theorem) *Suppose in the t-period problem, period t's demand is satisfied by an order placed in period $j^*(t)$ in an optimal solution. Then there exists an optimal solution for the $t + 1$-period problem in which period $t + 1$'s demand is satisfied by an order placed in period $j^*(t+1)$ such that $j^*(t+1) \geq j^*(t)$.*

How is the above theorem useful in developing the solution algorithm? In the algorithm, the iterations are indexed by time t. Thus, in each iteration, we consider a t-period problem. We assume that $x_{t+1} = 0$ in the optimal solution. Given this assumption, we identify the period in which an order is placed to satisfy period t's demand, that is, $j^*(t)$. In the subsequent iteration, we consider a $t + 1$-period problem. By virtue of the planning horizon theorem, we need to consider only $j^*(t), j^*(t) + 1, \ldots, t + 1$ (as opposed to all the prior periods) as possible periods in which an order could be placed to satisfy period $t + 1$'s demand. This observation may result in making substantially fewer calculations.

4.2 Wagelmans–Hoesel–Kolen (WHK) Algorithm

Recall that the Wagner–Whitin algorithm is based on a forward in time dynamic programming recursion. That is, we first consider a limited version of the problem consisting only of the first period while ignoring the rest of the periods in the horizon. Then we expand the problem to include period 2. Once again, we ignore the remaining periods and just solve the two-period problem. In general, we go *forward* gradually and include one additional period at a time.

An alternative to this procedure is to go backwards. That is, first consider only the Tth period and ignore periods 1 through $T - 1$. Then we go backwards and include period $T - 1$ as well. In general, we iteratively add one additional period at a time and solve the resulting problem.

As we stated earlier, the advantage of the WHK algorithm is that its worst-case computational performance is better than that of the WW algorithm. Whereas the WW algorithm's worst-case performance in terms of computational time is of the order of T^2, this algorithm's worst-case computational time is of the order of $T \log T$ for problems with time varying ordering and holding costs. When these costs are stationary, the worst-case computational time is of the order of T. Note that this does not mean that this algorithm will always perform better than the WW algorithm. The algorithm may not perform better even in terms of the average time taken when the two algorithms are run for a certain number of data instances. That WHK has better worst-case performance means that for large values of T, this algorithm will outperform the WW algorithm from a computational perspective. Remember, both algorithms find the optimal solution.

The outline for this section is as follows. First, we develop other formulations for the dynamic lot sizing problem. Then, we discuss the general approach behind the algorithm. Finally, we present the algorithm in detail for the stationary cost case and illustrate its steps with an example.

4.2.1 Model Formulation

Recall that when we discussed the Wagner–Whitin model and algorithm we let x_t represent the number of units of on-hand inventory at the beginning of period t prior to placing an order and y_t represent the on-hand inventory following the placement of an order in period t. We assumed that the lead time was zero in length. Furthermore, we let $z_t = y_t - x_t$, which equals the amount ordered in period t. We also let K_t, h_t, d_t, and C represent the fixed ordering cost, per-unit holding cost, demand, and per-unit purchase cost in period t.

Then the dynamic lot sizing problem is

$$\min \sum_{t=1}^{T} \left\{ K_t \delta_t + h_t \sum_{i=1}^{t} (z_i - d_i) \right\} + C \sum_{t=1}^{T} d_t \tag{4.4}$$

subject to

$$\sum_{i=1}^{t} z_i \geq \sum_{i=1}^{t} d_i, \quad t = 1, \ldots, T,$$

$$0 \leq z_t \leq \delta_t \sum_{i=t}^{T} d_i,$$

$$\delta_t \in \{0, 1\},$$

where δ_t indicates whether or not an order is placed in period t. The first constraint implies that the total supply must be at least as large as total demand through the end of each period. The second constraint implies that the amount purchased in a period cannot exceed the total demand in that and future periods in the planning horizon.

Now let $f_t = \sum_{i=t}^{T} h_i$, $t = 1, \ldots, T$. Then (4.4) can be written as

$$C \sum_{t=1}^{T} d_t + \min \left\{ \sum_{t=1}^{T} (f_t z_t + K_t \delta_t) - \sum_{t=1}^{T} h_t \sum_{i=1}^{t} d_i \right\}.$$

To see that this is the case, observe that

$$\sum_{t=1}^{T}\left\{f_t z_t - h_t \sum_{i=1}^{t} d_i\right\} = \sum_{t=1}^{T}\left\{z_t \sum_{i=t}^{T} h_i - h_t \sum_{i=1}^{t} d_i\right\}$$

and that

$$\sum_{t=1}^{T} z_t \sum_{i=t}^{T} h_i - \sum_{t=1}^{T} h_t \sum_{i=1}^{t} d_i = \sum_{t=1}^{T} h_t \sum_{i=1}^{t} z_i - \sum_{t=1}^{T} h_t \sum_{i=1}^{t} d_i$$

$$= \sum_{t=1}^{T} h_t \left(\sum_{i=1}^{t} (z_i - d_i)\right).$$

We have assumed, of course, that $x_1 = x_{T+1} = 0$.

Since $C\sum_{t=1}^{T} d_t$ is a constant, we ignore the variable procurement cost term in our optimization process. Also, $\sum_{t=1}^{T} h_t \sum_{i=1}^{t} d_i$ is a constant and this term can also be ignored in the model.

Next, let z_{ti} be the amount of period i's demand that is produced in period t. Then $z_t = \sum_{i=t}^{T} z_{ti}$. Combining the preceding observations and definitions, we rewrite the dynamic lot-sizing problem as

$$Z_1 = \min \sum_{t=1}^{T}\left\{K_t \delta_t + f_t \sum_{i=t}^{T} z_{ti}\right\} \tag{4.5}$$

subject to

$$\sum_{t=1}^{i} z_{ti} = d_i, \quad i = 1, \ldots, T,$$

$$d_i \delta_t \geq z_{ti}, \quad i = t, \ldots, T, \quad t = 1, \ldots, T,$$

$$z_{ti} \geq 0, \quad i, t = 1, \ldots, T,$$

$$\delta_t \in \{0, 1\}.$$

We know that in any optimal solution that $z_{ti} \in \{0, d_i\}$ and that if $z_t > 0$, $z_t \in \{d_t, d_t + d_{t+1}, \ldots, \sum_{i=t}^{T} d_i\}$.

4.2.2 An Order $T \log T$ Algorithm for Solving Problem (4.5)

We now will present the WHK algorithm for solving (4.5). As we will see, it is a backward dynamic-programming-like method.

To simplify our discussion, let us assume that $d_t > 0$ for all t, $t = 1, \ldots, T$. Let us define Z_t to be the cost of following an optimal policy over a horizon consisting of periods t through T when $x_t = 0$. Define $Z_{T+1} = 0$.

Since $z_t \in \left\{ d_t, d_t + d_{t+1}, \ldots, \sum_{i=t}^{T} d_i \right\}$ and $z_t > 0$ only if $x_t = 0$,

$$Z_t = \min_{t < s \leq T+1} \left\{ K_t + f_t \sum_{j=t}^{s-1} d_j + Z_s \right\}, \quad t = 1, \ldots, T. \tag{4.6}$$

Recursion (4.6) states that we procure an amount in period t to satisfy demands through the end of period $s - 1$ and then follow an optimal strategy thereafter. Thus we begin with $t = T$ and compute Z_T. Next, we determine Z_{T-1} using the value of Z_T. Using (4.6) we continue the process until we find Z_1 and an optimal procurement plan.

The recursion that we have described for computing Z_t requires comparing $T - t + 1$ expressions. If this process is executed in the usual manner, then the recursion results in an order T^2 algorithm, essentially the same as is required to carry out the steps of the Wagner–Whitin algorithm. However, as we will now discuss, we can calculate (4.6) in order $T \log T$ time rather than order T^2 time.

Let us define the notion of an *efficient period*. As we will see, efficient periods are important in the sense that in a t-period problem, the next order after period t can be placed only in one of these efficient periods. Suppose $t = 4$ and $T = 8$ and periods 5 and 7 are efficient periods. Then the next order after period 4 can only be placed in either period 5 or 7 but not in periods 6, 8 or 9 (an order in period 9 means no order is placed after period 4 until the end of the horizon). Thus, the number of periods that we have to consider when determining the next period in which a procurement order is placed is reduced when knowing what periods are efficient periods.

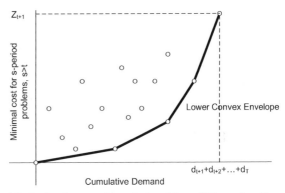

Fig. 4.1. Optimal costs of t-period problems (Z_t) as a function of cumulative demand computed backwards.

We now show how to determine efficient periods graphically. Suppose we know $Z_{t+1}, Z_{t+2}, \ldots, Z_T$. Suppose we plot these values in a graph against cumulative demands *computed backwards*. That is, we plot the points $(0,0) = (0, Z_{T+1}), (d_T, Z_T), (d_T + d_{T-1}, Z_{T-1}), \ldots, (d_T + d_{T-1} + \cdots + d_{t+1}, Z_{t+1})$. Figure 4.1 shows a sample plot. Think of these points as pins on the graph. Suppose we now take a cord and wrap it around the pins so that all the pins are above the cord and the cord is not slack. Efficient periods correspond to the pins that touch the cord. For instance, if a pin corresponding to the point (d_T, Z_T) touches the cord, then period T is efficient. Points $(0,0)$ and $(d_T + d_{T-1} + \cdots + d_{t+1}, Z_{t+1})$, being endpoints, are always touched by the cord. If we now draw a curve that imitates the cord, then this curve is called the *lower convex envelope* or the greatest convex minorant of the curve.

Suppose we have identified the efficient periods. Let s_t be the optimal value for s in (4.6), that is, the next period in which an order will be placed following period t. Observe that the lower convex envelope is a piecewise linear function. Between two consecutive efficient periods k and l, $k < l$, its slope is equal to $\frac{Z_k - Z_l}{d_k + d_{k+1} + \cdots + d_{l-1}}$. In addition, this slope is increasing as we move to the right. The value s_t in (4.6) is equal to the smallest efficient period k such that

$$\frac{Z_k - Z_l}{d_k + d_{k+1} + \cdots + d_{l-1}} > \sum_{j=t}^{T} h_j = f_t \ . \tag{4.7}$$

That is, s_t is the left endpoint of the last segment (as we go right) in the lower convex envelope whose slope is greater than f_t. With s_t known, we can compute Z_t.

Observe that the right-hand side of inequality (4.7) increases as t decreases. This means that the slope of the segment that determined s_{t+1} may not satisfy the inequality for the t-period problem. Therefore, $s_t \leq s_{t+1}$. This observation is helpful because when we search for s_t we need only consider efficient periods from $t+1$ through s_{t+1}.

We now discuss how to determine efficient periods algebraically. For the t-period problem, clearly period t is an efficient period. It turns out that other efficient periods for the t-period problem form a subset of efficient periods for the problem beginning in period $t+1$. In other words, to search for the efficient periods for the t-period problem, it suffices to consider the set of efficient periods for the $(t+1)$-period problem. Furthermore, only a certain number of consecutive periods that lie adjacent to t cease to be efficient as we go from the problem beginning in period $t+1$ to the problem beginning in period t. More precisely, if $t+1 = t_1 < t_2 < \cdots < t_r < \cdots < t_n = T+1$ are the efficient periods for the $(t+1)$-period problem, then the set of efficient periods for the t-period problem will consist of all of these periods (plus period t, of course) except $t_1, t_2, \ldots, t_{r-1}$ for some r. It is possible that all the periods that are efficient in the $(t+1)$-period problem continue to remain efficient for the t-period problem.

Suppose t_r is the smallest period such that

$$\frac{Z_t - Z_{t_r}}{d_t + d_{t+1} + \cdots + d_{t_r-1}} > \frac{Z_{t_r} - Z_{t_{r+1}}}{d_{t_r} + d_{t_r+1} + \cdots + d_{t_{r+1}-1}} \ . \tag{4.8}$$

Then periods $t_1, t_2, \ldots, t_{r-1}$ cease to be efficient as we go from the $(t+1)$-period problem to the t-period problem. If the above inequality holds for t_1, then all the efficient periods for the $(t+1)$-period problem remain efficient for the t-period problem. Another important fact that we state without proof is that $t_r \leq s_{t+1}$. This fact reduces the number of efficient periods that we need to consider to determine t_r.

Why is it that some periods cease to be efficient when we consider the t-period problem? Recall how we generated the efficient periods by wrapping a cord around the pins. Suppose we have the pins that correspond to periods $t+1$ through $T+1$ and we add another point $(d_t + d_{t+1} + \cdots + d_T, Z_t)$. If we wrap the cord around the pins once again, we now have to include the pin corresponding to period t. But in

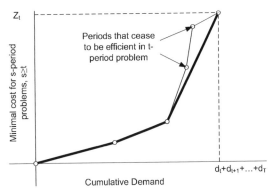

Fig. 4.2. Elimination of some efficient periods as we include period t.

doing so, as Figure 4.2 shows, some of the points that were touching the cord earlier do not do so anymore. Consequently, they are not part of the new lower convex envelope and are no longer efficient.

We now describe an iterative procedure based on these observations. The algorithm is described and illustrated with an example in the following subsection.

Before we present the details of the WHS algorithm let us discuss why it can be executed in the order of $T \log T$ time. Clearly f_t, K_t and $\sum_{i=t}^{T} d_i = \sum_{i=t+1}^{T} d_i + d_t$, $t = 1, \ldots, T$, can be found in time proportional to T.

In each of the T iterations of the algorithm we will calculate Z_t, $t = 1, \ldots, T$. The value of Z_t is found during iteration $T - t + 1$ since the algorithm is initiated with $t = T$. Let S be an ordered list of the efficient points at the time iteration $T - t + 1$ is executed. Then the value of t_r that satisfies (4.8) can be found using a binary search on S, which can be accomplished in the order of $\log T$ time. In each iteration, some efficient periods in S at the beginning of the iteration may need to be removed from S. That is, they are no longer efficient periods when the next iteration is executed. Each period can only be removed from the set of efficient periods once. Furthermore, at each comparison in an iteration we either conclude that we have found t_r or we move to the next period considered in the order list S and continue the binary search. Each such calculation is made for each of the T periods in the planning horizon. Combining these observations, we see that the total complexity associated with calculating Z_1 is of the order of $T \log T$.

Let us focus on a special case when $h_t = h$. Then $f_t = (T - t + 1)h$ and therefore $f_t > f_{t+1} > \cdots > f_T$. In this case, the algorithm, which we will state shortly, can be solved in linear time. In fact, this occurs in general when $f_1 \geq f_2 \geq \cdots \geq f_T$. This is the case for the Wagner–Whitin model. This occurs because the binary search process can be replaced by a search that requires only a number of comparisons that in total is of the order of T.

4.2.3 Algorithm

We will now state the algorithm for solving the economic lot size problem in order T time. We assume that the elements of S are ordered from smallest to largest, where S is the set of efficient periods.

Step 0: Let $S = \{T, T + 1\}$, $t = T$, $s_T = T + 1$, $Z_{T+1} = 0$, and compute $Z_T = K + h d_T$.

Step 1: $t = t - 1$. If $t = 0$, go to Step 4; otherwise, go to Step 2.

Step 2: Find the smallest $k \in S$, $k \leq s_{t+1}$, such that

$$\frac{Z_k - Z_l}{d_k + \cdots + d_{l-1}} < (T - t + 1)h,$$

where l is the next largest efficient period exceeding k. Set $s_t = k$.

Given k, compute $Z_t = K + (T - t + 1)h \sum_{i=t}^{k-1} d_i + Z_k$.

Step 3: Let $S = \{t_1, \ldots, t_q\}$, $t_1 < t_2 < \cdots < t_q$, be the set of efficient periods. Let $t_r \in S$, $t_r < s_t$, be the smallest efficient period for which

$$\frac{Z_t - Z_{t_r}}{\sum_{i=t}^{t_r-1} d_i} > \frac{Z_{t_r} - Z_{t_{r+1}}}{\sum_{i=t_r}^{t_{r+1}-1} d_i},$$

if one exists.

Let $S = \{t\} \cup (S \setminus \{t_1, \ldots, t_{r-1}\})$, the new set of efficient periods.

Return to Step 1.

Step 4: Compute the procurement policy:

(a) Set $t = 1$
(b) $z_t = \sum_{i=t}^{s_t-1} d_i$
(c) Set $t = s_t$ and go to Step 4(b)

Let us now illustrate the workings of the above algorithm using the data from the example in Section 4.1.2.

Iteration 1:

Step 0: $S = \{5,6\}$, $t = 5$, $s_5 = 6$, $Z_6 = 0$, and $Z_5 = 100 + (1)(110) = 210$.

Step 1: $t = 4$. Since $t > 0$, go to Step 2.

Step 2: s_4 must be either 5 or 6. (If it were 5, then a procurement order for d_4 units would be placed in period 4; otherwise, $d_4 + d_5$ units would be ordered.) Begin with $k = 5$. Since $\frac{Z_5 - Z_6}{d_5} = \frac{210}{110} = 1.91 < 2 = (5 - 4 + 1)(1)$, $s_4 = 5$. $Z_4 = K + (T - t + 1)hd_4 + Z_5 = 100 + 300 + 210 = 610$.

Step 3: Because $s_4 = 5$, no elements of S will be dropped. $S = \{4\} \cup \{5,6\} = \{4,5,6\}$. Return to Step 1.

Iteration 2:

Step 1: $t = 3 > 0$, so go to Step 2.

Step 2: Since $s_4 = 5$, we need only consider efficient periods 4 and 5 when computing s_3. With $k = 4$, $\frac{Z_4 - Z_5}{d_4} = 2.67 < 3$ and therefore $s_3 = 4$. Furthermore, $Z_3 = K + (T - t + 1)hd_3 + Z_4 = 755$.

Step 3: Since $s_3 = 4$, $S = \{3\} \cup \{4,5,6\} = \{3,4,5,6\}$. Return to Step 1.

Iteration 3:

Step 1: $t = 2 > 0$, and thus go to Step 2.

Step 2: Because $s_3 = 4$, we only consider efficient periods 3 and 4 when finding s_2. We begin with $k = 3$. Since $\frac{Z_3 - Z_4}{d_3} = 9.67 > 4$, $s_2 \neq 3$ and hence $s_2 = 4$, the only alternative. $Z_2 = K + (T - t + 1)h(d_2 + d_3) = 1010$.

Step 3: Since $s_2 = 4$, we need only consider $t_r = 3$ for elimination from S. Because $\frac{Z_2 - Z_3}{d_2} = 4.25 < 9.67 = \frac{Z_3 - Z_4}{d_3}$, period 3 is removed from S. After adding period 2 to S, $S = \{2,4,5,6\}$. Return to Step 1.

Iteration 4:

Step 1: $t = 1 > 0$, and hence go to Step 2.

Step 2: Because $s_2 = 4$, we must consider only efficient periods 2 and 4 when finding s_1. But when $k = 2$, $\frac{Z_2 - Z_4}{d_2 + d_3} = 5.33 > 5$ and therefore $s_1 \neq 2$. Consequently $s_1 = 4$ and $Z_1 = K + (T - t + 1)h(d_1 + d_2 + d_3) + Z_4 = 1135$.

Step 3: Period 2 is eliminated from S and $\{1\}$ is added to S. Hence, $S = \{1,4,5,6\}$. Return to Step 1.

Iteration 5:

Step 1: Since $t = 0$, go to Step 4.

Step 4: Set $t = 1$. Since $s_1 = 4$, $z_1 = d_1 + d_2 + d_3 = 85$. Next, set $t = s_1 = 4$. Because $s_4 = 5$, $z_4 = 150 = d_4$. Lastly $z_5 = 110$.

The example demonstrates why the algorithm in this case is a linear time algorithm.

4.3 Heuristic Methods

As we noted earlier, the execution of the Wagner–Whitin algorithm may require computation of cost terms for all production periods. Hence the algorithm may require a number of calculations proportional to T^2. Thus, algorithms and heuristics that require fewer potential computations are desirable. The Wagelmans–Hoesel–Kolen is one such algorithm. In this section, we present two such heuristics: the Silver–Meal heuristic and the least unit cost heuristic.

The heuristics imitate the algorithms we have discussed to this point in that they do not allow an order to include only part of the demand of a period. Specifically, they seek to determine the number of consecutive periods whose demands should be included in an order that is placed. However, they use the average cost criterion to identify the order quantity. They do not identify an optimal policy to minimize the total cost over the planning horizon; they merely employ the average cost criterion to identify a reasonable policy. In doing so, the heuristics use a rule to establish lot sizes that are consistent with the rule used to obtain an optimal lot sizing decision in the basic EOQ model.

4.3.1 Silver–Meal Heuristic

The first heuristic we will describe is the Silver–Meal heuristic. Define $\mathcal{C}(s,t)$ as the average cost over periods $s, s+1, \ldots, t$ such that an order is placed in period s to satisfy the demands of periods $s, s+1, \ldots, t$. We obtained an expression for the total cost incurred over periods $s, s+1, \ldots, t$ in (4.2). The average cost $\mathcal{C}(s,t)$ is equal to the total cost divided by the number of periods $t - s + 1$. Thus,

$$\mathcal{C}(s,t) = \frac{K + h(d_{s+1} + d_{s+2} + \cdots + d_t) + h(d_{s+2} + d_{s+3} + \cdots + d_t) + \cdots + h(d_t)}{t - s + 1}.$$

The basic idea on which the heuristic is constructed is as follows. Suppose the order placed in period s includes demands of periods $s, s+1, \ldots, t-1$ and we are considering whether to add period t's demand to the order as well. This decision depends on whether the average cost decreases or increases after the inclusion of period t's demand in the order. If the average cost decreases, that is, $\mathcal{C}(s,t-1) > \mathcal{C}(s,t)$, then the order is expanded to include period t's demand. Otherwise, if the average cost increases, the order quantity is equal to $d_s + d_{s+1} + \cdots + d_{t-1}$ and a fresh order is placed in period t. Therefore, the heuristic seeks to find an order quantity such that a local minimum is attained according to the average cost criterion. For this heuristic to find the optimal solution, the average cost calculations would have to possess the following property. Suppose for period s we find that t, $t > s$, is the first period for which $\mathcal{C}(s,t-1) < \mathcal{C}(s,t)$. This is the period that would be found using the Silver–Meal heuristic. Then for this period to be optimal, $\mathcal{C}(s,t-1)$ would have to satisfy $\mathcal{C}(s,t-1) \leq \mathcal{C}(s,j)$, $j > t$. This relation would have to hold for all s. This does not have to happen, however.

The steps of the heuristic are outlined below.

Silver–Meal Heuristic

Step 1: Set $s = 1$ and $t = 2$. Set $Q(s) = Q(1) = d_1$ as the current order quantity and compute $\mathcal{C}(1,1) = K$.

Step 2: If $\mathcal{C}(s,t) < \mathcal{C}(s,t-1)$, increase $Q(s)$ by d_t and go to Step 3.

Otherwise, $Q(s) = d_s + d_{s+1} + \cdots + d_{t-1}$. Set $s = t$, $Q(t) = d_t$, $\mathcal{C}(t,t) = K$ and go to Step 3.

Step 3: Set $t \leftarrow t + 1$. Stop if $t = T + 1$.

Otherwise, go to Step 2.

We demonstrate the heuristic using the data given in the example in Section 4.1.2.

Iteration 1:

Step 1: Set $s = 1$, $t = 2$, $Q(1) = d_1 = 10$, and $C(1,1) = 100$.

Step 2: We compute

$$\mathcal{C}(1,2) = \frac{K + hd_2}{2} = \frac{100 + 1(60)}{2} = 80,$$

which is less than $\mathcal{C}(1,1) = \frac{K}{1} = 100$. Hence, $Q(1)$ is increased by d_2 to 70.

Step 3: Set $t = 3$. Since $t < T + 1 = 6$, we proceed to the next iteration.

Iteration 2:

Step 2: We compute

$$\mathcal{C}(1,3) = \frac{K + h(d_2 + d_3) + hd_3}{3} = \frac{100 + 1(60 + 15) + 1(15)}{3} = 63.33,$$

which is less than $\mathcal{C}(1,2) = 80$ as computed in Iteration 1. Hence, $Q(1)$ is increased by d_3 to 85.

Step 3: Increase t by 1 to 4. Since $t < T + 1$, we proceed to the next iteration.

Iteration 3:

Step 2: We compute

$$\mathcal{C}(1,4) = \frac{K + h(d_2 + d_3 + d_4) + h(d_3 + d_4) + hd_4}{4}$$

$$= \frac{100 + 1(60 + 15 + 150) + 1(15 + 150) + 1(150)}{4} = 160,$$

which is greater than $\mathcal{C}(1,3) = 63.33$ computed in Iteration 2. As a result, period 4's demand will not be satisfied by the order placed in period 1. The size of period one's order is 85 units.

Another order is placed in period 4. We set $s = 4$, $Q(4) = d_4 = 150$, and $\mathcal{C}(4,4) = K = 100$.

Step 3: Increase t by 1 to 5. Since $t < T + 1$, we proceed to the next iteration.

Iteration 4:

Step 2: We compute

$$\mathcal{C}(4,5) = \frac{K + hd_5}{2} = \frac{100 + 1(110)}{2} = 105$$

which is greater than $\mathcal{C}(4,4) = 100$. As a result, the order placed in period 4 will not satisfy period 5's demand. The quantity ordered in period 4 is equal to its demand, 150 units.

Another order is placed in period 5. We set $s = 5$ and $Q(5) = d_5 = 110$.

Step 3: Increase t by 1 to 6. Since $t = T + 1$, we stop.

Thus, the policy proposed by the heuristic is to place orders for 85 units, 150 units, and 110 units in periods 1, 4, and 5, respectively. Observe that the same policy was prescribed by the Wagner–Whitin algorithm and hence it is optimal.

4.3.2 Least Unit Cost Heuristic

The least unit cost heuristic is almost identical to the Silver–Meal heuristic. The two heuristics differ in only one aspect: the criterion for placing a new order. Whereas in the Silver–Meal heuristic we use the average cost per unit time $\mathcal{C}(s,t)$ to determine when to place another order, in the least unit cost heuristic we use a slightly different measure to determine when to place the next order. The new one, which we denote by $\mathcal{C}_1(s,t)$, measures the average cost per unit demanded during an order cycle. That is,

$$\mathcal{C}_1(s,t) = \frac{K + h(d_{s+1} + d_{s+2} + \cdots + d_t) + h(d_{s+2} + d_{s+3} + \cdots + d_t) + \cdots + h(d_t)}{d_s + d_{s+1} + \cdots + d_t}.$$

(Note that the numerators of $\mathcal{C}(s,t)$ and $\mathcal{C}_1(s,t)$ are identical.) The application of this measure to determine when to place the next order is identical to the Silver–Meal heuristic. To see this, suppose that the order placed in period s includes demands of periods $s, s+1, \ldots, t-1$ and we are considering whether to add period t's demand to the order as well. No order is placed in period t if the cost per unit demand decreases, that is, $\mathcal{C}_1(s, t-1) < \mathcal{C}_1(s,t)$; the order in period s is expanded to include period t's demand. Otherwise, if the cost per unit demand increases, a new order is placed in period t. The steps of the algorithm can be obtained by replacing $\mathcal{C}$ by $\mathcal{C}_1$, and we omit the details. We next illustrate the heuristic using the data given in the example in Section 4.1.2.

Iteration 1:

Step 1: Set $s = 1$, $t = 2$, and $Q(1) = d_1 = 10$.

Step 2: We compute

$$\mathcal{C}_1(1,2) = \frac{K + hd_2}{d_1 + d_2} = \frac{100 + 1(60)}{10 + 60} = 2.29,$$

which is less than $\mathcal{C}_1(1,1) = \frac{K}{10} = 10$. Hence, $Q(1)$ is increased by d_2 to 70.

Step 3: Set $t = 3$. Since $t < T + 1 = 6$, we proceed to the next iteration.

Iteration 2:

Step 2: We compute

$$\mathcal{C}_1(1,3) = \frac{K + h(d_2 + d_3) + hd_3}{d_1 + d_2 + d_3} = \frac{100 + 1(60 + 15) + 1(15)}{10 + 60 + 15} = 2.24,$$

which is less than $\mathcal{C}_1(1,2) = 2.29$ computed in Iteration 1. Hence, $Q(1)$ is increased by d_3 to 85.

Step 3: Increase t by 1 to 4. Since $t < T + 1$, we proceed to the next iteration.

Iteration 3:

Step 2: We compute

$$\mathcal{C}(1,4) = \frac{K + h(d_2 + d_3 + d_4) + h(d_3 + d_4) + hd_4}{d_1 + d_2 + d_3 + d_4}$$

$$= \frac{100 + 1(60 + 15 + 150) + 1(15 + 150) + 1(150)}{235} = 2.72,$$

which is greater than $\mathcal{C}(1,3) = 2.24$. As a result, period 4's demand will not be satisfied by the order placed in period 1. The quantity of this order is equal to 85.

Another order is placed in period 4. We set $s = 4$, $Q(4) = d_4 = 150$, and $\mathcal{C}(4,4) = \frac{K}{d_4} = \frac{100}{150} = 0.67$.

Step 3: Increase t by 1 to 5. Since $t < T + 1$, we proceed to the next iteration.

Iteration 4:

Step 2: We compute

$$\mathcal{C}(4,5) = \frac{K + hd_5}{d_4 + d_5} = \frac{100 + 1(110)}{260} = 0.81$$

which is greater than $\mathcal{C}(4,4)$. As a result, the order placed in period 4 will not satisfy period 5's demand. The quantity ordered in period 4 is equal to its demand, 150 units.

Another order is placed in period 5. Thus, the policy proposed by the heuristic is to place orders for 85 units, 150 units, and 110 units in periods 1, 4, and 5, respectively. Observe that the same policy was prescribed by the Wagner–Whitin algorithm and Silver–Meal heuristic.

4.4 A Comment on the Planning Horizon

After studying the theorems and propositions in the preceding sections of this chapter, it would be easy to conclude that there will always be a horizon length such that period one's production plan will always be fixed at some amount no matter how many more periods are added to the planning horizon. Unfortunately, this is not always the case.

Let us consider the following example. Suppose $K = 150$, $h = 1$, and $d_1 = d_2 = 110$ and $d_n = 100$ for $n \geq 3$. It would appear that the best strategy would be to produce every other period.

Suppose the planning horizon consists of three periods. Then the optimal solution is to produce 110, 210, and 0 units in periods one, two and three, respectively. Suppose the planning horizon consists of four periods. Then the optimal solution is to produce 220, 0, 200, 0 in periods one through four, respectively. Note that increasing the length of the planning horizon alters the production quantity in period one. In fact, in general, the optimal policy when the planning horizon consists of an odd number of periods is to produce $110, 210, 0, 200, 0, \ldots$ units in periods one, two, and beyond. When the planning horizon consists of an even number of periods, the optimal policy is to produce $220, 0, 200, 0, 200, \ldots$ units in periods one, two, and beyond, respectively. Hence there does not exist a planning horizon length for which period one's production quantity is unaltered as the horizon length is extended by one period.

4.5 Exercises

4.1. The cost model that we developed charged holding costs based on the period ending on-hand inventory level. Suppose we consider charging holding costs continually through time. Thus if the inventory level is x at time t in period j, we would charge holding costs at that instant proportional to the quantity x. If h_j is the holding cost rate for period j and T_j is the length of period j, h_j/T_j is the instantaneous rate at which the holding cost is incurred. Prove that the optimal ordering policy is unaffected by the inclusion of these holding costs in an optimization model.

4.2. When developing the Wagner–Whitin model and establishing properties affecting the computational procedure for finding the optimal procurement policy, we assumed that costs were stationary through time. Suppose purchasing and holding costs are time-dependent. Develop a forward dynamic programming model and an algorithm for finding the optimal procurement policy in this case.

4.3. A manufacturer produces a variety of integrated circuits. The cost of setting up to produce a lot of one particular type of integrated circuit is $4,000. The variable production cost is estimated to be $8. By collaborating with its customers, the manufacturer estimates that the demand during the next 8 months will be as follows: 15,000; 24,000; 20,000; 10,000; 5,000; 5,000; 20,000; and 4,000 units. The holding cost rate is .24 on an annual basis. Assume production times are small compared to the length of a month. Determine the optimal periods in which lots should be produced and the size of the corresponding production batches. Suppose that there are 5000 circuits on hand initially and that it is desired to have at least 4,000 units on hand at the end of the eighth month. Solve the problem using the Wagner–Whitin algorithm, the Wagelmans–Hoesel–Kolen algorithm, the Silver–Meal heuristic, and the least unit cost heuristic.

4.4. A component used in a manufacturing facility is ordered from an outside supplier. The component is used in a variety of finished items and therefore the demand is high. Forecast demand (in thousands) over the next 8 weeks is:

Week	1	2	3	4	5	6	7	8
Demand	21	33	29	15	9	41	50	45

Each component costs 70 cents and the inventory carrying charge rate is 0.56 cents ($0.0056) per unit per week. The fixed order cost is $300. Assume instantaneous delivery.

(a) What is the optimal ordering policy? Obtain the optimal policy using both the Wagner–Whitin algorithm and the least unit cost heuristic.
(b) What is the ordering policy recommended by the Silver–Meal heuristic?

4.5. Repeat the above questions with the following data set.

Week	1	2	3	4	5	6	7	8	9	10	11	12
Demand	69	29	36	61	61	26	34	67	45	67	79	56
Setup Cost	85	102	102	101	98	114	105	86	119	110	98	114

Take $h = 1$. Note that the setup cost varies over time.

4.6. Consider the deterministic lot sizing problem.

(a) Show that in any optimal policy $x_t(y_t - x_t) = 0$ for all t.
(b) Prove that either the order quantity in period t is zero or it is equal to $d_t + d_{t+1} + \cdots + d_k$ for some $k \geq t$.

4.7. Prove the following statements in the context of a deterministic lot sizing problem.

(a) In the optimal policy, if d_t is satisfied by an order placed in period j, $j < t$, then $d_{j+1}, d_{j+2}, \ldots, d_{t-1}$ are also satisfied by the same order.
(b) Suppose the optimal policy is unique (this is just to keep the statement simple). If $x_t = 0$ in the optimal policy, then the optimal values of $y_j, j = 1, 2, \ldots, t-1$ in the T-period model are same as the corresponding optimal values in a $t-1$-period model. In a $t-1$-period model, the goal is to satisfy only the demands of periods 1 through $t-1$.
(c) The planning horizon theorem.
(d) In the dynamic lot sizing problem, suppose period t's demand is such that $hd_t > K$. Then an order is placed in period t in any optimal policy.

4.8. Consider the Wagner–Whitin algorithm. Recall that we used $j^*(t)$ to denote the period in which an order is placed to satisfy period t's demand. We know that $j^*(t + 1) \geq j^*(t)$. That is, the order for period $t + 1$ in the $t + 1$-period problem is placed in period $j^*(t)$ or later.

In the implementation of the algorithm, we consider $j^*(t), j^*(t) + 1, \ldots, t + 1$ as possible periods in which an order could be placed to satisfy period $t + 1$'s demand. Let k be one such period, that is, $k \in \{j^*(t), j^*(t) + 1, \ldots, t + 1\}$. Suppose k is such that the cost of holding d_{t+1} from period k to $t + 1$, $(t + 1 - k)hd_{t+1}$, is greater than the fixed cost K. Show that in the optimal policy, an order cannot be placed in period k to satisfy period $t + 1$'s demand.

Specifically, you have to show that it is *better* to place an order in period $t + 1$ itself than to place it in period k.

4.9. Consider a company facing a demand pattern and costs as follows:

Month	1	2	3	4	5	6
Demand	50	20	40	20	25	60

Take the single-item cost as $C = \$5$, the holding cost rate as $I = .20$, and the setup cost as $K = \$50$. Determine a replenishment strategy using the Silver–Meal heuristic and the least unit cost heuristic.

4.10. Mark the following statement as true or false and justify your answer.

Suppose the setup cost in the dynamic lot sizing problem is stationary over time. If we increase the setup cost, the total number of orders placed over the horizon is likely to decrease in the Wagner–Whitin algorithm.

4.11. Consider a computer making firm that uses a Materials Requirement Planning (MRP) system for desktops. To illustrate how the two optimization methods or the heuristics could be incorporated into an MRP system, consider the following simple product structure for a desktop.

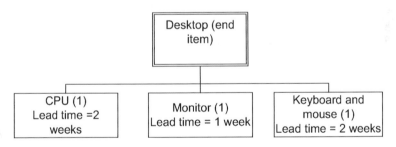

Fig. 4.3. Product structure for a desktop.

Assume that we are in week 1 of the current planning horizon. Suppose the projected net requirement of the desktops (weeks 3 through 9) are as follows:

Week	3	4	5	6	7	8	9
Demand	200	150	140	210	200	100	190

Using the product structure, the *planned order release* (that is, the requirement) for CPU units (given two-week lead time) can be generated as follows:

Week	1	2	3	4	5	6	7
Demand	200	150	140	210	200	100	190

Determine a production schedule for CPU units using the Wagner–Whitin algorithm or the least unit cost heuristic. Take the holding cost to be $1 per unit per period and the fixed cost of production to be $250.

4.12. Discuss in detail why the Wagner–Whitin algorithm requires calculations proportional to T^2 in the worst case.

4.13. Suppose the variable and fixed procurement costs and the per-unit per-period holding costs are time-dependent. Denote them by C_t, K_t, and h_t, respectively. As usual, let d_t represent the demand in period t, and assume $d_t > 0$ for all $t = 1, \ldots, T$. Generalize the Wagelmans–Hoesel–Kolen algorithm to this case.

5

Single-Period Models

The models and environments discussed in the preceeding chapters have all been based on the premise that the demand process is deterministic. In most real situations, this assumption is violated. Our goal in this, and in subsequent chapters, is to consider uncertainty directly in the decision models. In this chapter we will study the most basic models in which demand is described by a random variable. These models pertain to situations in which only a single procurement decision is made, and the effect of that decision is felt over a single period of finite duration. These models are often called one-shot or newsvendor models.

The newsvendor name for this type of model arises for the following reason. Suppose a newsvendor operates a corner newsstand. Each day the newsvendor places an order for newspapers which will be delivered to the newsstand the following morning. Only one order can be placed for the papers. Suppose there is a cost to purchase each newspaper and that there is a selling price as well. Demand for the newspapers occurs throughout the day. If this demand exceeds the quantity ordered, there are lost sales. There may be long-term consequences of lost sales since unsatisfied customers may look elsewhere for papers in the future. On the other hand, if the demand is less than the supply of papers, the newsvendor will incur the cost of disposing of the unsold papers. Day old papers have no value. The decision faced by the newsvendor is therefore how many papers should be purchased.

There are many situations to which single-period models may apply. For example, oftentimes stores will buy clothing to meet demand in a subsequent season. A single order is placed with the supplier for goods that will be sold throughout the fashion season. In the military, spare parts are placed on submarines prior to the beginning of a lengthy patrol. The types and quantities of each part type to stock on the submarines must be selected carefully to ensure the success of the mission.

J.A. Muckstadt and A. Sapra, *Principles of Inventory Management: When You Are Down to Four, Order More*, Springer Series in Operations Research and Financial Engineering, DOI 10.1007/978-0-387-68948-7_5, © Springer Science+Business Media, LLC 2010

In this chapter we will develop a basic philosophy for structuring models for environments in which uncertainty is present. We will demonstrate how to formulate models and will establish methods for obtaining optimal stock levels. We will start by examining a situation in which there is only a single item and then show how to extend the ideas to cases where there are many items present.

5.1   Making Decisions in the Presence of Uncertainty

Making inventory procurement decisions when uncertainty is present requires a perspective that differs from the ones we have discussed to this point. The factors and structural relationships that are the basis for the deterministic problems discussed in earlier chapters must be extended when uncertainty exists. Our modeling framework must now identify the sources and nature of uncertainty. We must determine how uncertainty can be modeled, and how the decision problem should be structured. While we may still want to determine when and how much to purchase, we need to state the objective that is desired in mathematical terms. Thus we must be able to ascertain the consequences of the decisions being made in operational and economic terms. As in earlier analyses, we must not only construct appropriate decision models, but we must establish approaches for finding the answers to these problems. Let us now examine a very simplified problem that captures some concepts that underlie the building of models in the presence of uncertainty.

5.2   An Example

Our goal is to determine a best stock level for a part that is integral to the operations of an offshore oil rig. Each week, this rig is resupplied from stocks on shore. Furthermore, any units of this part that remain unused on the rig at the time of replenishment of the part's stock are taken back to the onshore location where they are refurbished. Hence, there is no carryover of stock from one week to the next. Given this situation, a planner each week determines how much stock to place on the oil rig. When making this decision, the planner is cognisant of the fact that if this part fails during the week and there is no stock on hand on the rig, production ceases until a replacement unit is airlifted from the onshore facility to the rig. Unfortunately, the number of failures of this part during a week is unknown in advance, which makes the selection of the proper stocking level a difficult task.

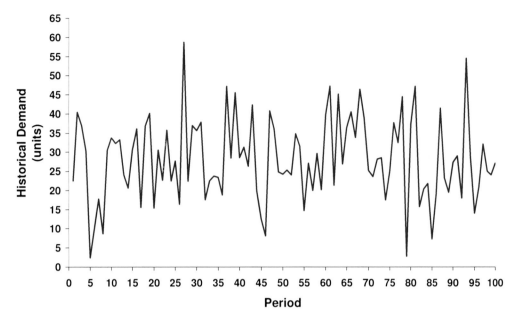

Fig. 5.1. Demand data for the past 100 weeks.

In order to make a stocking decision, we must construct a model. But what objective should be chosen? What data are required to build the model? How will the consequences of the decision be measured?

Clearly, there are two types of data that must be considered. First, we must have an idea of how much demand could occur. Consequently, we will need to examine failure data from past history. Second, we must have economic data, that is, the cost of refurbishing unused units at the end of a week and the impact of not having stock on hand on the rig when a failure occurs.

5.2.1 The Data

Fortunately, we have data pertaining to failures that have occurred during the past 100 weeks. These data are displayed in Figure 5.1. Before proceeding, we must ascertain whether or not the demand process is stationary and if there is any correlation in the demands from week to week.

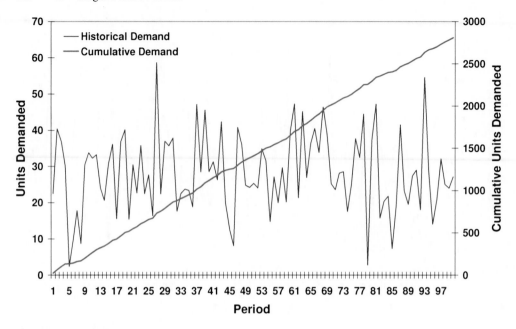

Fig. 5.2. Demand data.

Figure 5.2 contains a plot of the cumulative demand over the 100 weeks. A quick inspection of that graph indicates that demand is likely accumulating at a constant rate. Subsequent statistical tests suggest that there is no reason to assume otherwise.

Similarly, a statistical test for independence indicates that there is no justification for rejecting the hypothesis that demand is independent from week to week. Next, we use the demand data to construct a histogram of the weekly demand. This histogram will be used to hypothesize a model of the uncertain nature of the demand process. (Note in general we would use the distribution of forecast errors as the basis for modeling uncertainty rather than this histogram.) This histogram is displayed in Figure 5.3, along with a graph of the cumulative demand.

At this point we must understand how to interpret the data displayed in this histogram. Although the sample size (100 observations) is reasonably large, these data do not represent the future of the demand process perfectly. For example, will we never experience a week in which there are 39 failures? Is it really the case that it is six times as likely to observe 29 failures than it is to observe 30 failures in a week? In other words, are we willing to use these historical data as the sole predictor of future events? Rather than using these data in this way, we act as if these data are just random observations. We construct the cumulative frequency of these observations and see in Figure 5.3 that the resulting graph resembles one that corresponds to a normally distributed random

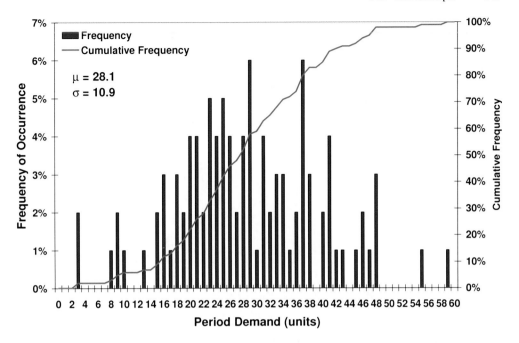

Fig. 5.3. Histogram of demand data.

variable. A statistical test indicates that we cannot reject this hypothesis. Hence we will use this representation of the demand process in our decision model.

We next turn to the cost data. Upon examining records, we estimate that it costs, on average, $1000 to refurbish an unused unit that is brought back to the onshore facility at the end of a week. Furthermore, on average, the cost incurred when there is insufficient stock on hand (a backorder occurs) is $9000. This cost includes the value of lost production and the cost of bringing a unit from shore to the oil rig using a contractor's helicopter. We assume in the following that sufficient quantities of the part are owned by the company and hence we are interested only in the incremental cost effect of stocking or not stocking the units on the oil rig.

5.2.2 The Decision Model

Each week's operation of the oil rig is identical to that of other weeks given our analysis of the demand (failure) and cost data. Hence it is reasonable to assume that the desired quantity to stock of our part is the same each week. Much more will be said about this

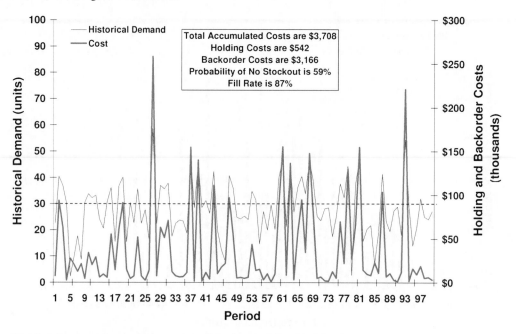

Fig. 5.4. Cost and demand time series when the stock level is 30 units.

topic in the next chapter. Suppose we had chosen to stock either 30, 40, or 100 units at the beginning of each week. Figures 5.4 through 5.6 contain graphs and data pertaining to the consequences of operating with these different stocking levels.

Each figure contains time series graphs of weekly demand and the corresponding weekly operating costs that would have resulted from using the particular stock level. Additionally, there are data indicating the total accumulated operating costs (in thousands of dollars), holding costs (in thousands of dollars), and backorder costs (in thousands of dollars) that would have been incurred over the 100-week horizon. Finally, there are data indicating the probability that all demand occurring during a week could be met from on-hand stock and the fraction of total demand met from on-hand stock over the 100-day horizon. We call the latter quantity the fill rate.

Let us compare the information displayed in these three figures. The average demand per week is about 30 units. Hence the data found in Figure 5.4 indicates how costs would be incurred if no safety stock were held. By adding 10 units of safety stock (a stock level of 40 units), we see that the operating costs would have been about $1.7 million lower. Note that holding costs increase and shortage costs decrease as safety stock is added. We also observe that the probability of having no stockout in a period is lower than the fill rate. The probability of no stockout measures the fraction of weeks during which no stockout occurred. When the stock level is 40, during 85 of the 100

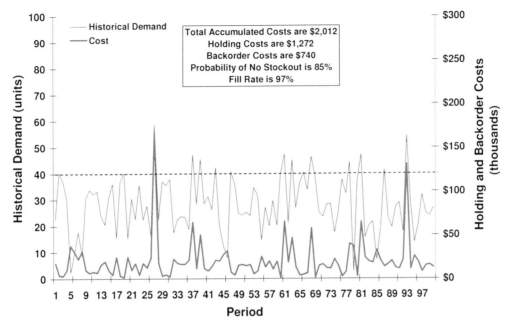

Fig. 5.5. Cost and demand time series when the stock level is 40 units.

weeks we see that demand was less than or equal to 40 units. In week 46 demand was 41 units and hence was one of the 15 weeks in which demand exceeded supply. But 40 of the 41 units of demand were satisfied from on-hand stock. While in general it is not true that the probability of no stockout in a period is less than the fill rate, it is the case for situations of the type we see in this example.

Note further that as the stock level increases, the average cost per week changes as well as the variance of the operating costs incurred per week. Observe that the variance reduces as the stock level increases.

The question remains as to how we should select the best stock level. What should be our objective when choosing this "best" level? Decisions of the sort we are making— that is, inventory decisions—are normally made so as to keep the average cost as low as possible. However, other measures could be used, and are in different contexts. Suppose we choose the stock level that results in the lowest average cost per week. This solution might result in a high variance in the weekly costs, which in some situations would not be tolerable by management. Observe in Figures 5.4 through 5.6 how the variance in the weekly operating costs changes as the stock level increases. We assume that our decision maker's utility function is such that the goal is to minimize the expected weekly operating costs.

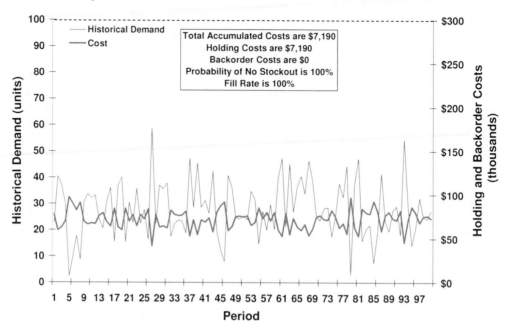

Fig. 5.6. Cost and demand time series when the stock level is 100 units.

Let us now construct a mathematical model that represents the environment we have discussed. We begin by defining nomenclature that we will use in the model. Let

D represent the demand random variable in each period,

s represent the target stock level,

h represent the holding (refurbishment) cost per unit per period,
     for unused units at the end of a week, and

b represent the economic cost per unit short.

Suppose we somehow know the amount of demand, D, in the coming week and suppose we stocked s units on the oil rig at the beginning of a week. Then the cost function for the week is given by

$$c(s,D) \equiv h\max\,(0,s-D) + b\max\,(0,D-s)$$
$$= h[s-D]^{+} + b[D-s]^{+},$$

which is displayed in Figure 5.7.

For the time being, suppose D is a discrete-valued random variable, and let $P(D)$ be the probability that demand in a week is equal to D.

Recall that our goal is to choose s so as to minimize the expected cost per week. We denote the expected cost per week given s by the function $G(s)$. Hence

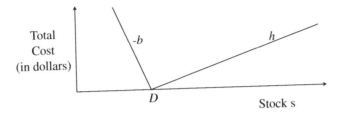

Fig. 5.7. Cost of stocking when demand is D units.

$$G(s) = E[c(s,D)] = \sum_{D} c(s,D)P(D),$$

where the sum is over all possible values of D, the weekly demand.

When D is a continuous random variable,

$$G(s) = \int_{a}^{b} c(s,x)f(x)\,dx,$$

where $f(x)$ is the density function describing demand $(D = x)$ in a week and a and b are the lower and upper bound limits on the weekly demand (i.e., $f(x) = 0$ when $x \notin [a,b]$).

When demand is normally distributed, as we have assumed in our oil rig problem, we can construct the function $G(s)$ as follows. In general,

$$G(s) = h\int_{-\infty}^{s}(s-x)f(x)\,dx + b\int_{s}^{\infty}(x-s)f(x)\,dx$$

when demand during the week is described by a continuous random variable having density function $f(x)$. This function can be rewritten as

$$\begin{aligned}
G(s) &= h\left[\int_{-\infty}^{\infty}(s-x)f(x)\,dx - \int_{s}^{\infty}(s-x)f(x)\,dx\right] \\
&\quad + b\int_{s}^{\infty}(x-s)f(x)\,dx \\
&= h(s-\mu) + (h+b)\int_{s}^{\infty}(x-s)f(x)\,dx,
\end{aligned} \tag{5.1}$$

where μ represents the expected weekly demand $(E[D] = \mu)$. When demand has a normal distribution, we can construct an exact expression for $\int_{s}^{\infty}(x-s)f(x)\,dx$.

Observe that

$$\int_{s}^{\infty}(x-s)f(x)\,dx = \int_{s}^{\infty}xf(x)\,dx - s\int_{s}^{\infty}f(x)\,dx.$$

Let $\bar{\Phi}(\cdot)$ represent the complementary cumulative distribution function for a standard normally distributed random variable. Then $s \int_s^\infty f(x)\,dx = s\,\bar{\Phi}\left(\frac{s-\mu}{\sigma}\right)$, where σ is the standard deviation of weekly demand. Next, let $\phi(\cdot)$ represent the density function for a standard normally distributed random variable. Then

$$\int_s^\infty x f(x)\,dx = \int_s^\infty \frac{x}{\sigma}\phi\left(\frac{x-\mu}{\sigma}\right)dx.$$

Letting $u = \frac{x-\mu}{\sigma}$,

$$\int_s^\infty \frac{x}{\sigma}\phi\left(\frac{x-\mu}{\sigma}\right)dx = \sigma\int_{\frac{s-\mu}{\sigma}}^\infty u\phi(u)\,du + \mu\int_{\frac{s-\mu}{\sigma}}^\infty \phi(u)\,du.$$

But

$$\int_{\frac{s-\mu}{\sigma}}^\infty u\phi(u)\,du = \frac{1}{\sqrt{2\pi}}\int_{\frac{s-\mu}{\sigma}}^\infty u e^{-u^2/2}du = \frac{1}{\sqrt{2\pi}}\int_{[(s-\mu)/\sigma]^2/2}^\infty e^{-v}\,dv$$

$$= \frac{1}{\sqrt{2\pi}}e^{-\frac{1}{2}\left(\frac{s-\mu}{\sigma}\right)^2} = \phi\left(\frac{s-\mu}{\sigma}\right).$$

Combining the above expressions, we see that

$$\int_s^\infty (x-s)f(x)\,dx = (\mu-s)\bar{\Phi}\left(\frac{s-\mu}{\sigma}\right) + \sigma\phi\left(\frac{s-\mu}{\sigma}\right),$$

which can be easily computed using today's spreadsheet tools.

Thus

$$G(s) = h(s-\mu) + (h+b)\left[(\mu-s)\bar{\Phi}\left(\frac{s-\mu}{\sigma}\right) + \sigma\phi\left(\frac{s-\mu}{\sigma}\right)\right]. \tag{5.2}$$

In our oil rig example, D has a normal distribution with $\mu = 30$ and $\sigma = 10$. The graph of $G(s)$ in this case can be found in Figure 5.8.

To find s^*, the optimal value of s, we compute $\frac{dG(s)}{ds}$ and find the value of s that satisfies

$$\frac{dG(s)}{ds} = 0. \tag{5.3}$$

As can be seen from Figure 5.8, $G(s)$ is a convex function, and therefore the value of s that satisfies (5.3) will yield the smallest value for (5.2). We will now show how to compute s^* and demonstrate that in general $G(s)$ is a strictly convex function when $f(x) > 0$ over the range $[a,b]$.

In general, using (5.1), we see that

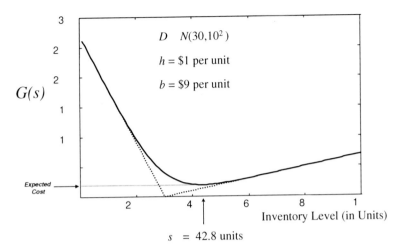

Fig. 5.8. Graph of $G(s)$.

$$\frac{dG(s)}{ds} = h + (h+b)\frac{d\left[\int_s^\infty (x-s)f(x)\,dx\right]}{ds}$$

$$= h + (h+b)\left[-\int_s^\infty f(x)\,dx\right]$$

so that

$$\bar{F}(s^*) = \frac{h}{h+b}, \tag{5.4}$$

where $\bar{F}(\cdot)$ is the complementary cumulative distribution function for the demand random variable. The expected cost function $G(s)$ is commonly called the newsvendor cost function and (5.4) is commonly called the newsvendor solution.

Observe that this newsvendor solution states that s^* is chosen so that the probability that demand exceeds supply is $h/(h+b)$. $G(s)$ is in general strictly convex when $f(x) > 0$ since $\frac{d^2 G(s)}{ds^2} = f(s) \cdot (h+b) > 0$. When D has a normal distribution, (5.4) becomes

$$\bar{\Phi}\left(\frac{s-\mu}{\sigma}\right) = \frac{h}{h+b}. \tag{5.5}$$

In our example problem, $h/(h+b) = .1$, which implies that $(s-\mu)/\sigma = 1.28$. Thus $s = 30 + (1.28)(10) = 42.8$, as indicated in Figure 5.8. The economic safety stock is $s - \mu = s - 30 = 12.8$ or about 13 units.

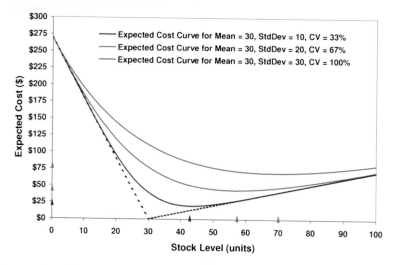

Fig. 5.9. Expected cost functions for various CV values.

Figure 5.9 contains graphs of $G(S)$ for three values of the per-week standard deviation of demand. Several observations can be made. The coefficient of variation (CV) is defined as follows: $CV = \sigma/\mu$. As the CV increases, that is, as the uncertainty in the weekly demand increases relative to the average weekly demand, we see that the optimal target inventory level and the expected weekly operating costs increase substantially. Note how flat $G(s)$ becomes in the region around s^* as the CV increases. This fact is clearly comforting since as the CV increases the accuracy of the parameter estimation decreases. For example, when the $CV = 1$, if we used a stock level of 60 rather than 70, our expected weekly operating costs would not differ substantially from the lowest possible value. Nonetheless, the greater the uncertainty, the higher the expected weekly costs.

5.3 Another Example

Let us examine another example environment. Suppose a buyer must choose how much of a high priced women's leather handbag to procure for the holiday season. The supplier of the handbag is in Italy. Since these handbags are custom made, the buyer must place a purchase order several months in advance of the holiday season. The buyer may place only one order with the supplier each season. The purchase cost per unit is c dollars and the selling price is q dollars per unit. Assume that demand is described by a discrete random variable. Let $p(x)$ represent the probability that the store will sell x units through the holiday season. If demand exceeds supply, the buyer believes there is a goodwill cost as well, which we denote by b dollars per unit. This cost is in addition to the lost revenue. The buyer believes that there is a goodwill cost since the handbag will be advertised, and, hence, arriving customers experiencing a stockout will be annoyed. On the other hand, if handbags remain on hand at the season's end, they will be sold at a price of ℓ dollars per unit, where $\ell < c$. The buyer wishes to establish the best quantity to purchase so as to maximize expected profit.

As in our earlier example, let s be the purchase quantity and s^* be the optimal quantity. Furthermore, let $G(s)$ now represent the expected profit for the season when s handbags are procured.

Then

$$G(s) = \text{expected revenue} - \text{expected cost}$$

$$= q \underbrace{\sum_{x \le s} x p(x)}_{\substack{\text{expected revenue} \\ \text{when demand does} \\ \text{not exceed supply}}} + q \cdot s \underbrace{\sum_{x > s} p(x)}_{\substack{\text{expected revenue} \\ \text{when demand} \\ \text{exceeds supply}}} + \ell \underbrace{\sum_{x \le s} (s - x) p(x)}_{\substack{\text{expected revenue of} \\ \text{after season sales when within} \\ \text{season sales are less than supply}}}$$

$$\underbrace{-cs}_{\substack{\text{purchase} \\ \text{cost}}} \underbrace{- b \sum_{x \ge s} (x - s) p(x)}_{\substack{\text{expected} \\ \text{goodwill costs}}}$$

$$= q \left(\sum_{x \ge 0} x\, p(x) - \sum_{x > s} x\, p(x) \right) + q \cdot s \sum_{x > s} p(x)$$

$$+ \ell \left\{ \sum_{x \ge 0} (s - x) p(x) - \sum_{x > s} (s - x) p(x) \right\}$$

$$- cs - b \sum_{x \ge s} (x - s) p(x)$$

$$= (q - \ell)\mu - (c - \ell)s - q \sum_{x > s} (x - s) p(x)$$

$$- \ell \sum_{x > s} (s - x) p(x) - b \sum_{x > s} (x - s) p(x)$$

$$= (q - \ell)\mu - (c - \ell)s - (q + b - \ell) \sum_{x > s} (x - s) p(x), \tag{5.6}$$

where μ represents the expected sales for the season.

First, we will show that $G(s)$ is a discretely concave function of s. We will accomplish this by demonstrating that the second difference is negative. The first difference, $\Delta G(s)$, is defined as follows:

$$\Delta G(s) = G(s+1) - G(s)$$

$$= -(c-\ell) - (q+b-\ell)\left(\sum_{x \geq s+1}(x-(s+1))p(x) - \sum_{x \geq s+1}(x-s)p(x)\right)$$

$$= -(c-\ell) + (q+b-\ell)\sum_{x \geq s+1}p(x).$$

The second difference is

$$\Delta^2 G(s) = \Delta G(s+1) - \Delta(G)$$
$$= -(q+b-\ell)p(s+1) < 0 \text{ since } q > \ell, \; b > 0, \text{ and } q > 0.$$

An optimal value of s is the smallest value for which $\Delta G(s) \leq 0$. Thus s^* is the smallest value for which

$$\sum_{x \geq s+1}p(x) \leq \frac{c-\ell}{q+b-\ell} \quad \text{or}$$

$$\sum_{x \leq s}p(x) \geq 1 - \frac{(c-\ell)}{q+b-\ell} = \frac{q+b-c}{q+b-\ell} = \frac{(q-c)+b}{(q-c)+b+(c-\ell)}. \qquad (5.7)$$

Observe that there may be more than one optimal solution. If $\Delta G(s^*) = 0$, then both s^* and $s^* + 1$ are optimal.

Observe that $c - \ell$ measures the loss incurred for any units remaining at the end of the season, $q - c$ represents the per-unit profit on units sold during the season, and $(q-c) + b$ is the cost that arises due to a stockout, the lost margin plus the goodwill cost.

Let us see how the solution is obtained for the following set of data. Suppose the purchase cost per handbag is \$250 and the selling price during the season is \$700. Any handbags left at the end of the season will be marked down to \$100 and are certain to sell at that price. The buyer estimates that the sales during the season are likely to be at least 100 but not more than 250. The buyer feels that it is equally likely that demand will assume values of 100 through 250 units. That is, demand is assumed to be uniformly distributed for sales during the season. In addition, the buyer feels that there is a goodwill cost of \$50 associated with stockouts. Hence $q = \$700$, $c = \$250$, $\ell = \$100$, $b = \$50$, and $p(x) = \frac{1}{151}$ for $x = 100, \ldots, 250$, and 0 otherwise.

The goal is to find the amount to stock that maximizes expected profit. This objective can be achieved by finding the smallest value of s for which

$$\sum_{x \leq s}p(x) \geq \frac{500}{650} = .7692.$$

This yields a value of $s^* = 217$ units.

An intern working with the buyer has looked at sales data for similar products sold in previous seasons that have been promoted in a manner similar to the one planned for the handbag. The intern found that the forecast errors were Poisson distributed for these products. In this case, the buyer estimates demand to be 175 units. The intern suggests to the buyer that this information be used to calculate the desired purchase quantity. The stock level computed by the intern is 185 units using the Poisson demand model rather than 217 units using the uniform demand model. We leave finding the incremental expected profit that is anticipated by using the intern's solution as an exercise.

5.4 Multiple Items

The single-period problems we have described to this point have focused on the management of one item type. However, in most cases there are many item types. The decisions made for each item type may not be independent because of constraints that link the items together, and therefore we must establish a new way to compute stock levels.

To illustrate how multiple item problems can be formulated and solved, let us return to the oil rig example. Suppose we have two types of parts that must be stocked on the oil rig. The storage space on the oil rig is limited for these two part types and hence setting stock levels for the parts independently may result in a space requirement that exceeds the available storage capacity. To simplify our analysis, let us assume that the parts are the same size. Thus the storage constraint limits the total number of units of the two part types that can be stocked. Suppose the distribution of weekly demand for each part follows a Poisson distribution. The probability mass functions for these parts are displayed in Figure 5.10 and Figure 5.11. The expected weekly demand differs between the parts. The mean weekly demand is 10 units for part type 1 and 20 units for part type 2.

Table 5.1 contains the individual probabilities and cumulative probabilities corresponding to weekly demands for the two part types.

Suppose the average refurbishing cost for the unused units is $1,000 for both part types. Additionally, assume the costs incurred when a stockout arises is $10,000 for either part type.

The goal is to select the stock levels for the two part types that minimize the expected weekly operating costs while satisfying the space constraint. Let $G_i(s_i)$ represent the expected weekly operating costs for part type i if the stock level for part type i is s_i units, $i = 1, 2$. Thus the problem is to

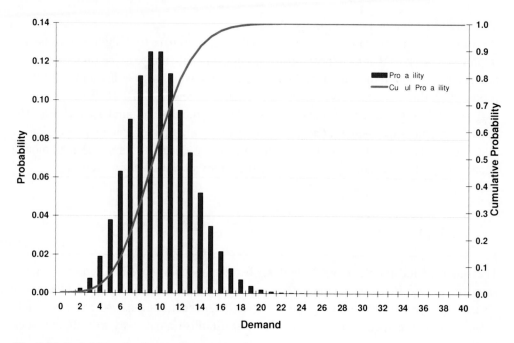

Fig. 5.10. Probability distribution for part type 1.

$$\text{minimize } G_1(s_1) + G_2(s_2)$$

subject to the constraints

$$s_1 + s_2 \leq V, \text{ (the available space)},$$
$$s_1, s_2 \geq 0 \text{ and integer}.$$

The graphs of the expected weekly costs for both part types are given in Figure 5.12. The expected weekly cost is lowest when $s_1 = 14$ and $s_2 = 26$. To achieve the lowest possible cost requires $V \geq 40$ units of storage space. Whenever $V < 40$, the stock levels must be adjusted to account for the lack of available storage capacity.

We will find the optimal stocking levels for a given value of V using a marginal analysis algorithm.

The expected weekly operating cost functions can be expressed as

$$G_i(s_i) = h_i(s_i - \mu_i) + (h_i + b_i) \sum_{x \geq s_i} (x - s_i) p_i(x),$$

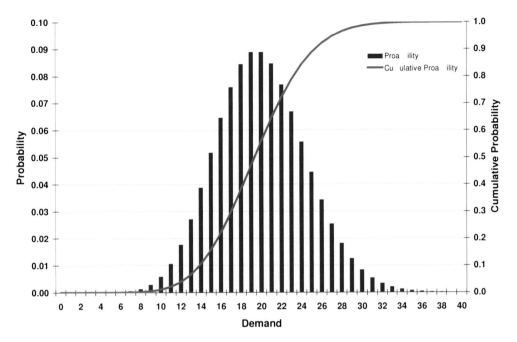

Fig. 5.11. Probability distribution for part type 2.

where $p_i(x) = e^{-\mu_i}\frac{\mu_i^x}{x!}$, where μ_i is the expected weekly demand for part type i. To find the optimal stock levels we will measure the impact on the expected weekly total cost associated with adding (or subtracting) a unit of stock of either type of part. This requires us to compute the first differences

$$\Delta G_i(S_i) = G_i(s_i + 1) - G_i(s_i)$$
$$= h_i - (h_i + b_i) \sum_{x \geq s_i + 1} p_i(x)$$
$$= h_i - (h_i + b_i) \left(1 - \sum_{x \leq s_i} p_i(x) \right)$$
$$= -b_i + (h_i + b_i) \sum_{x \leq s_i} p_i(x).$$

Recall that $p_i(x) = e^{-\mu_i}\frac{\mu_i^x}{x!} = p_i(x-1)\frac{\mu}{x}$ so that $\sum_{x \leq s_i} p_i(x) = \sum_{x \leq s_i - 1} p_i(x) + p_i(s_i - 1)\frac{s_i}{\mu}$.

Table 5.2 contains the values of $\Delta G_i(s_i)$, $i = 1, 2$, for our example (the values are in the thousands of dollars).

Table 5.1. Probability distributions in tabular form.

Demand	Part1 Probability	Part1 Cumulative Probability	Part 2 Probability	Part 2 Cumulative Probability
0	0.0000	0.0000	0.0000	0.0000
1	0.0005	0.0005	0.0000	0.0000
2	0.0023	0.0028	0.0000	0.0000
3	0.0076	0.0103	0.0000	0.0000
4	0.0189	0.0293	0.0000	0.0000
5	0.0378	0.0671	0.0001	0.0001
6	0.0631	0.1301	0.0002	0.0003
7	0.0901	0.2202	0.0005	0.0008
8	0.1126	0.3328	0.0013	0.0021
9	0.1251	0.4579	0.0029	0.0050
10	0.1251	0.5830	0.0058	0.0108
11	0.1137	0.6968	0.0106	0.0214
12	0.0948	0.7916	0.0176	0.0390
13	0.0729	0.8645	0.0271	0.0661
14	0.0521	0.9165	0.0387	0.1049
15	0.0347	0.9513	0.0516	0.1565
16	0.0217	0.9730	0.0646	0.2211
17	0.0128	0.9857	0.0760	0.2970
18	0.0071	0.9928	0.0844	0.3814
19	0.0037	0.9965	0.0888	0.4703
20	0.0019	0.9984	0.0888	0.5591
21	0.0009	0.9993	0.0846	0.6437
22	0.0004	0.9997	0.0769	0.7206
23	0.0002	0.9999	0.0669	0.7875
24	0.0001	1.0000	0.0557	0.8432
25	0.0000	1.0000	0.0446	0.8878
26	0.0000	1.0000	0.0343	0.9221
27	0.0000	1.0000	0.0254	0.9475
28	0.0000	1.0000	0.0181	0.9657
29	0.0000	1.0000	0.0125	0.9782
30	0.0000	1.0000	0.0083	0.9865
31	0.0000	1.0000	0.0054	0.9919
32	0.0000	1.0000	0.0034	0.9953
33	0.0000	1.0000	0.0020	0.9973
34	0.0000	1.0000	0.0012	0.9985
35	0.0000	1.0000	0.0007	0.9992
36	0.0000	1.0000	0.0004	0.9996
37	0.0000	1.0000	0.0002	0.9998
38	0.0000	1.0000	0.0001	0.9999
39	0.0000	1.0000	0.0001	0.9999
40	0.0000	1.0000	0.0000	1.0000

To find the solution to the example problem, we employ the following algorithm when the capacity constraint is binding.

Marginal Analysis Algorithm

Step 0: Set $A = 0$, $s_1 = s_2 = 0$.

Step 1: Compute $\Delta G_i(s_i)$, $i = 1, 2$.

Step 2: While $A \leq V$.

(a) Select i^* such that ΔG_{i^*} is smallest (most negative).
(b) Set $s_i^* = s_i^* + 1, A = A + 1$.
(c) Compute $\Delta G_{i^*}(s_i^*)$.
(d) Repeat.

In each iteration of Step 2 we select the part that reduces the expected weekly cost by the largest amount.

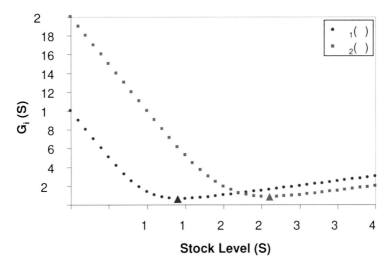

Fig. 5.12. Expected cost functions.

Earlier we stated that when $V \geq 40$, $s_1^* = 14$ and $s_2^* = 26$. Again refer to Table 5.2. Observe that $\Delta G_1(14) > 0$ and $\Delta G_2(26) > 0$, which indicates that the expected weekly cost would increase if $s_1 > s_1^*$ or $s_2 > s_2^*$.

Another way to solve this problem is to start with the unconstrained solution $(s_1^* + s_2^* = 40)$. Suppose $V = 30$ units. Then we could use the following marginal analysis algorithm to obtain the solution.

An Alternative Marginal Analysis Algorithm

Step 0: Set $A = (s_1^* + s_2^*) - V$ (s_1^*, s_2^* being the solutions to the unconstrained single item part problems) and $s_1 = s_1^*$, $s_2 = s_2^*$.

Step 1: Compute $\Delta G_1(s_1 - 1)$, $\Delta G_2(s_2 - 1)$.

Step 2: While $A > 0$.

(a) Select i^* such that $\Delta G_{i^*}(s_i - 1)$ is the smallest increase in the expected cost.
(b) Set $s_{i^*} = s_{i^*} - 1, A = A - 1$.
(c) Recompute $\Delta G_{i^*}(s_{i^*} - 1)$.
(d) Repeat.

The implementation of this algorithm is illustrated in Table 5.2. The numbers in the circles indicate the iteration of Step 2 that resulted in lowering the stock level of that part by one unit.

Table 5.2. First differences for the expected cost functions.

y	$\triangle G_1(S_1)$	$\triangle G_2(S_2)$
0	-10.000	-10.000
1	-9.995	-10.000
2	-9.970	-10.000
3	-9.886	-10.000
4	-9.678	-10.000
5	-9.262	-9.999
6	-8.568	-9.997
7	-7.578	-9.991
8	-6.339	-9.977
9	-4.963	-9.945
10	-3.587	-9.881
11	-2.335	-9.765
12	-1.293	-9.571
13	-0.491	-9.273
14	0.082	-8.846
15	0.464	-8.278
16	0.703	-7.568
17	0.843	-6.733
18	0.921	-5.804
19	0.962	-4.827
20	0.983	-3.850
21	0.992	-2.919
22	0.997	-2.073
23	0.999	-1.338
24	0.999	-0.724
25	1.000	-0.234
26	1.000	0.143
27	1.000	0.423
28	1.000	0.622
29	1.000	0.760
30	1.000	0.852

S_1 (30) — 10

S_1 (0) — 14

S_2 (30) — 20

S_2 (0) — 26

Note that in many real problems the value of V is not always known. Hence, a graph is constructed that measures the total expected weekly operating cost as a function of V. The data in Table 5.2 can be used to construct such a graph.

5.4.1 A General Model

The example discussed in the previous section illustrates how multiple-item, single-period problems can be formulated and solved. Let us now turn to a more general problem.

Suppose there are I part types, and the goal is to find the stock levels, s_i^*, for each type of part so as to minimize the sum of the expected single-period cost functions, $G_i(s_i)$, while satisfying a constraint. Suppose each unit of part type i consumes v_i units of a resource, and suppose in total there are V units of the resource available. Then the general problem is to

$$\text{minimize} \quad \sum_{i=1}^{I} G_i(s_i),$$

subject to

$$\sum_{i=1}^{I} v_i s_i \le V,$$

$$s_i \ge 0 \quad \text{(and possibly integer).} \tag{5.8}$$

This problem is a knapsack problem and hence can be solved using a dynamic-programming-based algorithm. Another way to obtain the stock levels is to employ a Lagrangian-relaxation-based approach. Let us describe how this approach can be used since we will be using it in subsequent chapters.

Let us assume that the constraint is active. Otherwise, the problem can be solved one part at a time.

Let us assume that the decision variables are continuous. (We will later assume that each s_i must assume an integer value.) Let us construct a Lagrangian relaxation of (5.8) as

$$\text{minimize} \sum_{i=1}^{I} G_i(s_i) + \theta \left(\sum_{i=1}^{I} v_i s_i - V \right), \tag{5.9}$$

where θ is the Lagrangian multiplier. Assuming the $G_i(s_i)$ are convex and differentiable (as they are in realistic problems), the solution for part i occurs where

$$\frac{dG_i(s_i)}{ds_i} + \theta v_i = 0.$$

Suppose $s_i(\theta)$ is the value of s_i obtained when solving (5.9). Recall that θ measures the rate of change of the objective function as the amount of available resource, V, changes. Now compute $\sum_{i=1}^{I} v_i s_i(\theta)$. If this quantity exceeds V, then the value of θ is too small; if $\sum_{i=1}^{I} v_i s_i(\theta) < V$, then θ is too large. In the former case, θ must be increased and in the latter it must be decreased. Once adjustments have been made to θ, problem (5.9) is re-solved. In practice, a bisection method is often employed to generate the sequence of the values of θ. We note that this approach may converge only after an infinite number of steps and therefore is normally stopped when $|\sum s_i(\theta) \cdot v_i - V| < \varepsilon$ for some tolerance $\varepsilon > 0$.

Oftentimes, decision makers want to know how the total expected cost $\left(\sum_{i=1}^{I} G_i(s_i) \right)$ depends on V. For example, suppose $G_i(s_i)$ measures the expected number of shortages that might be incurred over the time period for part i given s_i, v_i is the unit cost for part i, and V is the budget for all the parts. The decision maker would want to know what the tradeoff is between investment and expected total shortages. In this case, such a tradeoff relationship could be constructed by solving (5.9) for a wide range of values of θ.

Now suppose that the values of s_i^* will be relatively small and integer-valued. Let us also assume that there are many part types and that the maximum value of v_i/V is small (say .05 or less). This is the case in almost all spare parts problems found in construction, automotive, and military applications. In these cases we would employ a marginal analysis algorithm to find a solution. As earlier, let $\Delta G_i(s_i) = G_i(s_i+1) - G_i(s_i)$. We could then use the following algorithm to find a solution.

A Marginal Analysis Algorithm

Step 0: Set $A = 0$, $s_i = 0$ for all i.

Step 1: Compute $\Delta G_i(s_i)$ for all i.

Step 2: While $A \leq V$.

(a) Select i^* such that $\left\{\frac{\Delta G_i}{v_i}\right\}$ yields the largest decrease in the value of the objective function per unit of resource invested.
(b) Set $s_{i^*} = s_{i^*} + 1$, $A = A + v_i$.
(c) Recompute $\Delta G_{i^*}(s_{i^*})$.
(d) Repeat.

The solution generated by the above algorithm will likely not be feasible, since the last iteration will likely increment the resource requirement above the available quantity. Several alternatives exist to adjust the solution. We could revert to the solution found in the previous iteration, which is feasible, and accept that as the solution. We could modify the algorithm so that Step 2(a) considers only parts for which $A + v_i$ remains feasible. Or we could ignore the fact that the solution generated by the algorithm is infeasible. This can safely be done in many economic applications when V is large and the incremental value of resource usage (v_i) is insignificant compared with V. If V is an investment budget that is in tens of millions of dollars and the values of the v_i do not exceed, say, $100,000 or so, then the significance of the infeasibility is minimal. In automotive applications, investments are in the billions of dollars and the maximum value of parts is on the order of $10,000, of which there are few in number out of the hundreds of thousands of parts managed in such systems.

Marginal-analysis-based algorithms are quite useful since solutions can be computed very quickly. The application of marginal analysis algorithms will be discussed further in later chapters.

5.4.2 Multiple Constraints

In the previous section we demonstrated how certain single-period, multi-item stocking problems can be represented and solved when there is a single constraint that links the items together. Let us now extend that discussion to certain cases in which multiple constraints exist that link the procurement decisions.

Suppose our buyer will purchase not only handbags but many other high fashion items for the holiday season. Let I represent the number of items that may be purchased. Let s_i, q_i, ℓ_i, μ_i, c_i, and b_i represent the quantity purchased, selling price per unit during the season, markdown price at the end of the season, the expected season's demand, the unit cost, and the lost goodwill cost for item i, $i = 1, \ldots, I$. As shown in Section 5.3, the expected profit for item i is given by

$$G_i(s_i) = (q_i - \ell_i)\mu_i - (c_i - \ell_i)s_i - (q_i + b_i - \ell_i) \sum_{x > s_i} (x - s_i)p_i(x), \qquad (5.10)$$

where $p_i(x)$ measures the probability that x units will be demanded for item i during the holiday season.

Let us assume that several constraints exist that will force a tradeoff to be made when purchasing the inventory. For example, there may be a budget limit for all the items that constrains the total amount purchased. There may be storage constraints, too. Let us assume that there are m linear resource constraints that limit purchases, and that a_{ij} represents the amount of resource j consumed per unit purchased of item i. Then we have m constraints of the form

$$\sum_i a_{ij}s_i \leq v_j, \quad j = 1, \ldots, m, \qquad (5.11)$$

where v_j is the amount of resource j that is available.

Observe that (5.10) is a non-linear function of s_i. We will now illustrate how this non-linear function can be replaced by a linear one. The non-linearity arises from the expected shortage function $\sum_{x > s_i} (x - s_i)p_i(x) = B_i(s_i)$. When $s_i = 0, B_i(0) = \mu_i$. Define $\Delta B_i(s_i) = B_i(s_i + 1) - B_i(s_i)$. As we saw earlier, $\Delta B_i(s_i) = -\sum_{x \geq s_i + 1} p_i(x)$. Let $\delta_{ik} = \Delta B_i(k)$, and

$$y_{ik} = \begin{cases} 1, & \text{when } s_i \geq k, \\ 0, & \text{otherwise .} \end{cases}$$

Then

$$s_i = \sum_k y_{ik} \text{ and } B_i(s_i) = \mu_i + \sum_k \delta_{ik}y_{ik}. \qquad (5.12)$$

Suppose we remove s_i from both the constraints and the objective function using (5.12). Then we can write the procurement problem as follows:

$$\text{maximize} \quad \sum_{i=1}^{I} \left\{ (q_i - \ell_i)\mu_i - (c_i - \ell_i) \sum_k y_{ik} \right.$$

$$\left. -(q_i + b_i - \ell_i) \left[\mu_i + \sum_k \delta_{ik} y_{ik} \right] \right\}$$

subject to

$$\sum_i a_{ij} \sum_k y_{ik} \leq v_j, \quad j = 1, \ldots, m,$$

$$y_{ik} = 0 \text{ or } 1. \tag{5.13}$$

Note that since $\delta_{ik} < \delta_{i,k-1}$, y_{ik} cannot equal one unless $y_{i,k-1}$ is equal to one. Because this is the case, we would likely replace the constraints $y_{ik} = 0$ or 1 with $0 \leq y_{ik} \leq 1$, which would make the problem a linear program. There would be a maximum of one fractional value for a y_{ik} for each item. A feasible solution could be found by setting each variable that assumes a fractional value in the solution to the linear program to 0. This assumes that $a_{ij} \geq 0$ for all i and j. Other rules could be employed to obtain a solution, too. If the constraints are "soft," that is the values of the v_i can be altered slightly, then the fractional values can be rounded upwards so that all the y_{ik} variables assume a value of 0 or 1. When the a_{ij} are small relative to the value of v_j, such adjustments are very often possible in inventory planning problems.

We have illustrated how a particular multi-item, single-period procurement problem can be formulated as a linear program. Many single-stage, probabilistic, production and inventory planning problems can be expressed as linear programs by replacing variables with others in the manner we have shown.

5.5 Exercises

5.1. A distributor of seasonal items wants to establish the best stock level to purchase. The distributor estimates that the demand for an item during the season will be normally distributed with a mean of 100 units and a standard deviation of 10 units. The items will sell for $25.00 and will cost the distributor $10.00. Units unsold at the end of the season will be sold to a third party for $5.00 a unit. How many units should be purchased to maximize expected profit?

5.2. Re-solve Problem 5.1 under the assumption that demand is Poisson distributed with a mean of 100 units. Now assume demand is uniformly distributed with the mean and standard deviation as stated in Problem 5.1. Again, re-solve the problem.

5.3. A manufacturer of a part, purchased by Llenroc Electronics (LE) for its airline customers, has just informed LE that it is going to terminate production of the part type. The manufacturer has informed LE that it has one last opportunity to obtain the part at the normal production rate price. The manufacturer is willing to sell units to LE for $2,000 at this point in time. However, if LE wishes to procure the part type at a later point in time the unit price will likely increase to $25,000. Another factor that LE faces is the remaining length of time the airlines might use the part. Suppose the remaining life is gamma distributed with a mean of 5 years and a standard deviation of 1.25 years. The demand for the part is Poisson distributed with a rate of 2 units per year. There is no residual value for these parts if they are not used. What quantity of this part would you recommend LE to purchase from the manufacturer? Assume LE has no stock on hand or on order.

5.4. In Problem 5.3 we assumed that the demand rate per year for the part is a constant. Suppose that this is not the case. Let $\Lambda(t)$ represent the expected remaining cumulative demand for the part given the remaining useful life of the part is t. Then the probability that n remaining units are demanded, given $\Lambda(t)$, is $e^{-\Lambda(t)}\frac{\Lambda(t)^n}{n!}$. Assume that the remaining lifetime is t years with probability $p(t)$. Suppose $p(3) = .1$, $p(4) = .2$, $p(5) = .4$, $p(6) = .2$, $p(7) = .1$, and $\Lambda(t) = 2\sqrt{t}$. Using the other data from Exercise 5.3, what quantity of the part should LE procure?

5.5. In Exercises 5.3 and 5.4 we assumed that the subsequent purchase cost of the part is $25,000, independent of the year in which such an order might be placed. Suppose the per-unit subsequent costs are estimated to be $25,000, $26,000, $28,000, $30,000, $33,000, $35,000, and $40,000 for years 1 through 7, respectively. These costs represent their discounted values. Thus, if a shortage for a unit occurs in year 5, the unit procurement cost would be $33,000. Additionally, there is a holding cost charged per unit per year. The holding cost of a particular unit is charged at a rate of .15 of the purchase cost of that unit. Using the demand and lifetime data given in Problem 5.4 and an initial procurement cost of $2,000 per unit, find LEs optimal procurement quantity.

5.6. Suppose the unit procurement cost is a function of the quantity purchased. In particular, suppose that there is a set of values $0 = a_0 < a_1 < \cdots < a_n$ such that the purchase price per unit is c_i if the quantity ordered q satisfies $a_{i-1} < q \leq a_i$, with $c_1 > c_2 > \cdots > c_n$. Assuming the selling price remains constant, extend the analysis developed in Section 5.2.2 to represent this situation.

5.7. Use the method developed in Problem 5.6 to solve Problem 5.1 when the unit price is $12.00 if 125 or fewer units are purchased but is $8.00 per unit if more than 125 units are purchased.

5.8. Each week Brian visits the fish market in Boston and procures quantities of a variety of fish that he will take to sell in Ithaca, NY. He has capacity to store only 1000 pounds of fish in his truck. Suppose he concentrates on four types of fish that appear to be best choices today. Suppose demand is normally distributed for these fish with expected sales of 100, 300, 400, and 100 pounds, respectively. The standard deviations for these demands are correspondingly 20, 30, 90, and 25 pounds. Unsold fish are worthless. The profits per pound are $1, $1.5, $1.7, and $1.8 respectively. How much of each type of fish should be procured? Solve this problem using a Lagrangian method.

5.9. A submarine is being stocked for a lengthy mission. Certain parts are carried on the vessel in case repairs are required while at sea. Three such critical parts are stored in a compartment that has limited space. The three part types all have very similar geometries. The capacity of the compartment is 10 units. The demand for these parts is negative binomially distributed on the basis of past history. They have expected demands of 2.5, 3, and 2 for the duration of the mission. The variances of demand are 3, 5, and 6, respectively. What quantity of each item should be stocked so as to maximize the probability that there will be no shortages? Demands are independent among the three part types. Suppose the objective is to minimize the expected number of shortages. What is the optimal solution in this case? Use marginal analysis to compute the answer.

5.10. Llenroc Electronics repairs equipment at customer locations. Customers call and report a failure of a system (e.g., copy machine) along with information about the nature of the failure. On the basis of this information, Llenroc dispatches a repair technician to the customer location along with a collection of parts. The type of repair, and hence the set of parts required to complete the repair, is not known with certainty at the time the technician is dispatched. Associate with each repair type i a collection of part quantities $R_i = \{a_{ij} : j = 1, \ldots, P\}$, where P represents the number of part types, and $a_{ij} \in \{0, 1, \ldots\}$ represents the quantity of part type j required to complete a repair of type i. Suppose p_i represents the probability that a repair of type i will be required. Suppose v_j represents the physical volume of part type j. Develop a model and a solution methodology for finding the mix of parts that should be carried by the technician if the goal is to minimize the chance that the technician does not have the set of parts needed to complete the repair, subject to a constraint on the total physical volume of parts that the technician can carry to the job site. Let V be the available volume. Assume $\sum_j a_{ij} \ll V$ for all i but $\sum_i \sum_j a_{ij} > V$. Develop an exact model and solution approach.

Also, construct a heuristic that can be used to determine the contents of the technician's repair parts kit.

5.11. Each day technicians at Llenroc Electronics repair equipment located at customer sites. Since these repair actions are often accomplished by removing and replacing defective parts, the technicians are required to take parts with them to complete the repairs in a timely manner. In Exercise 5.10 we assumed that there was an unlimited supply of parts available that could be selected by the technician and that there was only one technician. Suppose now that there are N technicians. Suppose also that there are b_j units of part type j on hand that are available to distribute to the N technicians. Technician n cannot carry a volume of parts greater than V_n. Assume each technician's demand for parts occurs in the manner described in Exercise 5.10. Construct an appropriate model and solution procedure that can be employed so that the expected number of customer repairs that can be completed using the allocation of parts to technicians is maximized considering part availability and volume constraints.

5.12. A buyer for a wine shop visits Italy every year and places an order with producers for a wide range of wines. Some are quite expensive while others are moderately expensive. However, the buyer has only this one chance to buy these wines since other buyers will likely purchase the remaining available supply. Suppose there are N types of wines the buyer is considering to purchase. Let c_i be the per-bottle cost for wine of type i and q_i the average selling price for a bottle of wine of type i, $i = 1, \ldots, M$. Wine of type i not sold after a year will be discounted and sold for ℓ_i dollars per bottle. Demand for wines of type i is normally distributed with mean μ_i and standard deviation σ_i. The buyer also has a constraint on the dollar amount he can purchase. The buyer also wants to have a mix of wines. The buyer wants to purchase at least a_i but no more than b_i bottles of type i wines, $i = 1, \ldots, M$. Construct a model that can be used to maximize the buyers expected profit subject to the constraints that limit the procurement decisions. Construct a Lagrangian type of solution method that can be used to calculate the best mix of wines to purchase. Next, show how a marginal analysis type of algorithm could be used to find the optimal purchase plan, assuming the demand distribution for each wine type is negative binomially distributed. Finally, show how to represent this decision problem as a linear program.

5.13. In Exercise 5.12, suppose a_i and b_i are measured in cases, where a case contains 12 bottles. Suppose further that purchases are made in case quantities. Assuming demand for each wine type is negative binomially distributed, construct a marginal analysis type of procedure that can be employed to find the optimal procurement strategy.

5.14. A fashion item must be purchased several months prior to the selling season. Purchases of the item can occur at only this time. Suppose the purchase price is c dollars per unit and the in-season selling price is q dollars per unit. Suppose s represents the procurement quantity and x the number of units sold during the season. When supply exceeds the in-season demand, that is, when $s - x > 0$, the remaining units are sold at a discounted price. The discount depends on the magnitude of $s - x$. Suppose when $s - x > 0$, the total revenue for these units is

$$q(s - x) - a(s - x)^2, \quad \text{where } a > 0.$$

Suppose the distribution of demand has a symmetric triangular distribution with lower and upper limits α and β, respectively, with $\alpha \geq 0$. Construct an algorithm for finding the optimal procurement quantity.

5.15. A buyer of clothing items must purchase these items well in advance of the selling season. The store's manager recognizes that sales volume depends on the advertised selling price. If the manager sets the advertised selling price to be q_i, then the probability distribution of the sales has density $f_i(x)$, with mean μ_i and standard deviation σ_i. Suppose I possible sales prices are considered. The store manager has noted that as q_i increases μ_i decreases and furthermore σ_i/μ_i increases as q_i increases. Suppose the buyer informs the manager that the unit procurement cost is c dollars. All clothing remaining at the season's end is sold for ℓ dollars per unit, where $\ell < c$. Additionally, the manager believes there is a goodwill cost incurred when there is insufficient stock on hand to meet customer demand. The goodwill cost depends on the selling price. The higher the selling price, the higher the goodwill cost per unit short. Construct a model and solution methodologies that the manager and buyer can use to find the optimal procurement quantity.

5.16. In Section 5.3 we discussed an example in which a buyer had to determine how many handbags to purchase for a holiday season. Using the data in that example, find the optimal order quantity assuming demand is Poisson distributed with mean of 175 units rather uniformly distributed over the range 100 to 250 units. What is the expected incremental profit that would be obtained if this solution were used rather than the one obtained assuming demand was uniformly distributed?

6

Inventory Planning over Multiple Time Periods: Linear-Cost Case

The discussion in the previous chapter focused on decision making when the planning horizon consisted of a single period. The assumption was made that decisions in one period would not affect those made in other periods. But what happens when a decision at one point in time does affect decisions in subsequent time periods? Hence, when is it necessary to consider other time periods when making these inventory decisions? Additionally, what policies are best for managing inventories in a dynamic environment? The focus of this chapter is on answering these and other related questions. Specifically, we will again restrict ourselves to the case where the fixed cost of placing an order is negligible. We will begin by showing that a base-stock or order-up-to policy is optimal in a number of different environments. Next we will develop a dynamic programming model that can be employed to find optimal stocking decisions. On the basis of this model, we will then establish properties that optimal solutions possess. The discussion of how capacities on production affect stocking decisions follows. The chapter concludes with a discussion of multi-echelon systems, and how the role of each location within a supply chain differs.

6.1 Optimal Policies

To begin, we will concentrate on establishing the form of an optimal policy for managing a single item. We will assume that time is divided into periods. Initially, we will assume the planning horizon is finite in length and that lead times are deterministic. Costs incurred in each period are (i) linear purchasing and shipping costs between stages in the more general case, (ii) linear holding and backorder costs at the demand location, and (iii) linear holding costs for all other locations in a supply chain. We will

J.A. Muckstadt and A. Sapra, *Principles of Inventory Management: When You Are Down to Four, Order More*, Springer Series in Operations Research and Financial Engineering, DOI 10.1007/978-0-387-68948-7_6, © Springer Science+Business Media, LLC 2010

then show how the analysis can be extended to consider random lead times, multiple stages, a more general demand model, limited production capacities in each period in serial systems, and how to consider an infinite planning horizon.

6.1.1 The Single-Unit, Single-Customer Approach: Single-Location Case

We will first examine a single-location system in detail. This system operates as follows.

At the beginning of each period, an order is placed on an outside supplier and arrives exactly $m-1$ periods in the future. Subsequent to the time the order is placed, the order due in that period is received from the supplier; customer demands are then observed. Demands in each period are assumed to be independent from period to period. All excess demand is backordered. At the end of each period, both holding and backorder costs are incurred. The approach we will use to establish the form of an optimal policy is called the "single-item, single-customer" approach that was recently developed by Muharremoglu and Tsitsiklis [250].

6.1.1.1 Notation and Definitions

We begin our analysis by presenting the notation and key definitions. As stated, we initially assume the planning horizon of the system consists of N periods, numbered $n = 1, 2, \ldots, N$, in that order. We assume that the probability distribution of D_n, the demand in period n, is known and is independent among periods.

We consider each unit of demand as an individual customer. Suppose at the beginning of period 1 there are v_0 customers waiting to have their demand satisfied. We index these customers 1, 2, ..., v_0 in any order. All subsequent customers are indexed $v_0 + 1$, $v_0 + 2$, ... in the order of the period of their arrivals, arbitrarily breaking ties among customers that arrive in the same period.

Next, we define the concept of the distance of a customer at the beginning of any period. See Figure 6.1. Every customer who has been served is at distance 0; every customer who has arrived, placed an actual order, but who has not yet received inventory, is at distance 1; all customers arriving in subsequent periods are said to be at distances 2, 3, ... corresponding to the sequence in which they will arrive. Distances are assigned to customers that arrive in the same period in the same order as their indices. This ensures that customers with higher indices are always at "higher" distances.

Next, we define the concept of a location for a unit. Again see Figure 6.1. There are $m + 2$ possible locations at which a unit can exist. If the unit has been used to satisfy a customer's order, the location of this unit is 0. If it is part of the inventory on hand, it is in location 1. If the unit has not been ordered from the supplier, it is in location $m+1$.

At the beginning of period 1, we assign an index to all units in a serial manner, starting with units at location 1, then location 2, ..., location $m+1$, and arbitrarily assign an order to units present at the same location. We assume a countably infinite number of units is available at the supplier, that is, location $m+1$, at all times.

We will use indices j and k to denote the indices of units as well as customers. We define y_{jn} to be the distance of customer j at the beginning of period n and z_{jn} to be the location of unit j at the beginning of period n.

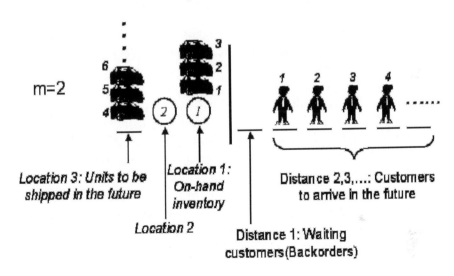

Fig. 6.1. Locations of units and distances of customers.

We define the state of the system at the beginning of period n to be the vector $x_n = ((z_{1n}, y_{1n}), (z_{2n}, y_{2n}), \ldots)$.

Next, we explain the sequence of events in period n in detail.

1. (z_{jn}, y_{jn}) is known for all j, $j = 0, 1, 2, \ldots$.

2. An order is placed, which we denote by q_n, where q_n is a non-negative integer. All units in locations $j = 2, 3, \ldots, m$ move to the next location prior to placing the order, that is, location $j - 1$. The q_n units move from location $m + 1$ to location m. If a unit was ordered from the supplier ℓ periods ago ($1 \leq \ell < m$), it is in location $m - \ell$.

3. Demand D_n is realized and these new customers arrive and are at distance 1. That is, customers at distances $2, 3, \ldots, 2 + D_n - 1$ all arrive and are, by definition, now at distance 1. All customers at distances $2 + D_n, 3 + D_n, \ldots$ at the beginning of the period move D_n steps towards distance 1.

4. Units on hand and waiting customers are matched to the extent possible. That is, as many waiting customers are satisfied as possible and as many units on hand are consumed as possible.

5. h dollars are charged per unit of inventory remaining on hand (at location 1) and b dollars are charged per waiting customer (at distance 1). Clearly, only one of these costs will be incurred in any period. We assume $b > h$. This ensures that if the inventory position is negative in some period, then the optimal policy will be to increase the inventory position to some non-negative level.

The performance measure under consideration in our initial discussion is the expected sum (discounted or undiscounted) of costs over the N period planning horizon.

Next, we define a policy. Let $u_{jn} \in \{\text{Release}, \text{Hold}\}$ denote the decision made in period n for unit j. By Release, we mean that an order is placed for a unit and that it enters the supplier's production/distribution system. However, note that the only units over which we have control are the units at the supplier, that is, at location $m + 1$. The movement of all other units is governed by the lead time and demand processes as defined previously. A policy is a function that maps every possible realization of x_n to a vector of Release/Hold actions for each unit at location $m + 1$.

Observe, however, that when a decision is made to release q units from location $m + 1$, it does not matter which q units are released. Consequently, we can consider a class of policies that always releases the units with the smallest indices from $m + 1$. We call such a policy a monotone policy. A class of policies is said to be optimal if it contains at least one optimal policy. The class of monotone policies is clearly optimal. A monotone policy releases unit j before or at the same time as unit $j + 1$ but never in any period following the one in which unit $j + 1$ is released. Similarly, we define a monotone state to be one where lower-indexed units are in the same or lower-indexed locations. That is, $z_{kn} \leq z_{jn}$ if $k \leq j$. Since lead times are the same for all units and we started period 1 in a monotone state, the system is always in a monotone state when a monotone policy is followed. Furthermore, by definition, the customers also arrive in the order of their indices. We define a policy as a committed policy if it ensures that the only customer that the jth unit can satisfy is customer j's demand and that the only unit that customer j can receive is the jth unit. Assume that the units at location 1 that are picked to satisfy customers at distance 1, as well as the customers at distance 1 picked to consume units at location 1, are those with the lowest indices. Hence, every monotone policy is also a committed policy. Consequently, the class of committed policies is also optimal. Since this is an important fact, we state this as a lemma.

Lemma 6.1. *The class of monotone policies is optimal. Furthermore, every monotone policy is a committed policy and hence the class of committed policies is also optimal.*

In the next section, we develop a proof of the optimality of base-stock policies for periodic review, single-stage uncapacitated systems of the type we have described.

6.1.1.2 Optimality of Base-Stock Policies

In this section, we first show that the system can be decomposed into a collection of countably infinite subsystems, each having a single unit and a single customer. Subsequently, we prove that each subsystem can be managed optimally by using a policy we call a "critical distance" policy. We prove that when the same "critical distance" policy is used to manage each subsystem, the system follows a base-stock policy.

6.1.1.2.1 Decomposition of the System into Subsystems

Let us first outline the proof technique. First, we observe that the cost of the system is the sum of the costs incurred for each unit–customer pair. Second, we show that each of these pairs can be controlled independently and optimally and that the resulting policy is optimal for the entire system. Third, we examine the individual unit–customer problem and show that the optimal policy is a "critical distance" policy: Release a unit if and only if the corresponding customer is closer than a critical distance. Last, we observe that operating each unit–customer pair using a critical distance policy produces an echelon base-stock policy in the original system.

Let us now precisely define the concepts of the system, the subsystems, and the sets of constraints that govern these systems and subsystems.

Definition 6.1. Let S refer to the entire system with all the units and all the customers. Subsystem w, represented by S_w, $1 \leq w$, refers to the unit–customer pair with index w.

Definition 6.2. Constraints on Monotone and Committed Policies in S:
   Monotonicity: Unit j ($j = 1, 2, \dots$) cannot be released before unit $j - 1$.
   Commitment: Unit j ($j = 1, 2, \dots$) serves customer j.

Definition 6.3. Constraint on Committed Policies in S_w:
   Commitment: Unit w serves customer w.

We will now show that the optimal cost for the system S is equal to the sum of the optimal costs for the subsystems S_w. We will prove this fact by demonstrating that every monotone and committed policy for system S corresponds to a set of monotone and committed policies for the subsystems, S_w, and that any set of monotone and committed policies for the subsystems yields a feasible policy for the system S. We will also show that when the individual subsystems are managed "independently and optimally," the resulting policy for the system S is optimal.

From now on, we will use $\tilde{S}$ to denote the group of all subsystems, that is, $\tilde{S} = (S_1, S_2, \ldots)$. When only monotone and committed policies are considered, the constraints in Definition 6.2 apply to S while the constraint in Definition 6.3 applies to $\tilde{S}$. When we say "the (optimal) expected cost for $\tilde{S}$," we mean the sum of the (optimal) expected subsystem costs.

We have assumed so far that x_n is the state information available to us while managing the entire system S or any subsystem S_w. However, observe that the subsystems are "operationally independent" in the sense that each subsystem can be managed independently without being affected by the policies used to manage the other subsystems. Consequently, we can find an optimal policy for managing S_w that uses only those parts of the state vector x_n that pertain to unit w and customer w. We define $x_n^w \overset{\text{def}}{=} (z_{wn}, y_{wn})$. Thus, x_n^w is a sufficient state descriptor for S_w. This means that an optimal policy for $\tilde{S}$ can be found by managing the subsystems independently. A subtle point to be noted here is that the subsystems, though operationally independent, are stochastically dependent through the demand process.

We are now ready to state and prove the results relating the optimal costs and policies for the system S and the subsystems S_w, $w = 1, 2, \ldots$.

Theorem 6.1. *For any starting state x_1 in period 1, the optimal expected discounted (undiscounted) cost in periods 1, 2, ..., N for system S equals the optimal expected discounted (undiscounted) cost in periods 1, 2, ..., N for the group of subsystems $\tilde{S}$. Furthermore, when each subsystem w is managed independently and optimally using the state vector x_n^w in every period n, the resulting policy is optimal for the entire system S.*

Proof. First, observe that the cost incurred by S is the sum of the costs incurred by every unit and every customer, since the holding and backorder costs are linear.

Second, observe that every monotone and committed policy for S produces a set of committed policies, one for each subsystem S_w. Consequently, the optimal expected cost for $\tilde{S}$ is a lower bound on the optimal expected cost for S over any number of periods, since a monotone and committed policy is optimal for S.

Third, observe that operating each subsystem independently using any committed policy is a feasible policy for S. Consequently, the optimal expected cost for S is a lower bound on the optimal expected cost for $\tilde{S}$.

Combining the two lower-bound arguments above proves that the optimal expected costs for S and $\tilde{S}$ are equal. The earlier discussion about the "operational independence" of the subsystems and this equality result show that when each subsystem w is managed independently and optimally using the state vector x_n^w, the resulting policy for the entire system, S, is optimal. $\square$

Next, we show the existence of an optimal policy with a very special structure for every subsystem.

6.1.1.2.2   Optimal Policy Structure for a Subsystem

Before examining an individual subsystem, we first observe that all subsystems are identical in the sense that (i) they have identical cost structures and (ii) given a state (x_n^w) and a fixed operating policy for a subsystem, the stochastic evolution of the subsystem is independent of the index w. Consequently, the optimal policies are identical across all subsystems.

We define $R_n^*(y) \subseteq \{\text{Release}, \text{Hold}\}$ to be the set of optimal decisions for subsystem w at time n if y_{wn} is y and if z_{wn} is $m+1$.

Next, we show that there is a "critical distance" policy that is optimal for a subsystem. We need the following lemma to prove this fact. The lemma states that if it is *uniquely* optimal for subsystem w to release unit w (if it is at location $m+1$) in period n when customer w is at a distance $y+1$, then it would be optimal to release it if the customer were any closer and the unit were at location $m+1$.

Lemma 6.2. $R_n^*(y+1) = \{Release\}$ *implies that* $R_n^*(y) \supseteq \{Release\}$.

Proof. The proof is by contradiction. Assume the statement is not true. That is, there exist n and y such that $R_n^*(y+1) = \{\text{Release}\}$ and $R_n^*(y) = \{\text{Hold}\}$. Another way of saying this is as follows: it is suboptimal for a subsystem to hold unit w if customer w is at a distance $y+1$ while it is suboptimal for a subsystem to release unit w if customer w were at a distance y.

Consider some monotone and committed policy for S. Assume in period n that we can find subsystems w and $w+1$ such that y_{wn} is y and $y_{(w+1)n}$ is $y+1$. Monotonicity implies that this policy would choose one of the following three pairs of actions for units w and $w+1$: (a) release both w and $w+1$, (b) hold both w and $w+1$, and (c) release w and hold $w+1$.

Cases (a) and (c) are suboptimal for subsystem w, while cases (b) and (c) are suboptimal for subsystem $w+1$ owing to our initial assumption. This implies that any monotone and committed policy for S is suboptimal for at least one of subsystems w and $w+1$. So, any monotone and committed policy for S has a higher expected cost than the optimal cost for $\tilde{S}$ from period n onwards, which is the same as the optimal cost for $\mathcal{S}$. This implies that no monotone and committed policy can be optimal for $\mathcal{S}$, which contradicts our earlier assertion about the optimality of some such policy. Therefore, our assumption about $R_n^*(y)$ and $R_n^*(y+1)$ is invalid.   □

We use this lemma to develop the notion of a "critical distance" policy. Let us define

$$y^*(n) \overset{\text{def}}{=} \max\{\, y \,:\, R_n^*(y) \supseteq \{Release\} \,\}.$$

$y^*(n)$ is defined in such a way that it is optimal to release unit w if and only if customer w in period n is at a distance of $y^*(n)$ or closer. This distance $y^*(n)$ is the "critical distance" in period n for every subsystem.

Consider the policy

$$R_n(y) \;=\; \{Release\} \text{ if and only if } y \leq y^*(n).$$

Policy R_n is an optimal policy for every subsystem. The next observation we make is that when policy R_n is used in period n for every subsystem, the resulting policy for the original system $\mathcal{S}$ is an order-up-to policy. This can be shown either using an algebraic proof or using a more intuitive argument, which we now provide.

Theorem 6.2. *The optimal policy for $\mathcal{S}$ is to release as many units as necessary to raise the inventory position to $y^*(n) - 1$ in period n when the planning horizon consists of N periods. That is, a state-dependent order-up-to or base-stock policy is optimal for the entire system when the planning horizon is finite.*

Proof. We know that policy R_n is optimal for every subsystem. It can be seen that if R_1, ..., R_{n-1} are the policies used on each of the subsystems in periods $1, \ldots, n-1$, we will start period n in a state where the units that are in location $m+1$ bear consecutive labels. Consequently, the corresponding customers who have not arrived are in consecutive distances. Among these customers, those in locations $2, 3, \ldots, y^*(n)$ are all within the critical distance $y^*(n)$. All backordered customers are also within the critical distance. The policy R_n dictated that we should release the waiting unit in just the right number of subsystems in period n so that all waiting customers and all future customers within the critical distance can be satisfied with the units on hand or on order. That is equivalent to saying we would raise the inventory position to $y^*(n) - 1$.   □

This concludes the proof of the finite-horizon optimality result for uncapacitated single-stage systems with constant lead times. In the next two sections, we will discuss how this approach can be extended to more general situations.

6.1.1.3 Stochastic Lead Times

So far, we have assumed that the lead time is exactly $m - 1$ periods. Let us now relax this assumption by allowing stochastic lead times subject to the restriction that orders cannot cross, that is, the sequence in which orders are received from the supplier corresponds to the sequence in which orders were placed on the supplier. We permit the lead time distribution to be governed by the lead time model described below.

The lead time process evolves as follows. There is a random variable ρ_n, whose distribution specifies the least "age" of orders that will be delivered in period n to location 1. This means all outstanding orders placed in period $n - \rho_n$ or earlier are delivered in period n. We assume that the sample space of the random variable ρ_n is $\{0, 1, 2, \ldots, m - 1\}$ and consequently, the maximum lead time of an order is $m - 1$ periods.

The sequence of events in a period as described in Section 6.1 is now modified slightly. Because of the possibility of more than one period's orders arriving at location 1 in a period, we include the following event just prior to observing the demand in period n. Then ρ_n is realized; if $\rho_n \leq m - 2$, all units in locations 2 through $m - \rho_n$ arrive from the supplier and are at location 1. If $\rho_n = m - 1$, then no units arrive at location 1.

It is easy to verify that all the analysis and results that we presented for the "deterministic lead time model" hold for the stochastic lead time model described above. Therefore, echelon base-stock policies are optimal for the single-stage system even when lead times are stochastic and non-crossing.

6.1.1.4 The Serial Systems Case

The next extension to our analysis is the case of serial systems. In the single-echelon case, the only location from which a unit could be released using a control policy was location $m + 1$. In the multiple echelon case, there are more "physical locations" from which a unit can be released using the control policy. These physical locations correspond to stages in the production/distribution system. In addition, there may be many "artificial locations" between successive stages. The number of these artificial locations corresponds to the maximum possible lead time between these stages. We still assume that orders do not cross. The cost model is the same as discussed in the beginning of Section 6.1. The optimality of state-dependent echelon base-stock policies can be verified by repeating the following arguments, which we used in the analysis of the single-stage system.

First, the cost for system $\mathcal{S}$ is still the sum of the costs for the subsystems because of the linear cost structure. Second, monotone and committed policies are still optimal.

Third, each subsystem can be operated independently and optimally, which results in an optimal policy for system $\mathcal{S}$. Fourth, the optimal policy for a subsystem should be such that if it is optimal to release a unit from a stage or physical location when the corresponding customer is at a distance y, then it would also be optimal to release the unit from that stage if the customer were any closer. Consequently, an appropriately defined critical distance policy is optimal for every subsystem. Now there is a critical distance corresponding to each stage. When this policy is used for every subsystem, the resulting policy is a state-dependent echelon base-stock policy for the system $\mathcal{S}$. The proofs of all these results are identical to the proofs for the corresponding results for the single-stage system. Consequently, it is clear that echelon base-stock policies are optimal for serial systems having non-crossing lead times.

In the next section, we discuss how the single-unit, single-customer approach can be extended to consider a more general demand model and state description.

6.1.1.5 Generalized Demand Model

In the preceding discussion, we assumed that demand was governed by a process in which demands were independent from period to period. A careful analysis of the arguments indicate that this assumption is unnecessarily restrictive, but was made only to simplify notation and exposition.

Now suppose that demands are governed by a Markov modulated process. That is, demand in each period is governed by an exogenous, stationary ergodic Markov chain, s_n. This state of the system is observed at the beginning of each period n. The transition probabilities for the Markov chain are assumed known. Once s_n is given, the probability distribution of the period n demand random variable, D_n is known. The state of the system at the beginning of period n is now the vector $x_n = (s_n, (z_{1n}, y_{1n}), (z_{2n}, y_{2n}), \dots)$. The vectors x_n^w are defined similarly.

The optimal release strategy is now redefined to include s_n. That is $R_n^*(s_n, y) \subseteq \{\text{Release, Hold}\}$ is the set of optimal decisions for subsystem w at time n when the state of the exogenous Markov chain is s_n and y_{wn} is y and z_{wn} is $m+1$. The "critical distance" is also a function of s_n. Let

$$y^*(n, s_n) = \max\{y : R_n^*(s_n, y) \supseteq \{\text{Release}\}\}.$$

Then $y^*(n, s_n)$ is defined so that it is optimal to release unit w if and only if customer w is at a distance of $y^*(n, s_n)$ or closer. Hence the optimal policy is an order-up-to policy, where in each period we would raise the inventory position to $y^*(n, s_n) - 1$. Net inventory is on-hand minus backorders and the inventory position is the net inventory plus on-order inventory.

6.1.1.6   Capacity Limitations

Janakiraman and Muckstadt [188] extended the ideas developed earlier to consider capacity constraints on the order quantities in each period and echelon.

For serial systems consisting of an external supplier, a distribution center and a retailer in which the supplier and distribution center can ship only a maximum of C units in each period, they show the following results.

First, the optimization problem can be decomposed into C problems, each of which represents a subsystem consisting of the serial system with a single unit of capacity at each stage. Second, when the supplier-to-distribution-center and distribution-center-to-retailer lead times are one period, then the optimal policy is a modified base-stock policy at the distribution center and retail levels. That is, order up to the target inventory position if there is sufficient capacity. Otherwise, order C units. Third, when the two lead times are now two periods in length, then the optimal policy is called a "two-tier base-stock policy" at the distribution center and retailer.

This type of policy requires knowing more about the contents and the locations of the units in the pipeline. The ordering decision then depends on these values. A detailed description of these results can be found in Janakiraman and Muckstadt [188].

Finally, the same decomposition technique used for the two-echelon system we have described can be employed to control an M-echelon serial system in which lead times are of arbitrary length and capacities are identical at all locations. As is the case for the two-echelon problem, more parameters are required to describe the optimal policy.

Thus complicated extensions to the simple order-up-to policies are required to manage capacitated systems optimally.

6.2   Finding Optimal Stock Levels

Now that we have identified the form of an optimal policy when there are no fixed ordering costs, let us turn our attention to finding optimal stock levels. The method we will discuss in some detail is based on creating a dynamic programming representation of the decision problem.

6.2.1   Finite Planning Horizon Analysis

Let us assume that our goal is to find an optimal order-up-to level for a single item managed at a single location over a finite planning horizon consisting of N periods. We denote the order-up-to level by s_n. To begin, we will assume that the order lead time, τ, is zero. This implies that an order placed in a period is available for use in the same period. We choose to let $\tau = 0$ only to simplify the discussion. We will show how to extend the analysis to the more general case where τ is positive. We will see that the analysis is virtually the same as in the zero lead time case.

Suppose period n events evolve in the following manner. First, we observe the net inventory (which equals the inventory position since the lead time is zero). Next, we place an order on an external supplier, which raises the inventory position to s_n. Then the inventory arrives, which raises the net inventory level to s_n. Following the arrival of the inventory we observe the demand that arises for the item. If this demand is less than the supply, a holding cost of h dollars is charged per unsold unit; if demand exceeds supply, a shortage or backorder cost of b dollars is assessed per unit short.

Let c represent the unit cost. If the planning horizon is lengthy, we would likely want to set the values of s_n so as to minimize the expected total discounted cost over the planning horizon. Let α represent the discount factor.

Let D_n be the random variable representing the demand quantity in period n. We assume that these random variables are independent. We could consider more complicated demand models in our analysis. For example, we could have the process evolve over time as a Markov modulated process. This would make the values of the order-up-to levels depend on the state of the demand process as well as the net inventory. We choose to simplify the discussion by concentrating on this special case. Extentions to more complicated demand models will become obvious.

The solution procedure begins by considering the decision made in period N. This is now a single-period problem. The value of s_N^*, an optimal value for s_N, is selected to minimize the holding and backorder costs for the period plus purchase costs. We also assume that inventory remaining on hand at the end of the horizon is returned to the supplier, and a credit of c dollars per unit is given. If shortages exist at the end of period N, we assume we satisfy the excess demand through purchases from the supplier at a cost of c dollars per unit.

Let $g_N(y) =$ total expected discounted cost in period N when the inventory position (net inventory) is y units before placing the order in period N.

$=$ purchase cost + expected holding and backorder costs + expected discounted purchase or return costs.

Furthermore, let

$$g_{N+1}(y) = -cy. \tag{6.1}$$

Hence, if backorders exist at the end of period N ($y < 0$), we purchase $-y$ units at a cost of c dollars per unit; if $y > 0$ we return these units and receive a credit of c per returned unit.

Then

$$g_N(y) = \min_{s \geq y}\{c(s-y) + C_N(s) + \alpha E[g_{N+1}(s-D_N)]\}, \tag{6.2}$$

where

$$C_N(s) = hE[(s-D_N)^+] + bE[(D_N-s)^+]. \tag{6.3}$$

In general let

$g_n(y) = $ total expected discounted cost from period n through the end
of the planning horizon when the inventory position (net
inventory in this $\tau = 0$ case) is y units before an order is placed
in period n.

$$= \min_{s \geq y}\{c(s-y) + C_n(s) + \alpha E[g_{n+1}(s-D_n)]\} \tag{6.4}$$

where

$$C_n(s) = hE[(s-D_n)^+] + bE[(D_n-s)^+]. \tag{6.5}$$

We showed in the last chapter that $C_n(s)$ is a convex function of s, and hence $C_N(s)$ is convex. Clearly $E[g_{N+1}(s-D_N)]$ is also a convex function of s, as is $c(s-y)$.

Let

$$G_n(s_n) = cs_n + C_n(s_n) + \alpha E[g_{n+1}(s_n-D_n)]. \tag{6.6}$$

Suppose $n = N$. We can easily see that $G_N(s_N)$ is a convex function of s_N. Let s_N^* be the smallest optimal solution to $\min_{s_N} G_N(s_N)$, which exists and is non-negative for problems of practical interest.

Returning to the function $g_N(y)$, observe that it satisfies

$$g_N(y) = \begin{cases} -cy + G_N(s_N^*) & \text{if } s_N^* \geq y, \\ -cy + G_N(y) & \text{if } y > s_N^*. \end{cases} \tag{6.7}$$

It is easy to see that $g_N(y)$ is a convex function, as illustrated in Figure 6.2. Furthermore, by induction on n, we conclude that for

$$g_n(y) = \min_{s_n \geq y}\{c(s_n-y) + C_n(s_n) + \alpha E[g_{n+1}(s_n-D_n)]\} \tag{6.8}$$

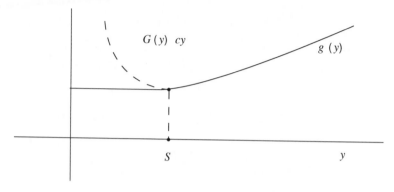

Fig. 6.2. Graphs of G_N and g_N.

$n = 1,\ldots,N$ is also convex in y. This follows immediately since the expectation of a convex function is convex as well.

Thus this set of relations represented by (6.1) and (6.8) establish a recursive method that can be used to find the optimal order-up-to values. These are called functional equations since $G_n(\cdot)$ and $g_n(\cdot)$ are unknown functions. The solution model we have described is called a dynamic programming representation of the decision problem.

Let us now return to the oil rig problem that we discussed in the previous chapter. Recall that in each period a decision must be made as to what amount of inventory to place on the rig. Let us now assume that all inventory that is unused at the end of a period remains on the rig, and is refurbished there rather than on shore. This represents a slight change in the system's operation from the one discussed in the previous chapter. Assuming all other data are the same as presented in the previous chapter, we want to know if the order-up-to policy, with $s = 43$, is optimal in every period.

When we posed the problem previously, we ignored the potential interaction of decisions among periods. The approach we took led to what is called a *myopic policy*. By implementing myopic policies, we focus only on the base-stock policy that minimizes the current period's cost while ignoring the impact of the current decision on future decisions and costs. An interesting question is under what conditions are myopic policies optimal.

Suppose we let $\bar{s}_n$ represent the myopic policy solution for period n and, as before, let s_n^* represent the optimal solution for period n. Let us first assume that the discount factor, α, is equal to 1. Recall $\bar{s}_n$ is obtained by finding the smallest value for which $P\{D_n \leq s_n\} \geq b/(b+h)$. Since we have assumed that the demand distribution is stationary, as well as that the costs are unchanging over time, we have $\bar{s}_1 = \bar{s}_2 = \cdots = \bar{s}_N$. By assumption, all demands over the N periods are ultimately satisfied, and no extra inventory is in the system after the end of the horizon, owing to our ability to return unused inventory. Hence the unit order costs do not come into play when making the inventory decision, and hence these unit costs can be assumed to be 0. This is a consequence of our assuming $\alpha = 1$. Thus $\sum_n C_n(\bar{s}_n)$ is a lower bound on $g_1(0)$, where we assume there is no inventory on hand initially and $c = 0$. But this solution provided by this myopic approach will also yield a feasible solution to the multi-period problem and hence is optimal.

But suppose $\alpha < 1$. In this case define a new, but related set of functions. Let

$$\bar{g}_n(y) = cy + g_n(y), \tag{6.9}$$

$$\bar{c} = c(1 - \alpha), \quad \text{and} \tag{6.10}$$

$$\bar{C}_n(s) = \alpha c \mu_n + \bar{c} s + C_n(s), \quad \text{where } \mu_n = E[D_n]. \tag{6.11}$$

Then

$$\bar{g}_{N+1}(y) = 0 \quad \text{and} \tag{6.12}$$

$$\bar{g}_n(y) = \min_{s \geq y} \{G_n(s)\}, \quad \text{where} \tag{6.13}$$

$$G_n(s) = \bar{C}_n(s) + \alpha E[\bar{g}_{n+1}(s - D_n)]. \tag{6.14}$$

Observe that (6.13) can be written as

$$\bar{g}_n(y) = G_n(\max(s_n^*, y)).$$

Let us now turn our attention to the function $\bar{C}_n(s)$, the single-period cost function in our revised model. Now let $\bar{s}_n$ represent the solution to this revised one-period problem. Then $\bar{s}_n$ is the smallest value of s for which

$$P\{D_n \leq s_n\} \geq \frac{b - \bar{c}}{b + h}.$$

An important question remains: how closely related are $\bar{s}_n$ and s_n^*? Let us now answer this question.

First, how does s_N^* compare with $\bar{s}_N$? As we will see, $s_N^* = \bar{s}_N$. Furthermore, in general $s_n^* \leq \bar{s}_n$. Let us see why this is the case. Let us assume that D_n is a continuous random variable so that $s_n^*, \bar{s}_n \in (-\infty, \infty)$. Observe that $\bar{g}_n(y) = \min_{s \geq y}\{G_n(s)\}$ is non-decreasing in y and therefore $E[\bar{g}_{n+1}(y - D_n)]$ is also non-decreasing in y. Recall that $\bar{s}_n$ satisfies $\bar{C}_n'(\bar{s}_n) = 0$. Thus $G_n'(\bar{s}_n) = \alpha E[\bar{g}_{n+1}'(\bar{s}_n - D_n)] \geq 0$, and, in general, $G_n'(s) \geq \bar{C}_n'(s)$. Consequently, $s_n^* \leq \bar{s}_n$. Now suppose $n = N$. Since $\bar{g}_{N+1}(y) = 0$, $E[\bar{g}_{N+1}(s - D_N)] = 0$, too, and $G_N'(s) = \bar{C}_N'(s)$. But $\bar{C}_N'(\bar{s}_N) = 0$ and therefore $G_N'(\bar{s}_N) = 0$ as well. Thus $\bar{s}_N = s_N^*$.

This result implies that if we consider the effect of the current decision on future decisions, we should never select a stock level that exceeds the level found by solving the corresponding one-period problem. Thus the myopic solutions are always upper bounds on the optimal solutions.

Second, suppose $\bar{s}_n \leq s^*_{n+1}$. Roughly speaking, this implies that the upper tail of the demand density in period $n+1$ falls to the right of that tail in period n ($\bar{s}_n \leq \bar{s}_{n+1}$). In this case, $s^*_n = \bar{s}_n$. But if $\bar{s}_n \geq \bar{s}_{n+1}$, $s^*_n \geq s^*_{n+1}$. To demonstrate that these relationships are true, recall that $\bar{g}_{n+1}(y) = G_{n+1}(\max(s^*_{n+1}, y))$, which is constant for all values of $y \leq s^*_{n+1}$. Thus for $y \leq s^*_{n+1}$, $E[\bar{g}_{n+1}(y - D_n)]$ is also a constant (assuming $D_n \geq 0$ with probability 1). Therefore $G'_n(y) = \bar{C}'_n(y)$, $y \leq s^*_{n+1}$. But $\bar{C}'_n(\bar{s}_n) = 0$ and hence $G'_n(\bar{s}_n) = 0$, too, which implies that $s^*_n = \bar{s}_n$.

When $\bar{s}_n \geq \bar{s}_{n+1}$, then $\bar{s}_n \geq s^*_{n+1}$, too. Thus $G'_n(y) = \bar{C}'_n(y) < 0$ when $y < s^*_{n+1}$ which implies $s^*_n \geq s^*_{n+1}$.

These observations demonstrate how the myopic and optimal order-up-to values tend to move together. The optimal value takes into account the risk of having too much inventory on hand when entering future time periods and hence s^*_n is bounded above by $\bar{s}_n$.

Third, $s^*_n \geq \min_{m \geq n}\{\bar{s}_m\}$. Thus the period-$n$ optimal stock level is also bounded below. Let us prove this result inductively. When $n = N$, the result is obvious. Suppose the result is true for all $m > n$. When $\bar{s}_n \leq \min_{m \geq n+1} \bar{s}_m \leq s^*_{n+1}$, then, as shown above, $s^*_n = \bar{s}_n = \min_{m \geq n} \bar{s}_m$. If $\bar{s}_n > \min_{m \geq n+1} \bar{s}_m$, $s^*_n \geq s^*_{n+1} \geq \min_{m \geq n+1} \bar{s}_m = \min_{m \geq n} \bar{s}_m$.

One important observation that is implied by these results is that if $\bar{s}_1 \leq \bar{s}_2 \leq \cdots \leq \bar{s}_N$, $\bar{s}_n = s^*_n$ for all $n = 1, \ldots, N$. Thus in this situation the myopic policy solution is an optimal solution. In fact, myopic policies are often very close, in an expected-cost sense, to the results obtained when employing the optimal solutions. The variation in the demand distributions between periods has to be substantial for the values of $\bar{s}_n$ and s^*_n to deviate dramatically.

Let us now illustrate the results of applying the methodology that we have discussed. To do this we will consider several examples. In each case the planning horizon will be 5 periods in length. Furthermore, we will assume that the unit cost is the same in all periods and hence we will ignore the variable procurement costs in our examples.

To begin, let us assume that demand in each period follows a uniform distribution. Assume that the backorder to holding cost ratio in each period is equal to 10.

In our first example, suppose demands are uniformly distributed as given in Table 6.1. Note that the expected demand is increasing over time.

Observe that the myopic order-up-to levels are increasing and therefore, as we have shown analytically, these levels are optimal.

Suppose the probability distribution for demands in each period are reversed, as indicated in Table 6.2.

In this case, the myopic order-up-to levels are strictly decreasing. Nonetheless, the optimal order-up-to levels are still equal to the myopic levels. Even changing the back-

Table 6.1. First example.

Demand Probability (Uniform Distribution)	Low	High
Period 1	5	20
Period 2	5	25
Period 3	7	28
Period 4	8	28
Period 5	9	30

	Myopic Order-Up-To Level	Optimal Order-Up-To Level
Period 1	19	19
Period 2	24	24
Period 3	26	26
Period 4	27	27
Period 5	28	28

Table 6.2. Second example.

Demand Probability (Uniform Distribution)	Low	High
Period 1	9	30
Period 2	8	28
Period 3	7	28
Period 4	5	25
Period 5	5	20

	Myopic Order-Up-To Level	Optimal Order-Up-To Level
Period 1	28	28
Period 2	27	27
Period 3	26	26
Period 4	24	24
Period 5	19	19

order to holding cost ratio does not alter this result. Thus the myopic solution remains optimal. Of course, this is not always the case.

Now suppose the demand distribution per period remains uniform but with ranges indicated in Table 6.3. The optimal solution now differs slightly from the myopic solution; however, the expected cost of implementing the myopic solution is very close to that of implementing the optimal solution.

Table 6.3. Third example.

Demand Probability (Uniform Distribution)	Low	High
Period 1	20	60
Period 2	55	85
Period 3	0	8
Period 4	50	100
Period 5	1	3

	Myopic Order-Up-To Level	Optimal Order-Up-To Level
Period 1	57	57
Period 2	83	81
Period 3	8	8
Period 4	96	92
Period 5	3	3

Suppose now that the probability distribution of demand per period is Poisson distributed. Does this change the results dramatically? Suppose the expected demand per period is given in Table 6.4. Even though the demand drops substantially between periods 3 and 4, we see that the myopic policy does not differ dramatically from the optimal policy. The difference in the cost of using the myopic rather than the optimal policy is 2.892%. This observation holds over a broad range of backorder to holding cost ratios. But, is this always the case?

Table 6.4. Fourth example.

Demand Probability (Poisson Distribution)	Mean
Period 1	30
Period 2	30
Period 3	30
Period 4	2
Period 5	2

	Myopic Order-Up-To Level	Optimal Order-Up-To Level
Period 1	37	37
Period 2	37	37
Period 3	37	35
Period 4	4	4
Period 5	4	4

Suppose now that the demand per period follows a geometric distribution. Geometric distributions have heavier upper tails than possessed by Poisson distributions having the same mean. Hence there is greater risk in stocking inventory when the demand distribution is a geometric distribution rather than a Poisson distribution.

First, observe in Table 6.5 that the myopic policy yields much larger stock levels when demand is geometrically distributed in each period. As mentioned, this is the case since the geometric has a heavier tail than the Poisson distribution having the same mean. Second, note that the optimal order-up-to levels differ substantially from the myopic order-up-to levels in periods 2 and 3. The expected cost of following a myopic policy rather than an optimal policy increases cost by 8.205%. Hence, in this case, we see that myopic policies are not as effective as in the other situations. Nonetheless, the cost difference is not too great.

Table 6.5. Fifth example.

Demand Probability (Geometric Distribution)	Mean
Period 1	30
Period 2	30
Period 3	30
Period 4	2
Period 5	2

	Myopic Order-Up-To Level	Optimal Order-Up-To Level
Period 1	73	71
Period 2	73	63
Period 3	73	46
Period 4	5	5
Period 5	5	5

6.2.2 Constant, Positive Lead Time Case

To this point we have assumed that $\tau = 0$. Suppose that τ is now some positive integral number of periods in length. In this case, there can be both on-order and on-hand inventory. Hence purchasing decisions must be based on the inventory position at the beginning of a period rather than on the amount of net inventory. Inventory purchased in period n arrives at the beginning of period $n + \tau$. Hence the inventory order placed in period n has no impact on costs incurred for holding and backorders in periods n

through $n + \tau - 1$. The optimal order-up-to level in a period must reflect the need for stock over periods n through $n + \tau$ plus any impact on periods following period $n + \tau$. The myopic policy approach would consider only demand over periods n through $n + \tau$, that is, over the subsequent $\tau + 1$ periods. The probability distributions used to calculate $\bar{s}_n$ and s_n^* are the distributions for the cumulative demand over the subsequent $\tau + 1$ periods. To be specific when $\tau = 0$ we saw that only period n's demand was needed to compute $\bar{C}_n(\cdot)$. When $\tau > 0$, we need to consider the total demand over the $\tau + 1$ periods, n through $n + \tau$, when determining $\bar{C}_n(\cdot)$. Thus the random variable D_n must capture total demand in periods n through $n + \tau$. With this change, the analysis we presented for the $\tau = 0$ case extends exactly to the $\tau > 0$ case.

We note that the planning horizon of length N corresponds to the periods in which procurement orders are placed for an item. The periods in which the effect of these decisions is observed are periods $\tau + 1$ through $\tau + N$ in the future, again a horizon of length N. Costs incurred from periods 1 through τ are not affected by the ordering decisions and hence can be ignored in the analysis. Of course, the inventory position at the beginning of period 1 must be taken into account when placing orders.

6.2.3   End-of-Horizon Effects

An important question that always arises is how large must N be. In our model we assumed that $g_{N+1}(y) = -cy$. Any assumption made about the functional form of $g_{N+1}(y)$ may affect the value of s_1^*, the value that tells us what to do now. In practice, procurement decisions do not terminate at the end of a finite horizon. N is chosen to be small because estimating demand (and costs) accurately over time is difficult to do. Computational requirements increase with N, too. Thus the dynamic program is not solved once and the values of the s_n^* implemented throughout the planning horizon. Rather the model is solved each period as time evolves. This so-called rolling planning horizon approach is employed to ensure future changes in the estimates of demand and costs are reflected in current decisions. While it is desirable to choose the planning horizon length so that current decisions are not sensitive to the end-of-horizon cost assumptions, the length is often dictated by the availability of accurate estimates of future costs and demands. More will be said about dealing with end-of-horizon effects at the end of the following section.

6.2.4 Infinite-Horizon Analysis

Now that we have examined the finite-horizon case, let us turn our attention to an infinite-horizon planning problem. Although we shall not prove it, an order-up-to policy is optimal in the infinite-horizon case, too. Assume, as we did in our study of deterministic models, that demand and cost parameters are stationary over an infinite horizon. In this case we assume that the demand random variables are independent and identically distributed across periods.

Let D represent the random variable for demand over $\tau + 1$ periods. Let us present some useful nomenclature. Let

$$\bar{c} = (1 - \alpha)c, \text{ where } c \text{ is the unit cost, } \alpha \text{ is the discount factor and}$$

$$\bar{C}(y) = \alpha c \mu + \bar{c}y + C(y),$$

where μ is the expected demand over $\tau + 1$ periods, and

$$C(y) = h \cdot E([y - D]^+) + b \cdot E([D - y]^+).$$

The parameter s represents the order-up-to level, or stock level, which is the inventory position achieved upon ordering in a period. The optimal value of s, which we denote by s^*, is the smallest value of s that minimizes $\bar{C}(y)$, that is,

$$P[D \leq s^*] = \frac{-\bar{c} + b}{b + h},$$

when D is a continuous random variable measuring demand over $\tau + 1$ periods, or s^* is the smallest value of s for which

$$P[D \leq s^*] \geq \frac{-\bar{c} + b}{b + h}$$

when D is a discrete random variable measuring demand over $\tau + 1$ periods. As discussed in the finite-horizon case, s^* is the myopic order-up-to or base stock level. The question then is how s^* compares with the optimal order-up-to level. We state the following theorem without proof.

Theorem 6.3. *When the costs and demand processes are stationary, s^* is also the optimal order-up-to level.*

Thus it is easy to calculate the optimal solution for this case. Furthermore, the optimal order quantity in a period is simply the demand in the previous period! Note that this is not the case when the planning horizon is finite and the demand process is not stationary.

Suppose $\alpha = 1$. This is called the average cost case. Then $\bar{c} = 0$ and $\bar{C}(y) = c\mu + C(y)$, and, in the continuous demand case, s^* is the smallest value of s for which

$$P[D \leq s] = \frac{b}{b+h}.$$

The value $\bar{C}(s^*)$ is the optimal average cost per period. In the case where $\alpha < 1$, the optimal expected discounted cost over the infinite planning horizon is $\bar{C}(s^*)/(1-\alpha)$, assuming the starting inventory level is less than or equal to s^*.

Suppose the demands and costs are not stationary over time. Again an order-up-to policy remains optimal.

In our discussion at the beginning of this section we noted that D represents demand over $\tau + 1$ periods. Suppose τ is three weeks in length and a period is one week in length. Then D represents the demand over the four-week horizon. Suppose that each week of this four-week horizon has a different distribution, although independent of the others. Let us assume, for simplicity, that a month consists of four weeks and that D_1, D_2, D_3, and D_4 are the random variables corresponding to the demand in each of the four weeks. Then $D = \sum_{i=1}^{4} D_i$.

Although the distribution of demand for each week individually may be different, the distribution of D is stationary. Hence the order-up-to values are also stationary.

In the preceding section we observed that the end-of-horizon cost function $g_{N+1}(y)$ can affect the choice of s_1^* as well as other values of s_n^*. One way to deal with this problem is as follows.

Suppose $\alpha < 1$. For periods 1 through N, solve the problem as discussed earlier with one exception. Estimate a distribution of lead time demand that is expected to occur in future periods. Use this distribution to compute the expected infinite-horizon discounted cost as a function of the ending inventory position. This provides us with a way to compute $g_{N+1}(y)$ that will not ignore the future beyond period N.

6.2.5 Lost Sales

Our analysis in this chapter has been based on the assumption that all demand in excess of supply is backordered. Clearly, in many, if not most, situations experienced in life this assumption is not satisfied. So why make this assumption? The answer to this question is simple. First, the form of the optimal operating policy is not trivial. It is not a simple order-up-to policy even when fixed ordering costs are negligible. Second, even if an approximate order-up-to policy is followed, computing these values is time-consuming except when lead time demand is low and lead times are relatively short.

In the complete backordering case, the inventory position in a period determines the order quantity. This occurs because the net inventory at the end of a period a lead time in the future is equal to $s_n - D_n$, where s_n is the inventory position following the placement of an order in period n and D_n is the total demand for periods n through $n + \tau$. But this need not be the case, and in the lost sales case is not likely to be the case, when the variance in the demand per period is substantial and demand in excess of supply is lost.

As a consequence, the construction of the dynamic program employed in the lost sales case to find the optimal order-up-to values (not the optimal order quantities) must take into account a more elaborate description of the state space than required in the backorder situation. In the backorder case only the inventory position must be known.

In the lost sales case we must know how much is on hand, what inventory will arrive in this period and throughout the remaining periods within the lead time. If all the inventory comprising the inventory position at the beginning of period n, call it y, is on hand, then the expected holding and lost sales cost incurred at the end of period $n + \tau$ is $h \cdot E[(y - D_n)^+] + b \cdot E[(D_n - y)^+]$. Here b measures the lost sales cost per unit of demand that is not satisfied and includes lost revenue margin as well as any other costs associated with losing the sale. On the other hand, if all y units are ordered in period n and there is nothing on hand or that will be received over the subsequent $\tau - 1$ periods, then the expected cost is quite different. All demands in the next $\tau - 1$ periods are lost. The net inventory at the end of period $n + \tau$ would equal y minus only the demand occuring in period $n + \tau$.

When lead times are lengthy and the amount ordered in each period can vary substantially the size of the state space (the current net inventory and the amounts due to arrive in each subsequent period) becomes enormous. Thus solving the resulting dynamic program is computationally intractable since the size of the state space determines the amount of required computation. The explosion of the size of the state space and its impact on the computational requirements is called the curse of dimensionality.

We leave the construction of a dynamic programming model for lost sales environments as an exercise.

6.3 Capacity Limited Systems

The analyses presented earlier in this chapter were based on the premise that the resupply systems had a reliable source of material (constant lead times).

We will now study a situation in which resupply is limited by a capacitated resource. The presence of capacity alters the method of analysis substantially.

6.3.1   The Shortfall Distribution

Let us introduce several important ideas by examining a simple capacity limited production facility. Production decisions are made in each period of an infinite planning horizon. There is only one item being produced and demands for this item are independent and identically distributed from period to period.

Assume that the system operates in the following manner. At the start of a period we observe the demand for the single item. On the basis of that period's demand and the item's current net inventory level, a production quantity is determined. Production then takes place. The production quantity is limited to a maximum of C units, the facility's per period production capacity. Whenever the demand exceeds C plus the on-hand stock, we assume the unsatisfied demand is backordered. On the other hand, if the demand is less than C plus the on-hand stock, then all customers are satisfied. After this production occurs, shipments are made to customers. The final act in a period is to charge holding and backorder costs of h and b dollars per unit held or backordered, respectively.

The optimal policy for managing this system is a modified version of the order-up-to policy or base-stock policy, as discussed earlier in this chapter. Suppose s is our target inventory level. The modified base-stock policy states that we should produce to have s units on hand after satisfying demand if capacity permits; otherwise, produce C units, the system's production capacity. As long as there is a positive probability that demand in a period can exceed C units, there is a positive probability that the production capacity in a period is not sufficient to raise the period ending inventory to s. The amount by which s exceeds the actual period ending inventory level is called the shortfall. The shortfall is a random variable. Our first goal is to show how to compute its distribution.

Then we will show how this distribution is used to develop a model for making production decisions. We will first show some general properties of this shortfall random variable and then focus on the case where demand is described by a discrete demand distribution. Note that if $C = \infty$, then there is never any shortfall. In this case, given the timing of events, $s = 0$ would be optimal.

6.3.1.1   General Properties

Let V_n represent the random variable for the shortfall in period n. We assume that the expected per period demand is strictly less than C. If this is not the case, then the backorder quantity will grow without bound as $n \to \infty$ with probability 1. Suppose now that D_n measures the demand in period n. Since $E[D_n] < C$ for all n, and the random variables D_n are independent and identically distributed, a stationary distribution exists for the shortfall process. Let V represent this random variable. Thus, in steady state, $V = s - I$, where I is the period ending net inventory level.

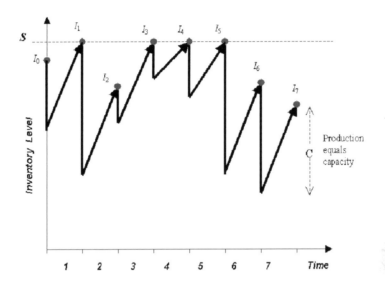

Fig. 6.3. System evolution of net inventory.

Recall the sequence of events that occur in period n. The initial net inventory I_n is equal to $s - V_{n-1}$. Demand is observed, that is, D_n is observed. A production quantity is then determined which equals $\min\{C, V_{n-1} + D_n\}$. Note that if $V_{n-1} > s$, then $(V_{n-1} - s)$ backorders exist at the beginning of period n. At the end of period n, the net inventory is $s - V_n$. If this quantity is positive, then there is stock on hand and is charged a holding cost of h dollars per unit; if negative, then there are backorders which are charged at a cost of b dollars per unit backordered. Figure 6.3 illustrates the evolution of the net inventory random variable.

Observe that

$$V_n = [V_{n-1} + D_n - C]^+ . \tag{6.15}$$

That is, if the capacity is large enough to satisfy both the entering shortfall plus the current period's demand, then $V_n = 0$; otherwise, V_n equals the difference between the total requirement $(V_{n-1} + D_n)$ and the production capacity (C).

We observe that V_n is independent of the target stock level s. Equation (6.15) describes the period-to-period dynamic behavior of the shortfall random variable. This behavior of V_n is illustrated in Figure 6.4.

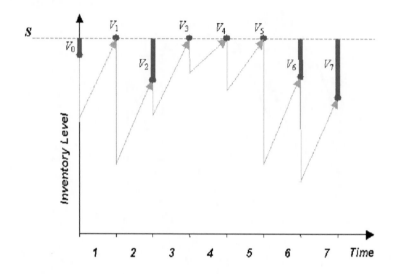

Fig. 6.4. Inventory shortfall.

6.3.2 Discrete Demand Case

Assume $V_0 = 0$. Observe from (6.15) that V_n depends only on V_{n-1} and D_n. That is, it does not depend on $V_0, \ldots, V_{n-2}$. Hence we can model the transitions of the shortfall process as a Markov chain. Specifically the transition probabilities for this chain are as follows:

$$p_{ij} \equiv P\{V_n = j | V_{n-1} = i\} = \begin{cases} P\{D \leq C - i\}, & j = 0 \text{ and } i \leq C, \\ P\{D = C + (j - i)\}, & j > 0, i \leq C + j, \\ 0, & \text{otherwise.} \end{cases}$$

Let $\mathcal{P} = [p_{ij}]$ be the matrix of transition probabilities. Since we assume that $E[D] < C$, a steady state distribution exists for the random variable, V, since the chain is ergodic. Let the stationary distribution that $V = i$ be denoted by π_i, and π the vector whose ith component is π_i. Then π solves

$$\pi \mathcal{P} = \pi,$$
$$\sum \pi_i = 1,$$
$$\pi_i \geq 0.$$

For practical situations, the matrix $\mathcal{P}$ and the corresponding vector π can be truncated to yield a finite system of equations. Some testing needs to be done to insure that accuracy is not sacrificed. The truncation process will depend on the difference between C and $E(D)$ and the variance of the demand process.

		Capacity Per Period (Utilization)			
		120 (83.3%)	**110** (90.9%)	**105** (95.2%)	**101** (99.0%)
Demand Variance Per Period (Coefficient of Variation)	**101** (0.10)	E(V) = 0.11 StDev = 0.93	1.36 3.83	5.51 9.44	45.04 50.59
	200 (0.14)	E(V) = 0.76 StDev = 3.22	4.26 9.06	13.32 19.83	92.47 100.72
	500 (0.22)	E(V) = 4.48 StDev = 11.48	15.36 25.48	39.41 51.56	238.60 252.68
	1000 (0.32)	E(V) = 12.96 StDev = 26.18	36.34 53.26	85.45 104.63	484.69 505.96
	2000 (0.45)	E(V) = 32.91 StDev = 56.36	81.39 109.17	180.57 210.92	979.88 1,012.55

Table 6.6. Expected shortfall and standard deviation of shortfall for various combinations of capacity utilization and demand variation.

Suppose C varies from period to period. Specifically, suppose C is a random variable that is independent and identically distributed from period to period. We assume that $E[D] < E[C]$. In this case, we can again represent the transitions of the shortfall process as a Markov chain. The process remains ergodic so that a stationary distribution will exist. In this case, the transition probabilities are given by

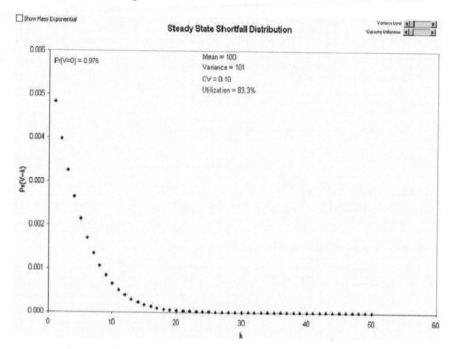

Fig. 6.5. Steady state shortfall distribution: Exact.

$$p_{ij} \equiv P\{V_n = j | V_{n-1} = i\} =$$
$$\begin{cases} \sum_{a \geq i} P\{D \leq a - i\} \cdot P\{C = a\}, & j = 0, i \leq C, \\ \sum_{a \geq i-j} P\{D = a + (j-i)\} P\{C = a\}, & j > 0, i \leq C + j, \\ 0, & \text{otherwise}. \end{cases} \quad (6.16)$$

In many cases the capacity is a random variable. If capacity can vary significantly from period to period, then the target stock level can increase substantially.

Let us now illustrate how the steady state shortfall distribution of V behaves for different levels of demand variation and available capacity. We assume that the expected demand per period is 100 units in all cases. We further assume the capacity C does not vary from period to period. Table 6.6 shows how both the expected value of V and standard deviation of V change for 20 combinations of the per period variance of demand and the amount of available per period capacity.

These data show how sensitive the mean and standard deviations of V are to changes in these values, and hence how inventory requirements will also depend on these values.

Additionally, Figures 6.5–6.10 provide the probability distributions for V for certain cases. Figures 6.5–6.7 contain plots of the distribution of V when the variance of the demand is set to 101 and when the capacity utilization rate assumes three different values:

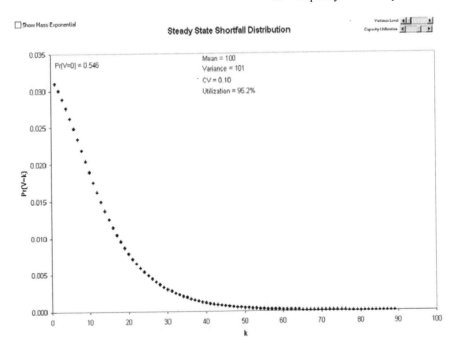

Fig. 6.6. Steady state shortfall distribution: Exact.

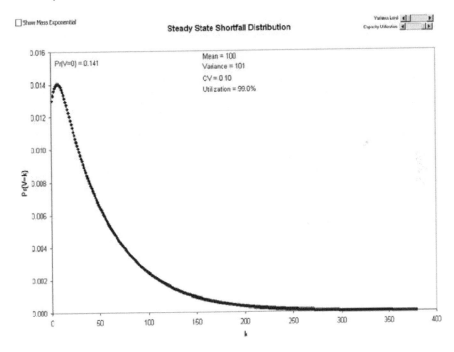

Fig. 6.7. Steady state shortfall distribution: Exact.

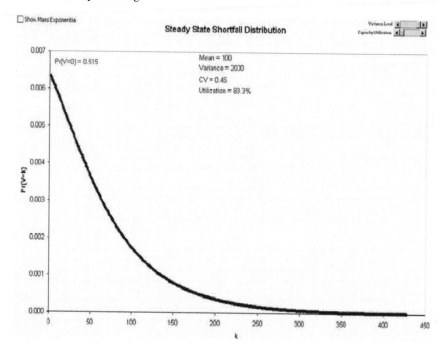

Fig. 6.8. Steady state shortfall distribution: Exact.

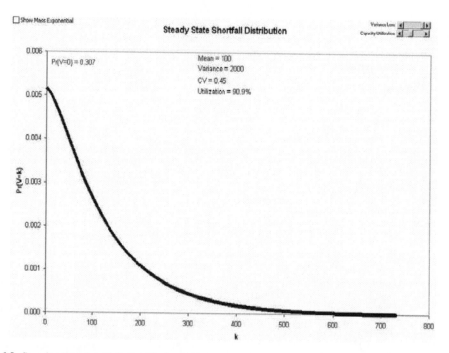

Fig. 6.9. Steady state shortfall distribution: Exact.

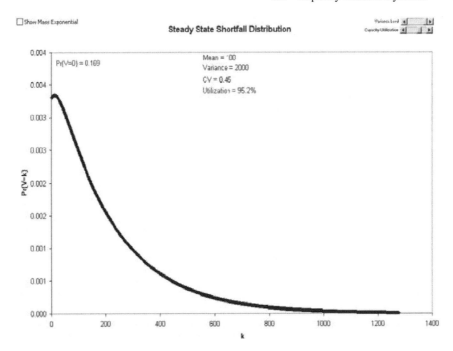

Fig. 6.10. Steady state shortfall distribution: Exact.

.833, .952, and .99. Similarly, Figures 6.8–6.10 show the probability distributions for
V when the variance of demand per period is 2000, again for the same three different
amounts of available capacity. Note how the shapes of the distributions are affected by
the variance of the demand process and the utilization rate. Furthermore, note how the
tails of the distributions behave.

6.3.3 An Example

Let us return to our oil rig example. Let us assume that regular replenishment of the
item to the oil rig is limited to C units. In addition, assume that each unused unit held
in inventory at the end of a period is maintained on the oil rig and the cost of doing so
plus other holding costs are estimated to be h dollars per unit. Each unit short incurs a
shortage cost estimated to be b dollars. Now there is no emergency resupply available
so the only replenishment of stock is the regular replenishment that occurs each period.

The system operates as follows. At the beginning of a period we request a shipment
from the on shore stock. The goal is to raise the on rig stock to s units. If there is not

sufficient space to send all the desired units, then only C units are placed on the vessel used to resupply the oil rig.

Note that this system differs from the one discussed in the preceding section since the sequence of events is not the same. Here we must make a decision on how much to order prior to knowing the amount of the demand in the period.

Assuming $C > E[D]$, where D is the random variable describing demand in each period, then a stationary distribution exists for the shortfall random variable. In this case, the shortfall in period n satisfies

$$V_n = [V_{n-1} + D_{n-1} - C]^+,$$

where V_n is the shortfall prior to the demand occuring in period n. Since the distribution of demand in periods $n-1$ and n are identical, the stationary distribution is identical to the one developed in the last section.

Our goal is to find s^*, the optimal base stock level, given the capacity limitations of C units. Suppose y units are on hand on the oil rig after the resupply delivery is received. Then the expected cost will be

$$hE[(y-D)^+] + bE[(D-y)^+].$$

But the value of y depends on s and the shortfall. Suppose the shortfall is i units. Then $y = s - i$. Since the shortfall random variable assumes the value of i with probability π_i, the expected cost per period is given by

$$F(s) = \sum_i \{hE[(s-i-D)^+] + bE[(D-(s-i))^+]\}\pi_i.$$

To find the optimal value of s we first note that $F(s)$ is a convex function in s, whose proof we leave as an exercise. Then, when we have a discrete random variable describing demand, s^* is the smallest value of s for which

$$\Delta F(s) = F(s+1) - F(s) \geq 0.$$

In this case

$$F(s) = \sum_i \left\{ h \sum_{d \leq s-i} (s-i-d)P[D=d] + b \sum_{d \geq s+1-i} (d-(s-i))P[D=d] \right\} \pi_i$$

$$= h(s-1-\mu) + (b+h)\sum_i \left\{ \sum_{d \geq s+1-i} (d-(s-i))P[D=d] \right\} \pi_i$$

and, after some algebra, s^* is the smallest value of s for which

$$\sum_i \left\{ \sum_{d \le s-i} P(D=d) \right\} \pi_i \ge \frac{b}{b+h}.$$

Observe that if $\pi_0 = 1$, then the above is

$$\sum_{d \le s} P[D=d] \ge \frac{b}{b+h},$$

which is the myopic solution found earlier.

6.4   A Serial System

Again let us consider the oil rig problem. But this time we will consider the interaction between the onshore stocking facility and the oil rig. The problem scenario is as follows.

Each period both the oil rig and onshore facility will place orders. The oil rig requests inventory from the onshore facility and the onshore facility from an external supplier. The external supplier is assumed to have an infinite amount of stock available. The replenishment lead time to the oil rig from the onshore facility is assumed to be negligible, that is, this lead time is zero. The lead time to resupply from the external supplier is assumed to be τ periods in length.

We will study two scenarios. In the first scenario, we assume that the demand process on the oil rig is non-stationary, but independent from period to period. We will also assume that $\tau = 0$. In the second scenario we assume demand on the oil rig is independent and identically distributed from period to period. Furthermore, we will assume τ is positive.

Our goals are to demonstrate the difference between stocking inventory at the two facilities on system cost and service, to show how to construct appropriate cost models in each case, and to present algorithms for calculating optimal ordering policies. Earlier in this chapter we discussed the optimality of order-up-to or base-stock policies for this type of system when there are no fixed ordering costs. Hence we assume such policies are followed at each location.

For ease of exposition, let us denote the onshore facility as facility 2 and the oil rig as facility 1. Subscripts on costs and performance measures will be either 1 or 2, corresponding to the appropriate location.

Let us begin by discussing the costs. Carrying costs are incurred on the basis of on-hand inventory at locations 1 and 2. We have assumed that the delivery time from the onshore facility to the oil rig is very short. In general, if the delivery time were lengthy, we would also charge holding costs to the in-transit stock between these locations.

Let h_1 and h_2 measure the holding cost per unit incurred at the end of a period at the two locations.

Note that shortages at each location play different roles in the system. Shortages on the oil rig imply that there is a true economic penalty incurred, which we assume is proportional to the number of non-functioning units (backorders).

On the other hand, a shortage at the onshore facility incurs no direct shortage cost. A shortage there implies that the order placed by the oil rig is not satisfied in its entirety. Hence if the number of backorders at facility 2 is x units, then these x units were ordered by 1 but not shipped or received on time. The implication of 2's not fulfilling 1's order is that facility 1 will incur expected shortage costs incremental to those that would have been anticipated if the order had been satisfied completely on time. Let b represent the shortage cost incurred per unit at facility 1, which is charged at the end of a period.

We will conduct our analysis of this system using echelon stock concepts, which were first introduced in Chapter 3. There are two possible and related views of system control mechanisms. In the first view, we consider stock levels and costs at each location as if the system operates in a decentralized manner. In the second, we consider a centralized view based on echelon stocks. In our case, the oil rig echelon inventory position measures its on-hand plus on-order less backorder position. The echelon inventory position for facility 2 considers this stock plus what it has on hand plus on order.

The echelon holding costs are related to installation-based costs. The cost h'_2, the echelon holding cost for facility 2, is charged on all units in the entire system, as it was in Chapter 3. Note that $h'_2 = h_2$. But h'_1, the echelon holding cost at facility 1, measures the incremental holding cost due to the placement of the unit on the oil rig rather than on shore. Note that $h'_1 = h_1 - h'_2 = h_1 - h_2$.

We will proceed as follows. After some introductory comments, we will construct a cost model along with a computational procedure that can be employed to find the optimal stocking policy.

6.4.1 An Echelon-Based Approach for Managing Inventories in Serial Systems

The system we will examine first consists of two facilities, the onshore facility and the oil rig. Our goal is to find the echelon order-up-to or echelon base stock levels for these locations in each of N time periods so that the total expected cost over the planning horizon is minimized.

To simplify our notation and to make the discussion easier to understand, we will assume that the lead time from the external supplier to the onshore facility is negligible. That is, we assume that $\tau = 0$.

As stated, the planning horizon is finite in length. We assume that at the end of the planning horizon any on-hand inventory can be returned to the external supplier and that the inventory can be returned to the external supplier and that the supplier will refund the original purchase price for each returned unit. If backorders exist at the end of the planning horizon, the external supplier will immediately provide inventory to fill the previously unsatisfied demand. Furthermore, these units will be purchased at the normal cost per unit. We will also not discount cost over time. On the basis of these assumptions, we can assume that the unit purchase cost is zero.

The costs considered when making the procurement and allocation decisions are holding and backorder costs. Thus, the goal is to minimize the total expected holding and backorder costs for the two facilities over the planning horizon.

As mentioned, demand on the oil rig will be non-stationary over time. We let D_n represent the random variable for oil rig demand in period n, $n = 1, \ldots, N$.

We will assume events in period n occur in the following sequence. First, each facility places an order on its supplier. The onshore facility orders a quantity q_{2n} from the external supplier. The oil rig orders q_{1n} units from the onshore facility. Next, each facility receives the stock it ordered from its supplier. None of these q_{2n} units received at the onshore facility is available for shipment to the oil rig in period n, however.

Finally demand occurs during period n at the oil rig. Holding and/or backorder costs are incurred, depending on the amount of net inventory at each facility at the end of period n.

6.4.1.1   A Decision Model

Let us introduce some additional notation. Let I_{in} measure the echelon inventory position at facility i, $i = 1, 2$, at the beginning of period n. Since the lead times are zero in length, I_{in} also measures the echelon net inventory at facility i at the beginning of period n. Since we are measuring echelon inventory position,

$$I_{1n} = \text{on-hand inventory on the oil rig} - \text{backorders on the oil rig}$$

and

$$I_{2n} = \text{on-hand inventory at the onshore facility} + I_{1n}.$$

Note that if lead times are positive, then in-transit inventories to each facility would be added to these echelon inventory positions.

Also, observe that the net on-hand inventory at the onshore facility at the beginning of period n is $I_{2n} - I_{1n}$, which is a non-negative quantity. The dynamics of the echelon inventory are as follows:

$$I_{1,n+1} = I_{1n} + q_{1n} - D_n$$

and

$$I_{2,n+1} = I_{2n} + q_{2n} - D_n.$$

Hence, I_{in} is a random variable, and the costs incurred over time are not known with certainty. As mentioned earlier, our goal is to establish a policy that minimizes the total expected costs of holding inventory and responding to shortages over the N period planning horizon.

These expected costs incurred in period n are expressed as follows:

$$h_1 E[(I_{1n} + q_{1n} - D_n)^+] + bE[(D_n - I_{1n} - q_{1n})^+] + h_2(I_{2n} + q_{2n} - I_{1n} - q_{1n}).$$

Let us define two cost expressions. The reason for defining them as we do will become clearer as we proceed.

First, let

$$C_{2n}(y) = h_2 y$$

and

$$C_{1n}(y) = h_1 E[(y - D_n)^+] + bE[(D_n - y)^+] - h_2 y.$$

Next, let

$$y_{in} = I_{in} + q_{in}, \qquad i = 1, 2.$$

Then the period n expected cost expression can be written as

$$C_{1n}(y_{1n}) + h_2 y_{1n} + h_2(y_{2n} - y_{1n}) = C_{1n}(y_{1n}) + C_{2n}(y_{2n}).$$

We will now show that this representation of the expected one-period costs provides the basis for finding the optimal ordering policy.

6.4.1.2   A Dynamic Programming Formulation of the Decision Problem

To obtain the optimal ordering policy, we will formulate the decision problem as a dynamic programming problem. Using this formulation, we will establish a key result that was first proven by Clark and Scarf [70].

Let $V_n(I_{1n}, I_{2n})$ represent the minimum expected cost incurred from period n through the end of the planning horizon, given that I_{1n} and I_{2n} are the echelon inventory positions at the beginning of period n for the onshore and oil rig facilities, respectively. Then

$$V_n(I_{1n}, I_{2n}) = \min C_{1n}(y_{1n}) + C_{2n}(y_{2n}) + E[V_{n+1}(y_{1n} - D_n, y_{2n} - D_n)]$$

subject to

$$I_{2n} \leq y_{2n}$$
$$I_{1n} \leq y_{1n} \leq I_{2n}. \tag{6.17}$$

What we will now show is that there exist functions f_{1n} and f_{2n} such that

$$V_n(I_{1n}, I_{2n}) = f_{1n}(I_{1n}) + f_{2n}(I_{2n}). \tag{6.18}$$

As a consequence, we will show that it is possible to independently find the optimal ordering policy for each facility. That is, we can find the best policy for the oil rig and onshore facilities by solving two independent problems, one for each facility.

Let

$$f_{2n}(I_{2n}) = \min_{y_{2n} \geq I_{2n}} C_{2n}(y_{2n}) + E[f_{2,n+1}(y_{2n} - D_n)] + \Delta_{2n}(I_{2n}), \tag{6.19}$$

for $n = 1, \ldots, N$. If I_{2n} is too small, then y_{1n} may be smaller than the oil rig would desire it to be, and hence additional expected costs may be incurred. The function $\Delta_{2n}(I_{2n})$ measures the expected incremental costs induced on the oil rig as a consequence of the value of I_{2n}.

Let

$$f_{1n}(I_{1n}) = \min_{y_{1n} \geq I_{1n}} C_{1n}(y_{1n}) + E[f_{1,n+1}(y_{1n} - D_n)], \tag{6.20}$$

for $n = 1, \ldots, N$. Observe that problem (6.20) is exactly the type studied in Section 6.2. Suppose y_{1n}^* is the optimal value for y_{1n}. Furthermore, let $\hat{y}_{1n} = \arg\min_{y_{1n}} \{C_{1n}(y_{1n}) + E[f_{1,n+1}(y_{1n} - D_n)]\}$. Given $\hat{y}_{1n}$, we can compute the value of $\Delta_{2n}(I_{2n})$. If $I_{2n} \geq \hat{y}_{1n}$, then the oil rig is able to raise its on-hand inventory to the level that minimizes its expected current and future costs. However, when $\hat{y}_{1n} > I_{2n}$, the oil rig's inventory will be equal to I_{2n} because $f_{1n}(\cdot)$ is convex. When $\hat{y}_{1n} > I_{2n}$, the expected incremental cost is equal to the difference between the expected cost incurred at the oil rig when $y_{1n} = I_{2n}$ and the expected cost when $y_{1n} = \hat{y}_{1n}$, that is,

$$\{C_{1n}(I_{2n}) + E[f_{1,n+1}(I_{2n} - D_n)]\} - \{C_{1n}(\hat{y}_{1n}) + E[f_{1,n+1}(\hat{y}_{1n} - D_n)]\}.$$

In summary, the expected incremental cost function is

$$\Delta_{2n}(I_{2n}) = \begin{cases} 0 & \text{if } \hat{y}_{1n} \leq I_{2n}, \\ C_{1n}(I_{2n}) + E[f_{1,n+1}(I_{2n} - D_n)] \\ \quad -C_{1n}(\hat{y}_{1n}) - E[f_{1,n+1}(\hat{y}_{1n} - D_n)] & \text{otherwise.} \end{cases} \tag{6.21}$$

We will now use an induction argument to prove that

$$V_n(I_{1n}, I_{2n}) = f_{1n}(I_{1n}) + f_{2n}(I_{2n}),$$

where $f_{1n}(I_{1n})$ and $f_{2n}(I_{2n})$ are defined in (6.19) and (6.20).

Recall that we assumed that all excess inventory at the end of the planning horizon would be returned to an external supplier and that the supplier will reimburse the company at the original purchase price per unit for units returned to it. Similarly, any shortages existing at the end of period N will be satisfied with purchases made at the normal price per backordered unit. As a consequence,

$$V_{N+1}(\cdot, \cdot) = 0.$$

Thus,

$$V_N(I_{1N}, I_{2N}) = \min\{C_{1N}(y_{1N}) + C_{2N}(y_{2N}) : I_{2N} \leq y_{2N}, I_{1N} \leq y_{1N} \leq I_{2N}\}$$
$$= \min_{y_{2N} \geq I_{2N}} C_{2N}(y_{2N}) + \Delta_{2n}(I_{2n}) + \min_{y_{1N} \geq I_{1N}} C_{1N}(y_{1N}),$$

where we assume that we have computed $\hat{y}_{1n}$ separately so that $\Delta_{2n}(I_{2n})$ is well-defined. But

$$f_{2N}(I_{2N}) = \min_{y_{2N} \geq I_{2N}} C_{2N}(y_{2N}) + \Delta_{2n}(I_{2n})$$

and

$$f_{1N}(I_{1N}) = \min_{y_{1N} \geq I_{1N}} C_{1N}(y_{1N})$$

so we have

$$V_N(I_{1N}, I_{2N}) = f_{1N}(I_{1N}) + f_{2N}(I_{2N})$$

and the decomposition exists for period N.

Assume, now, that $V_{n+1}(I_{1,n+1}, I_{2,n+1}) = f_{1,n+1}(I_{1,n+1}) + f_{2,n+1}(I_{2,n+1})$. We want to show that we can express $V_n(I_{1n}, I_{2n})$ similarly. For period n, we have

$$V_n(I_{1n}, I_{2n}) = \min \left\{ \sum_{i=1}^{2} C_{in}(y_{in}) + E\left[V_{n+1}(y_{1n} - D_n, y_{2n} - D_n)\right] \right.$$

$$\left. : y_{2n} \geq I_{2n}, I_{1n} \leq y_{1n} \leq I_{2n} \right\}$$

$$= \min \left\{ \sum_{i=1}^{2} C_{in}(y_{in}) + \sum_{i=1}^{2} E\left[f_{i,n+1}(y_{in} - D_n)\right] \right.$$

$$\left. : y_{2n} \geq I_{2n}, I_{1n} \leq y_{1n} \leq I_{2n} \right\}$$

$$= \min \left\{ \sum_{i=1}^{2} C_{in}(y_{in}) + \sum_{i=1}^{2} E\left[f_{i,n+1}(y_{in} - D_n)\right] \right.$$

$$\left. : I_{in} \leq y_{in} \leq I_{i+1,n} \right\}$$

where $I_{3n} = \infty$.

Let us consider two cases. In case one, we suppose that $\hat{y}_{1n} \leq I_{2n}$. Clearly, we have $\Delta_{2n}(I_{2n}) = 0$. The expected cost incurred on the oil rig is not affected by the constraint $y_{1n} \leq I_{2n}$. However, for case two, $\hat{y}_{1n} > I_{2n}$, and so $\Delta_{2n}(I_{2n}) \neq 0$. We have

$$V_n(I_{1n}, I_{2n}) = \min_{y_{2n} \geq I_{2n}} C_{2n}(y_{2n}) + E\left[f_{2,n+1}(y_{2n} - D_n)\right] + \Delta_{2n}(I_{2n})$$

$$+ \min_{y_{1n} \geq I_{1n}} C_{1n}(y_{1n}) + E\left[f_{1,n+1}(y_{1n} - D_n)\right]$$

$$= \sum_{i=1}^{2} f_{in}(y_{in}).$$

Thus, in both cases we have

$$V_n(I_{1n}, I_{2n}) = f_{1n}(I_{1n}) + f_{2n}(I_{2n}) . \tag{6.22}$$

This proof can be extended to serial systems containing more than two facilities. When there are M facilities, then $V_n(\bar{I}_n)$ can be written as the sum of M functions, $f_1(I_{1n}), \ldots, f_M(I_{Mn})$, where $\bar{I}_n$ is a vector of M echelon stock levels at the beginning of period N, $I_{jn}, j = 1, \ldots, M$. Each function corresponds to a facility upstream of the final facility, that is, the one that serves customers. Each function $f_{jn}(I_{jm})$, for $j = 2, \ldots, M$ has the form of (6.19), in which the $\Delta_{jn}(I_{jn})$ are computed as in (6.21).

6.4.1.3   An Algorithm for Computing Optimal Echelon Stock Levels

We have demonstrated that $V_n(I_{1n}, I_{2n})$ can be expressed as the sum of two functions, each of which is a function of a single variable. When showing that this is possible, we also established the basis for an algorithm that can be used to find the optimal operating policy for each facility in the system. Let us now formally state this algorithm.

Algorithm for Finding the Optimal Operating Policy

Step 1: Begin with the oil rig (customer serving location). For all n, $n = 1, \ldots, N$, find $\hat{y}_{1n}$. Begin with period N and work backwards to find $\hat{y}_{11}$. When $I_{1n} > \hat{y}_{1n}$, $y^*_{1n} = I_{1n}$; otherwise, $y^*_{1n} = \hat{y}_{1n}$.

Step 2: Given the values of $\hat{y}_{1n}$, calculate $\Delta_{2n}(I_{2n})$, $n = 1, \ldots, N$, for all $I_{2n} \leq \hat{y}_{1n}$. (We do not need to consider $I_{2n} \geq \hat{y}_{1n}$ since $\Delta_{2n}(I_{2n}) = 0$ in these cases.) Again, begin with period N and work backwards until $\Delta_{21}(I_{21})$ is determined.

Step 3: Solve problem (6.19), $n = 1, \ldots, N$, given the values of $\Delta_{2n}(I_{2n})$ computed in Step 2. Begin with period N and work backward until the solution is obtained for period 1.

This algorithm could be extended for any serial system problem. We leave the development of this extension as an exercise.

6.4.1.4   Solving the Oil Rig Problem: The Stationary Demand Case

Finding an optimal echelon stocking policy for the onshore facility and oil rig is somewhat easier than we have described when the demand process at the oil rig is stationary. Here the goal is to find y^*_1 and y^*_2, which are the optimal echelon stock levels for the oil rig and onshore facility, respectively, for all N periods.

Recall that the lead time for the onshore facility is τ periods in length. In the previous sections, we assumed $\tau = 0$; we must modify the analysis we presented slightly to account for τ being positive. Also, since the demand process is stationary, a myopic policy is also the optimal policy. This observation greatly simplifies our analysis.

Let $C_2(y) = h_2 E(y - \bar{D})$, where $\bar{D}$ is the random variable corresponding to the demand over τ periods. Notice that this definition is different from the previous formulation of the problem; the reasons for the difference will become clear shortly. Let $C_1(y) = h_1 E[(y - D)^+] + b E[(D - y)^+] - h_2 y$, where D is the random variable for demand occurring in one period. Then

$$\{C_1(y_1)+h_2y_1\}+\{h_2E(y_2-\bar{D})-h_2y_1\}=C_1(y_1)+C_2(y_2)$$

is the expected cost incurred per period.

Suppose the onshore facility echelon stock is y_2. Then after $\tau+1$ periods, the echelon inventory position random variable, I_2, which measures the inventory available to supply the oil rig, is equal to y_2-D_1, where D_1 is the demand over the lead time plus one period. This is the case because we have assumed that inventory arriving to the onshore facility in a period is not available to send to the oil rig in that period.

The value of y_1^* is found by determining the $\text{argmin}\,C_1(y)$. Once this value is computed, we can find the value of $\Delta_2(I_2)=\Delta_2(y_2-d)$, where $d=y_2-I_2$ is the realized demand over $\tau+1$ periods. If $I_2 \geq y_1^*$, then $\Delta_2(I_2)=0$; otherwise, $\Delta_2(I_2)=C_1(I_2)-C_1(y_1^*)$.

To find y_2^*, we solve the following problem:

$$\min C_2(y)+E[\Delta(y-D_1)].$$

This formulation accounts for the incremental expected cost incurred by the oil rig as a consequence of the echelon stock policy followed at the onshore facility. In summary, the algorithm for finding the optimal echelon stock levels is as follows:

Algorithm for Determining Optimal Oil Rig and Onshore Echelon Stock Levels

Step 1: Find y_1^*, the smallest optimal solution to

$$\min C_1(y).$$

Step 2: For all $I_2 < y_1^*$, calculate

$$\Delta_2(I_2)=C_1(I_2)-C_1(y_1^*).$$

Step 3: Find y_2^*, the smallest optimal solution to

$$\min C_2(y)+E[\Delta_2(y-D_1)].$$

6.5 Exercises

6.1. In Section 6.2 we developed a dynamic programming approach for finding the optimal order-up-to values over a finite planning horizon of N periods when there are no fixed ordering costs. Suppose that $N=6$, and $h=1$ and $b=10$. Let $\alpha=1$ and assume the unit cost is constant over time. Suppose demand is Poisson distributed and independent from period to period.

(a) Suppose the demand rates in the 6 periods are 10, 20, 40, 40, 3, 1, respectively. Find $\bar{s}_i$ and s_i^* for each period, $i = 1,\ldots,6$.
(b) Suppose the demand rates for these 6 periods are 50, 50, 50, 20, 5, 1, respectively. Find $\bar{s}_i$ and s_i^*, $i = 1,\ldots,6$.

6.2. Assume the data are the same as in Exercise 6.1 except now the demand in each period has a negative binomial distribution with mean 5.

(a) Suppose the variances in each of the 6 periods are 25, 25, 15, 10, 6 and 6. Find the optimal values for $\bar{s}_i$ and s_i^*, $i = 1,\ldots,6$.
(b) Suppose the variances are 6, 6, 10, 15, 25 and 25 for the 6 periods. Now find the values of $\bar{s}_i$ and s_i^*, $i = 1,\ldots,6$.

6.3. The model developed in Section 6.2 was based on the assumption that the cost data were constant over time. Suppose $\tau = 1$ and the costs are period-dependent. Derive results similar to those found in Section 6.2 for this case. Develop the dynamic programming model and corresponding recursions.

6.4. Llenroc Electronics repairs components found in radar sets for the Air Force. One particular component in the radar set is costly and fails frequently. When Llenroc receives failed parts it is not capable of repairing all of them. Units awaiting repair are called reparables. By policy, Llenroc attempts to maintain a stock of serviceable units that can be sent to the Air Force when a defective part arrives to be repaired or scrapped. In each period, Llenroc can purchase new components from the manufacturer. To simplify the situation, assume both the lead time from the manufacturer and the repair lead times are negligible. In each period, Llenroc must decide how many units to purchase from the supplier, how many of the available reparables to repair, and how many serviceable and reparables to scrap. Assume demand for replacement components placed in each period follows a Poisson distribution whose mean can fluctuate over time. Further assume that the cost data imply that it is best to repair before purchasing and to scrap reparables before serviceable units. Let s_n be the on-hand serviceable stock level after the repair, procurement and scrapping decisions are made in period n of an N period planning horizon. In each period the beginning state of the system indicates the number of serviceables and reparables on hand. Demand arises at the beginning of each period followed by the repair, purchasing or scrapping decision. If demand exceeds supply, a backorder cost is incurred at the rate of b dollars per unit. Supply in excess of demand incurs holding costs of h_1 and h_2 for serviceables and reparables, respectively.

Let C be the unit procurement cost, r the unit repair cost, u_1 and u_2 the scrap values of serviceables and reparables, respectively. Assume the discount factor is equal to one.

Develop a dynamic programming model that can be used to find the optimal repair, purchase, and scrapping policy.

6.5. Let us consider the oil rig example discussed in Section 6.4. In that example, there are two stages, the onshore and oil rig facilities. Let us assume that an order placed by the material clerk on the oil rig at the beginning of a period nominally arrives at the beginning of the next period at the oil rig. We say nominally, because the quantity ordered can only be shipped if there is enough stock on hand at the onshore warehouse to satisfy the order. Orders placed by the onshore facility in the beginning of a period arrive there two periods later.

The cost to hold a unit at the end of a period at the onshore facility is $1. The installation holding cost for a unit held at the end of a period on the oil rig is $2. The backorder cost on the oil rig is $20 per unit per period.

The demand per period is either 0,1, or 2 units with probabilities .3, .5, .2 respectively. Demands are independent from period to period.

Determine the optimal stocking levels (echelon inventory position) for each facility.

6.6. In Section 6.3 we developed an expression for the expected cost per period, $F(s)$. Prove that $F(s)$ is a convex function.

6.7. In Section 6.2 we discussed the difference between the lost sales and the backorder cases. Suppose the lead time is two periods. Construct a dynamic programming model that can be used to find the optimal solution to the lost sales problem in this case. Use the notation found in Section 6.2 when developing your answer.

6.8. Suppose demands over an infinite planning horizon are independent and identically distributed from period to period. Assume the production and inventory system operates in the manner described in Section 6.3.1. Assume production capacity in each period is 2 units. The demand is either 0, 1, 2, or 3 units per period with probabilities .2, .3, .3, and .2, respectively. Construct the shortfall probability distribution.

6.9. Suppose in the previous problem that capacity is one unit with probability .1, two units with probability .8, and three units with probability .1. Construct the shortfall probability distribution.

6.10. Find the optimal stock level for the situation described in Problem 6.8, assuming a shortage cost of $1,000 and a holding cost of $100 per unit per period.

7

Background Concepts: An Introduction to the $(s-1,s)$ Policy under Poisson Demand

In the previous chapter we studied order-up-to policies when time was divided into periods. We will now discuss the implications of following a similar policy, called an $(s-1,s)$ inventory policy. In this chapter we assume inventories are reviewed continuously in time. Recall that the stock level, s, measures the amount of inventory on hand plus on order minus backorders, that is, the stock level represents the inventory position for a particular location. In certain situations, we will refer to the on-order quantity as the "in resupply" quantity. This "in resupply" terminology is often used in military and aviation applications in which items fail and are repaired or are procured from an external source. When an $(s-1,s)$ policy is followed in continuous review environments, an order is placed immediately whenever a demand occurs for one or more units of an item. The order quantity matches the size of the demand exactly. Hence, the inventory position is constant in the case where the demand process and costs are stationary over an infinite planning horizon, which is the one we will examine in some detail in this chapter.

A major objective in this chapter is to show how to compute the stationary probability distribution of the quantity of units in resupply. The amount in resupply at a random point in time is a key random variable in the study of the behavior of systems managed using an $(s-1,s)$ policy. Once its stationary distribution is known, we can easily determine the stationary distribution for on-hand and backordered inventory. We will focus primarily on the case where backorders are allowed, since the analysis is simpler. As a special case, however, we will also analyze a situation where excess demand over supply is lost.

We first show how to compute the distribution for the quantity in resupply in the backorder case when the replenishment lead times or equivalently the resupply times are independent and identically distributed. We will focus on the case where the demand

J.A. Muckstadt and A. Sapra, *Principles of Inventory Management: When You Are Down to Four, Order More*, Springer Series in Operations Research and Financial Engineering, DOI 10.1007/978-0-387-68948-7_7, © Springer Science+Business Media, LLC 2010

process is a Poisson process. We will generalize these results to the case where the demand process is a compound Poisson process.

After we show how to calculate the stationary distributions, we show how to determine key statistical measures of supply system performance. Lastly, we present optimization models and algorithms for computing stock levels when items are managed using an $(s-1,s)$ policy.

7.1 Steady State Distribution of the Number of Units in Resupply

The construction of the steady state distribution of the number of units in resupply follows from the properties of the underlying Poisson process generating the orders. Let us begin by reviewing some of these properties.

Let λ represent the demand rate of the customer order process. Suppose exactly one order occurs during the time interval $[0,t]$. Given that this order has occurred, let us establish the distribution of the time at which the order was placed. Intuitively, this distribution should be uniform since a Poisson process has stationary and independent increments. Let T be the time at which this event occurs, and let $N(t)$ represent the number of customer orders received in $[0,t]$. Then, for $s < t$,

$$
\begin{aligned}
P[T < s | N(t) = 1] &= \frac{P[T < s; N(t) = 1]}{P[N(t) = 1]} \\
&= \frac{P[N(s) = 1; N(t-s) = 0]}{P[N(t) = 1]} \\
&= \frac{P[N(s) = 1] \cdot P[N(t-s) = 0]}{P[N(t) = 1]} \\
&= \frac{\lambda s e^{-\lambda s} e^{-\lambda(t-s)}}{\lambda t e^{-\lambda t}} \\
&= \frac{s}{t}.
\end{aligned}
$$

Hence the time at which the customer arrival occurs is uniformly distributed over the interval $[0,t]$.

This result can be generalized as follows. Suppose $X_1, \ldots, X_n$ are n independent and identically distributed random variables. The random variables $X_{(1)}, \ldots, X_{(n)}$ are order statistics corresponding to $X_1, \ldots, X_n$ if $X_{(k)}$ corresponds to the kth smallest value among the random variables, $X_1, \ldots, X_n$. Let $f(x_i)$ represent the common density function for the X_i. Then the joint density function for the $X_{(i)}$ is

$$f_{X_{(1)},\dots,X_{(n)}}(x_1,\dots,x_n) = n! \prod_{i=1}^{n} f(x_i), \quad x_1 < \cdots < x_n. \tag{7.1}$$

The term $n!$ appears because there are that many permutations of $X_1,\dots,X_n$ that lead to the same order statistic.

Now suppose that $N(t) = n$ and suppose $X_1,\dots,X_n$ are the arrival times of the 1st, 2nd, $\dots$, nth customer orders, respectively. Then $X_1,\dots,X_n$ have the same distribution as do the order statistics corresponding to n independent random variables that have uniform distributions over the interval $[0,t]$. We can prove this fact in the following manner.

Suppose we have times $t_1,\dots,t_n$, where $0 < t_1 < t_2 < \cdots < t_n < t$, and Δ_i is small enough so that

$$t_i + \Delta_i < t_{i+1} \quad \text{and} \quad t_n + \Delta_n < t. \tag{7.2}$$

Then

$$P[t_1 \leq X_1 \leq t_1 + \Delta_1, \dots, t_n \leq X_n \leq t_n + \Delta_n | N(t) = n]$$

$$= \frac{P[\text{1 cust order is placed in } [t_i, t_i + \Delta_i], i=1,\dots,n, \text{ and no customer orders are placed elsewhere in } [0,t]]}{P[N(t)=n]}$$

$$= \frac{(\lambda \Delta_1 e^{-\lambda \Delta_1}) \cdots (\lambda \Delta_n) e^{-\lambda \Delta_n} \cdot \left(e^{-\lambda(t - \sum_{i=1}^{n} \Delta_i)}\right)}{e^{-\lambda t} \frac{(\lambda t)^n}{n!}}$$

$$= \frac{n!}{t^n} \prod_{i=1}^{n} \Delta_i,$$

and therefore

$$\frac{P[t_1 \leq X_1 \leq t_1 + \Delta_1, \dots, t_n \leq X_n \leq t_n + \Delta_n | N(t) = n]}{\Delta_1 \cdots \Delta_n} = \frac{n!}{t^n}.$$

Taking the limit of the left-hand side as $\Delta_i \to 0$ for all i, we obtain

$$f_{X_1,\dots,X_n}(t_1,\dots,t_n) = \frac{n!}{t^n}, \quad 0 < t_1 < t_2 < \cdots < t_n < t, \tag{7.3}$$

which is the desired result. Thus, we may conclude that if n customer orders are placed in $[0,t]$, then the times at which these orders are placed, considered as unordered times, are independent and uniformly distributed over the interval $[0,t]$.

7.1.1 Backorder Case

We are now ready to establish a remarkable result, which is a restatement of a theorem attributed to Palm [263].

Theorem 7.1. *Suppose s is the stock level for an item whose demands are generated by a Poisson process with rate λ. Suppose further that the resupply time random variables have density functions $g(\tau)$ with mean $\bar{\tau}$, and have distribution functions $G(\tau)$. Suppose further that the resupply times are independent and identically distributed from customer order to customer order. Then the steady state probability that x units are in resupply is given by*

$$e^{-\lambda\bar{\tau}}\frac{(\lambda\bar{\tau})^x}{x!}. \tag{7.4}$$

Proof. Suppose $N(t) = n$ customer orders have been placed in $[0,t]$. We know that

$$P[N(t) = n] = e^{-\lambda t}\frac{(\lambda t)^n}{n!}. \tag{7.5}$$

Since an $(s-1,s)$ policy is employed to manage the inventory, each customer order generates a corresponding request on the resupply system. Next, let

$$q_t(x|n) = P[x \text{ units are in resupply at time } t|N(t) = n]. \tag{7.6}$$

Consider any one of the n orders. As we just demonstrated, the time of its placement is uniformly distributed over the interval $[0,t]$. Suppose this order was placed at time $s \in [0,t]$. Then the probability that the corresponding unit remains in the resupply system at time t is $1 - G(t-s)$.

Let p be the common probability that any unit that arrives during $[0,t]$ remains in the resupply system at time t. Since $1 - G(t-s)$ measures the conditional probability that the unit entering the resupply system at time s remains unsatisfied at time t, the unconditional probability is given by

$$\begin{aligned}
p &= \int_0^t [1 - G(t-s)]\frac{ds}{t}\\
&= \frac{1}{t}\int_0^t [1 - G(t-s)]ds\\
&= -\frac{1}{t}\int_t^0 [1 - G(u)]du\\
&= \frac{1}{t}\int_0^t [1 - G(u)]du.
\end{aligned}$$

Since each arriving order in $[0,t]$ has a probability p that its corresponding resupply request is not satisfied by time t, the probability that x of the n arriving units in the resupply system remain unsatisfied at time t is given by

$$q_t(x|n) = \binom{n}{x} p^x (1-p)^{(n-x)}. \tag{7.7}$$

Hence the unconditional probability that x units remain in the resupply system at time t is

$$
\begin{aligned}
q_t(x) &= \sum_{n=x}^{\infty} q_t(x|n) \cdot P[N(t) = n] \\
&= \sum_{n=x}^{\infty} \binom{n}{x} p^x (1-p)^{n-x} e^{-\lambda t} \frac{(\lambda t)^n}{n!} \\
&= \sum_{n=x}^{\infty} \frac{n!}{(n-x)!x!} p^x (1-p)^{n-x} e^{-\lambda t} \frac{(\lambda t)^n}{n!} \\
&= \frac{e^{-\lambda t}(p\lambda t)^x}{x!} \sum_{n=0}^{\infty} \frac{[\lambda t(1-p)]^n}{n!} \\
&= \frac{e^{-\lambda t} e^{\lambda t - \lambda t p}(p\lambda t)^x}{x!} \\
&= e^{-\lambda t p} \frac{(\lambda t p)^x}{x!}.
\end{aligned}
$$

But $p = \frac{1}{t} \int_0^t [1 - G(u)]du$, so

$$q_t(x) = e^{-\lambda \int_0^t [1-G(u)]du} \frac{[\lambda \int_0^t [1-G(u)]du]^x}{x!}. \tag{7.8}$$

Let

$$q(x) = \lim_{t \to \infty} q_t(x). \tag{7.9}$$

Recall that

$$\lim_{t \to \infty} \int_0^t [1 - G(u)]du = \int_0^{\infty} [1 - G(u)]du = \bar{\tau}, \tag{7.10}$$

and therefore

$$q(x) = e^{-\lambda \bar{\tau}} \frac{(\lambda \bar{\tau})^x}{x!}. \tag{7.11}$$

Thus the probability that there are n units in the resupply system is Poisson distributed with mean $\lambda\bar{\tau}$; i.e., we do not need to know the density function for the resupply time, but only the mean of the resupply time distribution, $\bar{\tau}$.   □

This remarkable result has an important generalization. Suppose at each demand event the quantity required is not always of size one, but rather is a random variable. Assume the size of each order is independent from order to order. When the order event process is a Poisson process and the order size random variables are independent and identically distributed, the resulting demand process is called a compound Poisson process.

Let us now state, without proof, the generalization of Palm's theorem for the case of a compound Poisson process for the backorder case.

Theorem 7.2. *Suppose demands occur according to a compound Poisson process where λ is the customer order arrival rate. Suppose also that the resupply times are independent and identically distributed with density $g(\tau)$ with mean $\bar{\tau}$. Assume when a customer order is received, the resupply time for all units in the order is the same and is drawn from the resupply time distribution. The steady state probability of x units in resupply is given by the compound Poisson distribution with mean $\lambda\bar{\tau}\bar{u}$, where $\bar{u}$ is the average customer order size.*

7.1.2 Lost Sales Case

To this point we have assumed that all customer orders in excess of the supply s are backordered. Let us now assume that this is not the case; that is, when a customer order is placed and there is no on-hand inventory, then the order is lost. We will prove a version of Palm's theorem for a special case of the lost order situation. A general and complicated proof of the lost sales case when demand is compound Poisson distributed is given by Feeney and Sherbrooke [120]. We will focus on a relatively simple situation where the order lead times are exponentially distributed. Specifically, the theorem that we will prove is as follows.

Theorem 7.3. *Suppose customer orders arrive according to a Poisson process with arrival rate λ. Furthermore, suppose the stock level is s. Assume resupply times for accepted customer orders are independent and identically distributed with common density $g(\tau) = \beta e^{-\beta\tau}$, with mean $\bar{\tau} = 1/\beta$. Then the steady state probability that x units are in resupply in the lost order case is given by*

$$\frac{e^{-\frac{\lambda}{\beta}}\left(\frac{\lambda}{\beta}\right)^{x}/x!}{\sum_{n=0}^{s}e^{-\lambda/\beta}\frac{\left(\frac{\lambda}{\beta}\right)^{n}}{n!}} = \frac{e^{-\lambda\bar{\tau}}(\lambda\bar{\tau})^{x}/x!}{\sum_{n=0}^{s}e^{-\lambda\bar{\tau}}(\lambda\bar{\tau})^{n}/n!}.$$

Proof. When $g(\tau) = \beta e^{-\beta\tau}$, we can derive the desired result by an argument used when analyzing queuing systems. Let $P_j(t)$ represent the probability that j units are in resupply at time t. Note that if $j < 0$ or $j > s$, then $P_j(t) = 0$.

Since the order arrival process is a Poisson process and the resupply time distribution is exponential, for $0 \le j \le s$,

$$P_j(t + \Delta t) = [1 - (\lambda + j\beta)\Delta t]P_j(t) + \lambda\Delta t \cdot P_{j-1}(t) + (j+1)\beta\Delta t \cdot P_{j+1}(t) + o(\Delta t). \tag{7.12}$$

Then

$$P_j'(t) = \lim_{\Delta t \to 0}\frac{P_j(t + \Delta t) - P_j(t)}{\Delta t}$$
$$= -(\lambda + j\beta)P_j(t) + \lambda P_{j-1}(t) + (j+1)\beta P_{j+1}(t).$$

Passing to the limit $(t \to \infty)$, $P_j'(t) \to 0$.

Let π_j represent the steady state probability that j units are in the resupply system. Then

$$0 = -(\lambda + j\beta)\pi_j + \lambda\pi_{j-1} + (j+1)\beta\pi_{j+1}. \tag{7.13}$$

For $j = 0$, we have

$$\lambda\pi_0 = \beta\pi_1 \quad \text{or} \tag{7.14}$$

$$\pi_1 = \frac{\lambda}{\beta}\pi_0. \tag{7.15}$$

For $j = 1$, we have

$$\lambda\pi_0 + 2\beta\pi_2 = (\lambda + \beta)\pi_1 \tag{7.16}$$

or

$$2\beta\pi_2 = (\lambda + \beta)\frac{\lambda}{\beta}\pi_0 - \lambda\pi_0 \tag{7.17}$$

$$\pi_2 = \frac{\lambda^2}{2\beta^2}\pi_0, \tag{7.18}$$

and, as is easily shown, for $0 < j < s$,

$$\pi_j = \frac{1}{j!}\left(\frac{\lambda}{\beta}\right)^j \pi_0. \tag{7.19}$$

For the case where $j = s$,

$$s\beta \pi_s = \lambda \pi_{s-1} \tag{7.20}$$

or

$$\pi_s = \frac{\lambda}{s\beta}\left[\frac{1}{(s-1)!}\left(\frac{\lambda}{\beta}\right)^{s-1}\right]\pi_0$$

$$= \frac{1}{s!}\left(\frac{\lambda}{\beta}\right)^s \pi_0.$$

Since $\sum_{j=0}^{s}\pi_j = 1$,

$$\pi_0 \sum_{j=0}^{s}\frac{1}{j!}\left(\frac{\lambda}{\beta}\right)^j = 1 \quad \text{or} \tag{7.21}$$

$$\pi_0 = \left[\sum_{j=0}^{s}\frac{1}{j!}\left(\frac{\lambda}{\beta}\right)^j\right]^{-1}$$

$$= \frac{e^{-\lambda/\beta}}{\sum_{j=0}^{s}e^{-\lambda/\beta}\frac{(\lambda/\beta)^j}{j!}}$$

$$= \frac{e^{-\lambda\bar{\tau}}}{\sum_{j=0}^{s}e^{-\lambda\bar{\tau}}\frac{(\lambda\bar{\tau})^j}{j!}}.$$

Thus,

$$\pi_j = \frac{e^{-\lambda\bar{\tau}}(\lambda\bar{\tau})^j/j!}{\sum_{i=0}^{s}e^{-\lambda\bar{\tau}}(\lambda\bar{\tau})^i/i!}. \tag{7.22}$$

$\square$

In the general case, that is, where $g(\tau)$ is an arbitrary density with mean value $\bar{\tau}$, the steady state probability that j units are in the resupply system is also given by the above expression.

7.2   Performance Measures

To this point, we have developed the steady state probabilities for the number of units that are in the resupply system at a random point in time when the demand process is either a Poisson or compound Poisson process. On the basis of these probabilities, we can calculate different measures of system performance. These measures relate to performance at a single location. Subsequently, we will see how steady state probabilities and performance measures are computed in multi-echelon situations.

We will begin by considering several measures that are single-item measures, confining our discussion to the backorder case. The first performance measure we consider, the fill rate, is the most commonly used measure in practice, and is defined as follows. Given a stock level of s, the fill rate, $F(s)$, is the expected fraction of demands that can be satisfied immediately from on-hand stock. As is intuitively clear, as s increases the fill rate will increase. We will develop an explicit expression for $F(s)$ in this section and will discuss its properties in the next section of this chapter.

A second performance measure is called the ready rate corresponding to stock level s. The ready rate measures the probability that an item observed at a random point in time has no backorders, that is, its net inventory is non-negative. We denote the ready rate by $R(s)$. This is an all or nothing measure. Either there are backorders or there are no backorders at a random point in time.

Observe that when computing either a fill rate or ready rate we are not concerned with the duration of backorders when they occur. Thus, for example, a fill rate of say 95% implies that, on average, 95 of every 100 units that are ordered have that request satisfied immediately. But we are not measuring how long it takes to satisfy the other 5% of the units requested. Thus it is not always clear that a firm that maintains a high fill rate is truly satisfying its customers needs. This is particularly true when fill rates are calculated for a large number of item types. In this case, the fill rate would measure the fraction of demands satisfied immediately over all items. Thus some items could have nearly 100% of the demands satisfied immediately while others could have a 0% fill rate.

Note also that the ready rate is always at least as high as the fill rate. For example, when $s = 0, F(s) = 0$. But $R(s)$ could approach 1 if the demand rate is low and the lead time is very short. It is not unusual that the measures $F(s)$ and $R(s)$ are confused in practice.

A third single-item performance criterion measures the expected number of backorders outstanding at a random point in time, and is denoted by $B(s)$. This measure accounts for the length of time backorders exist. Hence, it is a response-time focused measure. Observe that $B(s)$ is equal to the demand rate times the average "waiting time" of a demand. This is a consequence of Little's law, $L = \lambda W$, where $B(s)$ is L, λ the demand rate, and W the average waiting time. We could also compute the conditional value of W, given that backorders exist.

Let us now see how these performance measures can be computed. Recall that, in the backorder case when the demand is a compound Poisson process, the steady state probability that x units are in resupply is given by

$$P\{X = x\} = p(x|\lambda\bar{\tau}) = \sum_{j=1}^{\infty} e^{-\lambda\bar{\tau}}\frac{(\lambda\bar{\tau})^j}{j!}u_x^{(j)}, x \geq 1,$$

$$P\{X = 0\} = e^{-\lambda\bar{\tau}},$$

where λ is the demand rate, $\bar{\tau}$ the average resupply time, and $u_x^{(j)}$ the probability that j customer orders generate a total demand of x units.

The ready rate is the probability that there are no backorders existing at a random point in time. This is the probability that the number of units in resupply is s or less. That is,

$$R(s) = \sum_{x=0}^{s} p(x|\lambda\bar{\tau}).$$

The computation of the fill rate is more difficult, but it is obtained from the steady state probabilities $p(x|\lambda\bar{\tau})$. Suppose a customer order is received. There will be one unit of the order satisfied if there are $s-1$ or fewer units in resupply. A second unit will be sent to the customer if the order is for two or more units and there are $s-2$ or fewer units in resupply. Remember that the timing and size of a customer order are independent of all past orders and resupply times. Hence, the expected number of units filled per customer order is given by

$$F_1(s) = \sum_{x \leq s-1} p(x|\lambda\bar{\tau}) + (1-u_1) \sum_{x \leq s-2} p(x|\lambda\bar{\tau})$$

$$+ (1-u_1-u_2)\sum_{x \leq s-3} p(x|\lambda\bar{\tau}) +$$

$$\cdots + \left(1 - \sum_{j \leq s-1} u_j\right) p(0|\lambda\bar{\tau}), \tag{7.23}$$

where, as before, u_j measures the probability that a customer order is for exactly j units. In the case of a simple Poisson demand process (that is, when $u_1 = 1$),

$$F(s) = F_1(s) = \sum_{x \leq s-1} p(x|\lambda\bar{\tau}). \quad \text{Hence, in this case,}$$

$$F_1(s) = F(s) = R(s) - p(s|\lambda\bar{\tau}) \text{ and } F(s) < R(s).$$

When the demand process is a compound Poisson process, $\lambda F_1(s)$ measures the expected number of units that can be shipped on time per day, when λ is the expected daily rate at which customers place orders. Furthermore, $\lambda \bar{u}$ measures the expected number of units demanded per day, where $\bar{u}$ is the expected number of units demanded per order. Thus

$$\frac{\lambda F_1(s)}{\lambda \bar{u}} = \frac{F_1(s)}{\bar{u}}$$

measures the fraction of the units ordered that are sent to customers on time. We let this quantity be defined as the fill rate, or

$$F(s) = \frac{F_1(s)}{\bar{u}}.$$

Next, we see that the expected number of units in a backorder status in steady state is

$$B(s) = \sum_{x>s} (x-s)p(x|\lambda \bar{\tau}).$$

That is, there are $x - s$ units backordered if and only if there are x units in resupply, $x > s$.

Suppose there are n item types in a system rather than just a single item. Then performance measures are computed somewhat differently.

First, the system fill rate is calculated by computing the conditional fill rate for the item type, multiplying by the probability that a demand was for a specific item type, and summing over item types. Let $\overline{F}(\bar{s})$ measure the system fill rate, where $\bar{s} = (s_1, \ldots, s_n)$ is a vector of item stock levels. If $F_i(s_i)$ measures the fill rate for item type i, then

$$\overline{F}(\bar{s}) = \sum_{i=1}^{n} \frac{\lambda_i}{\sum_{j=1}^{n} \lambda_j} \cdot F_i(s_i)$$

because $\lambda_i/\sum_{j=1}^{n} \lambda_j$ is the probability that a customer demand is for item type i when the customer order process is a Poisson process. This calculation is based on the assumption that an order is for a single item type.

The expected number of backorders at a random point in time for n items is simply

$$\sum_{i=1}^{n} B_i(s_i) = \sum_{i=1}^{n} \sum_{x>s_i} (x-s_i)p(x|\lambda_i \bar{\tau}_i).$$

When there is more than one item, the ready rate measure must be modified. The new measure is called the operational rate. We assume a system is operational if and only if all item types are operational. A particular item type will not be available, and hence the system will not be operational, if the number of units in resupply for that item exceeds its stock level. Assuming demand and resupply times are independent from item type to item type, the operational rate is given by

$$OR(\bar{s}) = \prod_{i=1}^{n} R_i(s_i).$$

Suppose there are many operating systems, say a fleet of aircraft. Furthermore, suppose part shortages can be consolidated into as few aircraft as possible. This process is often called "cannibalization" of the aircraft. Suppose there may be more than one unit of a particular type on an aircraft, say q_i units of type i. Then, assuming independence and cannibalization,

$$\prod_{i=1}^{n} R_i(s_i + q_i) = \text{probability that all aircraft}$$
$$\text{or one less than all aircraft}$$
$$\text{are operational.}$$

$$\text{and } \prod_{i=1}^{n} R_i(s_i + kq_i) = \text{probability that } k \text{ or}$$
$$\text{fewer aircraft are non-operational.}$$

Let Y be a random variable that measures the number of non-operational aircraft. Assuming cannibalization and independence,

$$P\{Y = 0\} = \prod_{i=1}^{n} R_i(s_j)$$

$$P\{Y \leq 1\} = \prod_{i=1}^{n} R_i(s_i + q_i)$$

$$P\{Y \leq k\} = \prod_{i=1}^{n} R_i(s_i + kq_i).$$

Consequently, the expected number of non-operational aircraft at a random point in time given cannibalization and independence is

$$E[Y] = \sum_{k \geq 1} k \cdot P\{Y = k\}$$
$$= \sum_{k \geq 1} P\{Y \geq k\}$$
$$= \sum_{k \geq 1} (1 - P\{Y \leq k - 1\})$$
$$= \sum_{k \geq 0} (1 - P\{Y \leq k\}).$$

Hence, if there are N aircraft, an approximation to the expected number of operational aircraft is

$$N - E[Y].$$

Since the demand process assumes an infinite population, this is a conservative estimate.

Another approximation for the expected number of operational aircraft can be developed as follows. Let us consider item type i. Recall that $B_i(s_i)$ measures the expected number of backorders for item type i at a random point in time. Suppose there is one unit of this item type per aircraft and there are N aircraft in the system. The probability that a random aircraft at a random point in time is missing a unit of item type i is $B_i(s_i)/N$, or $1 - B_i(s_i)/N$ is the probability that the aircraft is not missing a unit of type i. Assuming independence, the probability that a random aircraft is operational at a random point in time is

$$p = \prod_{i=1}^{n} (1 - B_i(s_i)/N).$$

The expected number of operational aircraft at a random point in time is given by

$$
\begin{aligned}
Np = N \prod_{i=1}^{n} (1 - B_i(s_i)/N) \\
= N \left(1 - \sum_{i=1}^{n} B_i(s_i)/N + \sum_{k \neq j} \frac{B_j(s_j)B_k(s_k)}{N^2} \right. \\
\left. - \sum_{i \neq j \neq k} \frac{B_i(s_i)B_j(s_j)B_k(s_k)}{N^3} + \cdots \right) \\
\cong N - \sum_{i=1}^{n} B_i(s_i),
\end{aligned}
$$

when $B_i(s_i)/N$ is small for all item types.

Thus there is a simple approximate correspondence between the expected number of backorders outstanding at a random point in time and the expected number of operational aircraft. The latter approximation of the expected number of operational aircraft is particularly useful from a computational viewpoint because of the mathematical properties of the functions $B_i(s_i)$, as we will now see.

7.3 Properties of the Performance Measures

Now that we have defined several key performance measures and have shown how to compute them, let us examine them more closely. We begin by studying the fill rate measure.

Let us assume, for simplicity, that the demand process is a simple Poisson process with rate λ. Furthermore, assume that resupply times for each order are independent and identically distributed with mean $\bar{\tau}$. As we have shown, the probability that x units are in the resupply system in steady state is given by

$$p(x|\lambda\bar{\tau}) = e^{-\lambda\bar{\tau}}\frac{(\lambda\bar{\tau})^x}{x!}.$$

Since the demand process is a simple Poisson process, the fill rate, given a stock level of s, is given by

$$F(s) = 1 - \sum_{x\geq s} p(x|\lambda\bar{\tau}) = \sum_{x<s} p(x|\lambda\bar{\tau}).$$

Perhaps our goal might be to choose stock levels for many items so that the average fill rate across items is maximized given some target investment level in inventory. This type of optimization problem would be easy to solve if $F(s)$ were a discretely concave function. Unfortunately, as we will now observe, it is not.

We know that if $F(s)$ were a discretely concave function in s, then its second difference must be non-positive for all $s \geq 0$. Let us now define both the first and second differences of $F(s)$. The first difference, $\Delta F(s)$, is given by

$$\Delta F(s) = F(s+1) - F(s),$$

and the second difference, $\Delta^2 F(s)$, is given by

$$\Delta^2 F(s) = \Delta F(s+1) - \Delta F(s).$$

Hence

$$\Delta F(s) = \sum_{x\leq s} p(x|\lambda\bar{\tau}) - \sum_{x\leq s-1} p(x|\lambda\bar{\tau})$$
$$= e^{-\lambda\bar{\tau}}\frac{(\lambda\bar{\tau})^s}{s!}$$

and

$$\Delta^2 F(s) = e^{-\lambda\bar{\tau}}\frac{(\lambda\bar{\tau})^{s+1}}{(s+1)!} - e^{-\lambda\bar{\tau}}\frac{(\lambda\bar{\tau})^s}{s!}$$

$$= e^{-\lambda\bar{\tau}}\frac{(\lambda\bar{\tau})^s}{s!}\left\{\frac{\lambda\bar{\tau}}{s+1} - 1\right\}.$$

When $\lambda\bar{\tau} > s+1$, then $\Delta^2 F(s) > 0$ and $F(s)$ is not concave in that region. In fact, when $s < \lambda\bar{\tau} - 1$, $F(s)$ is discretely convex. Hence $F(s)$ is discretely concave only when $s \geq \lfloor\lambda\bar{\tau}\rfloor$, when $\lambda\bar{\tau}$ is non-integer, and $s \geq \lambda\bar{\tau} - 1$, when $\lambda\bar{\tau}$ is an integer.

Graphs of $F(s)$ for two cases are given in Figures 7.1 and 7.2. In the first case $\lambda\bar{\tau} = 3.2$ and in the second case $\lambda\bar{\tau} = 3$. The graphs illustrate what we have proven. Tables 7.1 and 7.2 contain values for $F(s)$, $\Delta F(s)$ and $\Delta^2 F(s)$ for the two cases. The table values show that the concavity property holds when $\lfloor\lambda\bar{\tau}\rfloor \leq s$. Note that when $\lfloor\lambda\bar{\tau}\rfloor = \lambda\bar{\tau}$, that is, when $\lambda\bar{\tau}$ is an integer, $\Delta F(\lambda\bar{\tau}) = \Delta F(\lambda\bar{\tau} - 1)$.

Next, we observe immediately that the ready rate function, $R(s)$, is also not a concave function of s for all values of s.

Thus, neither $F(s)$ nor $R(s)$ possesses the mathematical property of concavity that is desirable when formulating and solving an optimization problem whenever we consider all values of $s \geq 0$. Hence, in practical cases, s is constrained to assume values that are greater than or equal to $\lfloor\lambda\bar{\tau}\rfloor$ to ensure that the fill rate or ready rate functions are

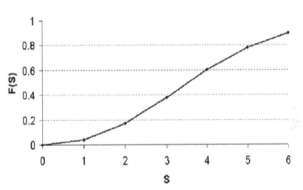

Fig. 7.1. Graph of fill rate vs. inventory (case 1).

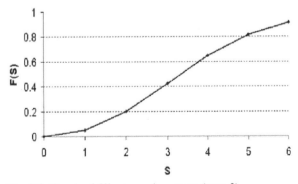

Fig. 7.2. Graph of fill rate vs. inventory (case 2).

Table 7.1. Fill rate vs. inventory tradeoff with Poisson demand.

	Mean Demand	**3.2**	
S	F(s) = P(D < S)	DF(S) = F(S + 1) − F(S)	D2F(S) = DF(S + 1) − DF(S)
0	0	0.040762204	0.089676849
1	0.040762004	0.130439053	0.078263432
2	0.171201257	0.208702484	0.013913499
3	0.379903741	0.222615983	-0.044523197
4	0.602519724	0.178092787	-0.064113403
5	0.780612511	0.113979383	-0.053190379
6	0.894591895	0.060789005	-0.032999745
7	0.955380899	0.027789259	-0.016673556
8	0.983170158	0.011115704	-0.007163453
9	0.994285862	0.00395225	-0.00268753
10	0.998238112	0.00126472	-0.000896801
11	0.999502832	0.000367919	-0.000269807
12	0.999870751	9.81116E-05	-7.39611E-05
13	0.999968862	2.41506E-05	-1.86304E-05
14	0.999993013	5.52013E-06	-4.3425E-06
15	0.999998533	1.17763E-06	-9.42102E-07
16	0.999999711	2.35525E-07	-1.91191E-07
17	0.999999946	4.43342E-08	-3.64526E-08
18	0.999999991	7.88163E-09	-6.5542E-09
19	0.999999998	1.32743E-09	-1.11504E-09

concave over the feasible region. Note that the operational rate and the first approximation for the expected number of operational systems are stated in terms of the ready rate. Hence, these performance measures are not easy to work with in optimization models unless $s \geq \lfloor \lambda \bar{\tau} \rfloor$.

The backorder function $B(s)$ does have very desirable mathematical properties, however. Recall that

$$B(s) = \sum_{x>s} (x - s) p(x | \lambda \bar{\tau}).$$

For $B(s)$ to be strictly discretely convex and strictly decreasing requires that

$$\Delta B(s) = B(s + 1) - B(s) < 0$$

and

$$\Delta^2 B(s) = \Delta B(s + 1) - \Delta B(s) > 0.$$

Table 7.2. Fill rate vs. inventory tradeoff with Poisson demand.

Mean Demand		3	
S	F(S) = P(D < S)	DF(S) = F(S + 1) − F(S)	D2F(S) = DF(S + 1) − DF(S)
0	0	0.049787068	0.099574137
1	0.049787068	0.149361205	0.074680603
2	0.199148273	0.224041808	0
3	0.423190081	0.224041808	-0.056010452
4	0.647231889	0.168031356	-0.067212542
5	0.815263245	0.100818813	-0.050409407
6	0.916082058	0.050409407	-0.028805375
7	0.966491465	0.021604031	-0.01350252
8	0.988095496	0.008101512	-0.005401008
9	0.996197008	0.002700504	-0.001890353
10	0.998897512	0.000810151	-0.000589201
11	0.999707663	0.00022095	-0.000165713
12	0.999928613	5.52376E-05	-4.24904E-05
13	0.999983851	1.27471E-05	-1.00156E-05
14	0.999996598	2.73153E-06	-2.18522E-06
15	0.99999933	5.46306E-07	-4.43873E-07
16	0.999999876	1.02432E-07	-8.4356E-08
17	0.999999978	1.80763E-08	-1.50636E-08
18	0.999999996	3.01272E-09	-2.53702E-09
19	0.999999999	4.75692E-10	-4.04338E-10

We see that

$$\Delta B(s) = \sum_{x \geq s+1} (x - (s+1))p(x|\lambda\bar{\tau}) - \sum_{x \geq s+1} (x - s)p(x|\lambda\bar{\tau})$$

$$= - \sum_{x \geq s+1} p(x|\lambda\bar{\tau}) = - \left(1 - \sum_{x \leq s} p(x|\lambda\bar{\tau}) \right)$$

and

$$\Delta^2 B(s) = - \sum_{x \geq s+2} p(x|\lambda\bar{\tau}) + \sum_{x \geq s+1} p(x|\lambda\bar{\tau})$$

$$= p(s+1|\lambda\bar{\tau}) > 0$$

and hence $B(s)$ is a strictly (discretely) convex function of s for all $s \geq 0$.

7.4   Finding Stock Levels in $(s-1,s)$ Policy Managed Systems: Optimization Problem Formulations and Solution Algorithms

Setting stock levels for items managed using an $(s-1,s)$ policy will depend on the objectives and constraints that are stipulated. For example, we could choose to minimize the average number of outstanding backorders across n item types subject to a constraint on investment in inventory. We could also select stock levels that minimize investment cost subject to an average fill rate constraint across items. Other optimization models could be formulated as well for complex resupply networks. We will study several such problems in the next chapter. In this chapter we will examine solution methods that will be employed subsequently for more general problems. The problem that we will study now is concerned with setting stock levels for many items at a single location.

One solution approach that we could use is to construct a Lagrangian relaxation of a particular optimization problem. We begin by solving the resulting relaxed problem for a given set of Lagrange multiplier values. We then adjust these multiplier values, and re-solve the relaxed problem. We continue in this manner until a stopping criterion of some sort is satisfied.

7.4.1   First Example: Minimize Expected Backorders Subject to an Inventory Investment Constraint

Suppose Llenroc Electronics manages a group of item types at a single location. The inventory policy followed for all items is an $(s-1,s)$ policy. Thus a replenishment order is placed on an external supplier whenever a unit is withdrawn from Llenroc's stock to satisfy a customer demand. The goal is to select the stock levels so that the average number of outstanding backorders is minimized subject to a constraint on the average investment in inventory. The demand process is assumed to be a stationary compound Poisson process for each of the n item types being managed. Order lead times for replenishment stock are assumed to be independent, identically distributed random variables for each item type and across item types.

Let

$\quad b \qquad$ represent the budget limit on the average value of on-hand inventory,

$\quad c_i \qquad$ be the unit cost for item type i,

$\quad s_i \qquad$ be the stock level for item type i,

$\quad \lambda_i \bar{\tau}_i \bar{u}_i$ be the expected demand over a lead time for item type i, and

$\quad B_i(s_i)$ be the expected number of backorders outstanding at a random point in time for item type i.

From Palm's theorem, we know that the steady state probability distribution for the number of units on order with the supplier has a compound Poisson distribution for each item type. Let us denote the probability that x units of item type i are on order with the supplier by $p(x|\lambda_i \bar{\tau}_i \bar{u}_i)$.

The inventory position when following an $(s-1,s)$ policy is a constant, s. In general, the inventory position is defined as

$$\text{inventory position} = \text{on hand} + \text{on order} - \text{backorders}.$$

In this case

$$s = E[\text{inventory position}] = E[\text{on hand}] + E[\text{on order}] - B(s).$$

The expected number of units on order for item type i is $\lambda_i \bar{\tau}_i \bar{u}_i$ from Little's law. Hence, for item type i,

$$E[\text{on hand}] = s_i - \lambda_i \bar{\tau}_i \bar{u}_i + B_i(s_i).$$

Let $\mu_i = \lambda_i \bar{\tau}_i \bar{u}_i$. Then the average investment in on-hand inventory for item i is $c_i[s_i - \mu_i + B_i(s_i)]$.

We are now in a position to state the optimization problem as

$$\text{minimize} \sum_{i=1}^{n} B_i(s_i)$$

subject to

$$\sum_{i=1}^{n} c_i[s_i - \mu_i + B_i(s_i)] \le b, \tag{7.24}$$

$$s_i = 0, 1, \ldots.$$

To solve this problem, we will use a Lagrangian relaxation method. Let θ represent the multiplier associated with the budget constraint that links the item stock level decisions.

The relaxation is

$$\min \sum_{i=1}^{n} B_i(s_i) + \theta \left[\sum_{i=1}^{n} c_i(s_i - \mu_i + B_i(s_i)) - b \right]$$

subject to $s_i = 0, 1, \ldots$

$$= \min_{s_i=0,1,\ldots} \sum_{i=1}^{n} [(1+\theta c_i)B_i(s_i)+\theta c_i s_i] - \left[\theta \sum_{i=1}^{n} c_i \mu_i + \theta b\right]$$

$$= -\theta \left[\sum_{i=1}^{n} c_i \mu_i + b\right] + \sum_{i=1}^{n} \min_{s_i=0,1,\ldots} [(1+\theta c_i)B_i(s_i)+\theta c_i s_i].$$

Thus, given a value of θ, the resulting relaxed optimization problem is separable by item type. The problem that must be solved for each item is of the same form, so we will temporarily drop the item subscript.

Let $f(s) = (1+\theta c)B(s) + \theta cs$. Since $B(s)$ is discretely strictly convex in s, $f(s)$ is convex, too. Define

$$\Delta f(s) = f(s+1) - f(s)$$
$$= (1+\theta c)\{B(s+1) - B(s)\} + \theta c.$$

Since we previously showed that

$$B(s+1) - B(s) = -\left(1 - \sum_{x \le s} p(x|\mu)\right),$$

$$\Delta f(s) = -(1+\theta c)\left(1 - \sum_{x \le s} p(x|\mu)\right) + \theta c.$$

Because of the convexity of $f(s)$, the optimal stock level, given θ, is the smallest non-negative integer, s^*, for which

$$\Delta f(s) \ge 0,$$

that is, the smallest value for which

$$(1+\theta c)\left(1 - \sum_{x \le s} p(x|\mu)\right) \le \theta c$$

or

$$\sum_{x \le s} p(x|\mu) \ge \frac{1}{1+\theta c}.$$

Clearly the value of s^* depends on the value of θ. Observe that as θ increases, s^* is non-increasing, and, similarly, as θ decreases, s^* is non-decreasing. Let

$$C(\theta) = \sum_{i=1}^{n} c_i [s_i^*(\theta) - \mu_i + B_i(s_i^*(\theta))].$$

It is clear that $C(\theta)$ is also non-increasing as θ increases and non-decreasing as θ decreases. A graph of this relationship is shown in Figure 7.3. The goal is to find a value of θ such that $C(\theta)$ is approximately equal to b. It is generally not possible to find a value of θ that yields $C(\theta) = b$, as illustrated in Figure 7.3. Hence, the goal is to construct the graph of the minimum expected backorders as a function of the average investment in on-hand inventory. Each value of θ yields a set of stock levels, a corresponding inventory investment, and a minimum number of average outstanding backorders. Thus it is obvious that by solving the relaxation corresponding to a set of multiplier values, $\theta_1 > \theta_2 > \cdots > \theta_M$, we can construct a graph of minimum expected backorders as a function of θ, as illustrated in Figure 7.4.

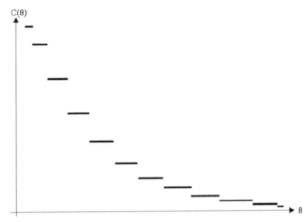

Fig. 7.3. Graph of $C(\theta)$ as a function of θ.

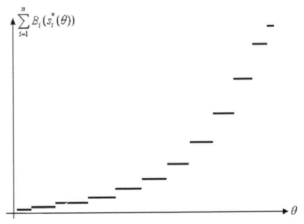

Fig. 7.4. Graph of minimum expected backorders as a function of θ.

Suppose we are given values $\theta_1 > \theta_2 > \cdots > \theta_M$, M Lagrange multiplier values. As stated earlier, since $\frac{1}{1+\theta_1 c} < \frac{1}{1+\theta_2 c} < \cdots < \frac{1}{1+\theta_M c}$, $s^*(\theta_1) \le s^*(\theta_2) \le \cdots \le s^*(\theta_M)$. To find $s^*(\theta)$, recall we find the smallest non-negative integer value of s for which

$$\sum_{x \le s} p(x|\mu) \ge \frac{1}{1 + \theta c}.$$

Thus to find $s^*(\theta_i)$ we know that

$$\sum_{x \leq s^*(\theta_{i-1})} p(x|\mu)$$

is a starting point for our calculation. Since we have already computed this value to determine $s^*(\theta_{i-1})$, the amount of computational effort required to find $s^*(\theta_i)$ may be reduced significantly.

Observe that there exists a $\theta > 0$ such that

$$p(0|\mu_i) = \frac{1}{1 + \theta c_i}, \text{ or}$$

$$\theta = \frac{1}{c_i} \left\{ \frac{1}{p(0|\mu_i)} - 1 \right\}.$$

Let $\theta_{max} = \max_i \frac{1}{c_i} \left\{ \frac{1}{p(0|\mu_i)} - 1 \right\}$. If $\theta = \theta_{max}$, then $s_i^*(\theta_{max}) = 0$ for all i.

Let us now state an algorithm that can be used to solve the original problem approximately using the Lagrangian method we have discussed.

Algorithm for Solving Problem (7.24)

Step 0: Set $\theta_{min} = 0$, $\theta_{max} = \max_i \frac{1}{c_i} \left\{ \frac{1}{p(0|\mu_i)} - 1 \right\}$, and $N = 0$.

Step 1: Compute $\theta = (\theta_{min} + \theta_{max})/2$, $N = N + 1$.

Step 2: For each item i, find the smallest value of s_i such that $\sum_{x \leq s_i} p(x|\mu_i) \geq \frac{1}{1 + \theta c_i}$ and call it $s_i^*(\theta)$.

Step 3: Calculate $A = \sum_{i=1}^n c_i [s_i^*(\theta) - \mu_i + B_i(s_i^*(\theta))]$.

If $|A - b| < \varepsilon$ or if $N > $ max iterations, stop; otherwise, if $A > b$, set $\theta_{min} = \theta$, and if $A < b$, set $\theta_{max} = \theta$. Return to Step 1.

Some of the ideas discussed in this section were described first in Fox and Landi [123] and later reviewed in Muckstadt [243].

7.4.2 Second Example: Maximize Expected System Average Fill Rate Subject to an Inventory Investment Constraint

As in our first example, suppose Llenroc Electronics manages n item types at a single location. We assume an $(s-1, s)$ policy is used for each of these items. We assume requests for these items are placed by customers on the firm. Each request that is made corresponds to a failure of a single unit of a particular part type. The failed parts are repaired at a repair facility. Repair times for an item type are independent and identically distributed; repair times across item types are also independent. We assume requests for serviceable parts for each item type i occur according to a Poisson process with rate λ_i.

In this case our goal is to find the stock levels that maximize the average expected system fill rate subject to an investment constraint. Here units do not leave the system, so that the investment corresponding to stock levels s_i is $\sum_{i=1}^{n} c_i s_i$.

We use the same notation as in the preceding example where appropriate. The function $F_i(s_i)$ measures the fill rate for item i given a stock level s_i.

In this situation, the probability that x units are in the repair process is given by

$$p(x|\lambda_i \bar{\tau}_i) = e^{-\lambda_i \bar{\tau}_i} \frac{(\lambda_i \bar{\tau}_i)^x}{x!},$$

from Palm's theorem.

The optimization problem can be stated as

$$\text{maximize} \sum_{i=1}^{n} \frac{\lambda_i}{\sum_{j=1}^{n} \lambda_j} F_i(s_i) \tag{7.25}$$

subject to

$$\sum_{i=1}^{n} c_i s_i \leq b,$$

$$s_i \geq \lfloor \lambda_i \bar{\tau}_i \rfloor \geq 0 \text{ and integral.}$$

Recall from our earlier discussion that

$$F_i(s_i) = \sum_{x < s_i} e^{-\lambda_i \bar{\tau}_i} \frac{(\lambda_i \bar{\tau}_i)^x}{x!}.$$

Recall also that $F_i(s_i)$ is concave in the region $s_i \geq \lfloor \lambda_i \bar{\tau}_i \rfloor$, and hence we have placed this constraint on s_i in our formulation of the inventory stocking problem.

We could obtain an answer to the problem using the Lagrangian relaxation method described earlier. However, we will use a simpler approach, marginal analysis. This greedy approach will produce an optimal solution for certain values of b and an approximately optimal solution for all other values of b, as we will see.

Define

$$\Delta_i(s_i) = \frac{\lambda_i}{\sum_{j=1}^{n} \lambda_j} \left\{ \frac{F_i(s_i + 1) - F_i(s_i)}{c_i} \right\},$$

which measures the increase in the expected system fill rate per incremental dollar invested in item i given the current stock level is s_i.

Suppose we have stock levels $s_i \geq \lfloor \lambda_i \bar{\tau}_i \rfloor$ and want to determine which item's stock level should be increased from s_i to $s_i + 1$. Since $\Delta_i(s_i)$ measures the change in perfor-

mance per incremental dollar invested, we would choose to increment the stock level
of item i^* if

$$i^* = \arg\max_i \Delta_i(s_i).$$

Initially set $s_i = \lfloor \lambda_i \bar{\tau}_i \rfloor$, and compute $\sum_{i=1}^{n} (\lambda_i / \sum_{j=1}^{n} \lambda_j) \cdot F_i(\lfloor \lambda_i \bar{\tau}_i \rfloor)$ and $\sum_{i=1}^{n} c_i \lfloor \lambda_i \bar{\tau}_i \rfloor$.
Next, compute $\Delta_i(\lfloor \lambda_i \bar{\tau}_i \rfloor)$ for all i and increment the stock level for the item having the
maximum value of $\Delta_i(s_i)$, say i^*. The solution

$$s_i = \lfloor \lambda_i \bar{\tau}_i \rfloor, \quad i \neq i^*,$$
$$s_{i^*} = \lfloor \lambda_{i^*} \bar{\tau}_{i^*} \rfloor + 1$$

is the optimal solution to (7.25) when

$$b = \sum_{i \neq i^*} c_i \lfloor \lambda_i \bar{\tau}_i \rfloor + c_{i^*} \left\{ \lfloor \lambda_{i^*} \bar{\tau}_{i^*} \rfloor + 1 \right\}.$$

Continuing in this manner, it is clear how we would construct a graph of the maximum average expected fill rate as a function of system investment in inventory. Thus
the proposed greedy algorithm will find the optimal solution for values of b that would
be generated sequentially as a consequence of constructing the solution as outlined.

7.5 Exercises

7.1. Suppose that time is divided into periods of equal length. Demand in each period
is Poisson distributed with a mean of λ units. An order-up-to policy is followed for
managing inventories. Replenishment lead times are assumed to be independent and
identically distributed with a mean of D periods. Prove that the number of units on
order (in resupply) in steady state has a Poisson distribution with mean λD. That is,
prove the discrete-time analogue of Palm's theorem.

7.2. Plot the probabilities for a random variable that has a negative binomial distribution
where its mean assumes values 1, 5, 25. For each mean, construct these plots when the
variance-to-mean ratios are 1.01, 3, 10. How do these probabilities compare with the
corresponding probabilities for a Poisson distributed random variable with the same
means?

7.3. Prove an extension to Palm's theorem when the arrival process is a compound
Poisson process and where every customer is willing to wait τ time units for delivery
of its order.

7.4. Plot the backorder, ready rate, and fill rate functions $(B(s), R(s), F(s))$ for a single-stage inventory system managed using a continuous review $(s-1, s)$ policy. Suppose the lead time demand is either Poisson distributed or negative binomially distributed with expected lead time demand being either 1, 5, or 10 units. In the case where lead time demand follows a negative binomial distribution, construct the plots for the variance-to-mean ratios of 1.01, 2, and 5. What do you observe? In particular, suppose the desired fill rate goal is .95. Find the required stock level in each case.

7.5. Suppose there are ten critical items on an aircraft. Compute the expected number of non-operational aircraft for several combinations of stock levels for these items assuming the demand process is Poisson. Expected demands over the lead time for these ten items are 10, 7, 2, 1, 0.7, 0.5, 0.3, 0.1, 0.04, and 0.01.

7.6. Llenroc Industries owns a fleet of aircraft and wants to determine how many spares of each of N components to buy. Aircraft failures occur according to a Poisson process. Each failure is due to the failure of a single component and the probability that it is due to component i is $\lambda_i / \sum_j \lambda_j$, where λ_i is the failure rate for component i. The average repair time for component i is t_i. The owner wants to set the spare parts' stock levels such that the average delay in completing the repair of an aircraft due to parts availability is minimized. There is a constraint on total investment in spare parts of C dollars. Assume each component of type i costs c_i dollars. An $(s-1, s)$ policy is used to manage the system.

(a) Develop a mathematical formulation of this problem. Construct a Lagrangian-based algorithm for finding the optimal stock levels.
(b) Use your algorithm to find the optimal solution when $N = 2, C = 6$,

$$c_1 = 1, \qquad\qquad c_2 = 2,$$
$$t_1 = 1/10, \qquad\qquad t_2 = 1/7,$$
$$\lambda_1 = 5, \qquad\qquad \lambda_2 = 7.$$

Plot the total investment in spare parts as a function of the Lagrange multiplier, and use this plot to obtain a solution for the problem.

7.7. Suppose a firm manages n item types at a single location. Assume an $(s-1, s)$ policy is used for each of these items. Requests for these items are placed by customers on the firm. Each request corresponds to a failure of a single unit of a particular part type. The failed parts are repaired at a repair facility. Repair times for an item type are independent and identically distributed; repair times across item types are also independent. Assume requests for serviceable parts for each item type i occur according to a Poisson process with rate λ_i. Construct a marginal analysis algorithm that determines stock levels that maximize the average expected system fill rate subject to an investment constraint. Use this algorithm to determine the optimal stock levels for the data given in the previous problem.

7.8. Suppose the demand process is a compound Poisson process where the order arrival rate is λ and where the order size is represented by a random variable, N, having a logarithmic distribution. That is, $P(N=n) = \frac{-(1-p)^n}{n}(\ln p)^{-1}$, where $0 < p < 1$, $n \geq 1$. Furthermore, suppose $\lambda = -\ln p$.

Show that the distribution of demand is described by a negative binomial distribution in this case. Do this by first constructing the generating function for the order size distribution and then by constructing the generating function for the compound Poisson distribution.

Using the generating function for the order size distribution, show how to compute the expectation and variance of the order size distribution, that is $E[N]$ and $\text{Var}(N)$.

8

A Tactical Planning Model for Managing Recoverable Items in Multi-Echelon Systems

There are different potential uses for inventory models. One of the uses is to create budgets that can then be employed to guide purchases of items. Another application is the setting of target stock levels (order-up-to levels) that are the basis for purchasing items that have long procurement lead times. Models that are used for these purposes are called tactical planning models. These models are extensively used in military environments. Our goals are to study one such tactical planning model and to demonstrate how echelons interact when continuous review $(s-1,s)$ policies are instituted.

In 1968, Sherbrooke [314]published a landmark paper in which he described a tactical planning model for the management of recoverable or repairable items called METRIC (Multi-Echelon Technique for Recoverable Item Control). The goal of the model was to provide a computationally tractable framework for planning budgets and procuring items for the U.S. Air Force. As we have discussed, the key to building models is the ability to calculate the distribution of the number of units in the resupply system. Unfortunately, establishing the exact probability distributions for these random variables in even two-echelon systems is not tractable for large numbers of items. Since these calculations are too computationally burdensome to be of practical use, Sherbrooke developed an approximation to this distribution that is easy to compute, and hence has been widely used in many applications.

We now summarize the key elements of Sherbrooke's ideas and important contributions and improvements to METRIC due to Graves [138], Sherbrooke [317] Muckstadt [239] and O'Malley [259].

J.A. Muckstadt and A. Sapra, *Principles of Inventory Management: When You Are Down to Four, Order More*, Springer Series in Operations Research and Financial Engineering, DOI 10.1007/978-0-387-68948-7_8, © Springer Science+Business Media, LLC 2010

8.1   The METRIC System

As stated, the system modelled by Sherbrooke [314] corresponds to one operated by the US Air Force, which consists of a set of bases, at which flying activity occurs, and a depot. Both the depot and bases stock inventory and repair defective parts. The parts removed from the aircraft requiring repair are called LRUs, or Line Replaceable Units. The system is assumed to operate as follows.

When an LRU fails at a base (removed from an aircraft), the following events occur. First, an LRU is withdrawn from base stock and placed on the aircraft, thereby returning the aircraft to an operational status. If serviceable stock is not available in base supply, then a backorder occurs. The failed unit is either repaired at the base or the depot.

The decision as to where the repair occurs depends only on the nature of the failure. Some types of LRU failures can only be diagnosed and repaired at the depot. Others can be diagnosed and repaired at the base level. We assume that whenever a failed unit can be repaired at a base, from a technical viewpoint, then it will be repaired there. Thus the choice of the repair location does not depend on the current on-hand stock at the base or the current workload in the base repair shop.

When the unit is shipped to the depot for repair, a request is made to have an LRU of the same type shipped to that base. If such a unit is on hand, it will be shipped immediately; otherwise, a backorder will occur at the depot. We assume two things. First, the depot meets demands on a first-come, first-serve basis. That is, no prioritization among bases occurs when making shipping decisions even though such prioritization may often be desirable. Thus the tactical planning model is conservative in that base level performance will be enhanced if proper prioritization practices are put into operation. Second, bases are not resupplied by other bases, that is, lateral resupply is not permitted in the model. Again this is a conservative assumption since lateral resupply will enhance base performance if executed appropriately.

Because the cost of each unit of an LRU is normally high and demand rates are usually low, the $(s-1,s)$ inventory policy is followed in practice at both the bases and the depot. We assume that all failed units can be repaired; however, if there are condemnations, then orders will be placed on an external supplier. In this case the orders may be for more than a single unit. In these cases, the $(s-1,s)$ policy is not followed and our subsequent discussion must be modified.

We will assume that failures of each LRU type occur according to a Poisson process. This assumption is necessary for some analytic reasons, as will become apparent as we proceed. However, we will indicate subsequently how this assumption is relaxed in practice.

8.1.1 System Operation and Definitions

As we discussed, the system operates as follows. Removals of LRU i at base j occur according to a Poisson process with rate λ_{ij}. With probability r_{ij} the unit will be repaired at the base, and with probability $(1 - r_{ij})$ will be repaired at the depot. Hence the arrival process to base maintenance for LRU i is a Poisson process with rate $r_{ij}\lambda_{ij}$. The arrival process to the depot maintenance activity for LRU type i is also a Poisson process with rate $\lambda_{i0} = \sum_j (1 - r_{ij})\lambda_{ij}$. That this process is a Poisson process is a consequence of the fact that failures occur at bases according to Poisson processes for LRU type i and that each such failure has a probability of r_{ij} of being repaired at the base and $1 - r_{ij}$ of being repaired at the depot. Furthermore, the superposition of the independent Poisson arrival processes from the bases corresponding to failures requiring depot repair for LRU type i is also a Poisson process.

We assume the depot repair cycle time for LRU i is denoted by D_i and is not dependent on the base from which the LRU is sent. The repair cycle time includes the packing, transportation, and actual repair times.

We let B_{ij} be the base repair cycle time for LRU i at base j and A_{ij} be the order, shipping and receiving time of LRU i at base j for shipments of LRU i received from the depot.

Let T_{ij} represent the average number of days that it takes to resupply base j's stock for LRU i once a unit enters the resupply system, that is, either depot or base repair cycles.

We assume λ_{ij} is measured in units per day and D_i, B_{ij}, and A_{ij} are measured in days. The flows of LRUs in this system are shown in Figure 8.1.

8.1.2 The Optimization Problem

Our goal is to develop a model that can be used to determine stock levels s_{ij} for LRU i at location j and to show how to compute these stock levels. The objective of the model is to minimize the total average number of outstanding backorders at the bases at a random point in time for a given level of investment. But why choose the backorder criterion as the performance measure of interest? The goal should be to maximize the expected number of operational aircraft at the bases. Recall that in the previous chapter, we showed, to a first-order approximation, that minimizing backorders at bases is equivalent to maximizing the expected number of operational aircraft at the base level. Thus maximizing the average number of available aircraft is approximately equal to minimizing the average number of outstanding base level LRU backorders at a random point in time.

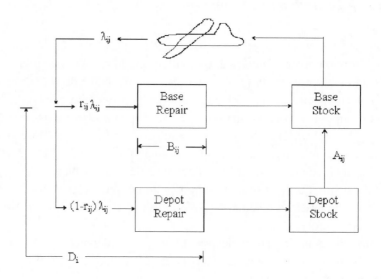

Fig. 8.1. The METRIC system.

Let

$$s_{ij} = \text{LRU } i \text{ stock level at base } j \text{ (or depot if } j = 0).$$

Before we state the optimization problem, we first develop some additional relationships.

The key equation that links the base and depot stock levels is the average resupply time equation for LRU i at base j. Let

$$T_{ij} = \text{average LRU } i \text{ resupply time at base } j$$
$$= r_{ij}B_{ij} + (1 - r_{ij})(A_{ij} + \text{ depot delay } (s_{i0})).$$

Thus the average resupply time is B_{ij} when the repair occurs at base j for LRU i and A_{ij} plus an expected waiting time (depot delay) due to the depot stock level when resupply comes from the depot. These average times are weighted by the probability that resupply occurs either from base maintenance or from depot stock. But how do we measure this expected waiting time?

Let

$\delta(s_{i0})$ = expected depot delay, or waiting time, given the depot
stock level is s_{i0} for LRU type i.

From Little's law,

$$\delta(s_{i0}) = \frac{\text{Average outstanding depot backorders } (s_{i0})}{\text{Depot demand rate } (\lambda_{i0})}.$$

Let $\mathcal{B}_D(s_{i0})$ = expected outstanding depot backorders for LRU i given s_{i0}. Hence

$$\delta(s_{i0}) = \frac{\mathcal{B}_D(s_{i0})}{\lambda_{i0}}.$$

Let us explore how $\delta(s_{i0})$ behaves. First, we note that $\delta(s_{i0}) \to 0$ quite rapidly as s_{i0} exceeds $\lambda_{i0}D_i$. To illustrate this observation, consider the cases where $\lambda_{i0}D_i$ is equal to 1, 5, 10, 50 and 100. We display the values of $\delta(s_{i0})$ for these cases in Table 8.1 through Table 8.5, where we assume $D_i = 1$.

As can be seen from the data in these tables, the expected depot delay becomes small quite quickly when $D_i = 1$. When A_{ij} is several days in length, the depot delay becomes a relatively inconsequential portion of the expected resupply time as s_{i0} exceeds $\lambda_{i0}D_i$.

As an example, suppose $r_{ij} = 0, A_{ij} = 5$, $\lambda_{ij} = 5$ and $\lambda_{i0} = 50$, with $D_i = 1$. Then the expected number of units in resupply for the base is $5 \times 5 + 5 \times \delta(s_{i0})$. If $s_{i0} = 50$, then $\delta(s_{i0}) = .0563$ and the expected number of units in the base's resupply system is 25.2815 units of which only .2815 are attributable to the depot delay in resupply of the base. In this example, we see that this minimal contribution to the expected number of units in the base resupply system occurred when the depot carried no safety stock. Thus as the depot demand rate increases and A_{ij} increases while r_{ij} decreases, there is little to be gained by having much depot safety stock.

Table 8.1. $\delta(s_{i0})$ when $\lambda_{i0} = 1$, $D_i = 1$.

s_{i0}	$\delta_i(s_{i0})$
0	1
1	.3679
2	.1036
3	.0233
4	.0043
5	.0007
6	.0001

Table 8.2. $\delta(s_{i0})$ when $\lambda_{i0} = 5, D_i = 1$.

s_{i0}	$\delta_i(s_{i0})$
4	.2874
5	.1755
6	.0987
7	.0511
8	.0244
9	.0108
10	.0044
11	.0017
12	.0006

Table 8.3. $\delta(s_{i0})$ when $\lambda_{i0} = 10, D_i = 1$.

s_{i0}	$\delta(s_{i0})$
9	.1793
10	.1251
11	.0834
12	.0531
13	.0322
14	.0187
15	.0103
16	.0055
17	.0028
18	.0013

Table 8.4. $\delta(s_{i0})$ when $\lambda_{i0} = 50, D_i = 1$.

s_{i0}	$\delta(s_{i0})$
49	.0667
50	.0563
51	.0471
52	.0389
53	.0318
54	.0258
55	.0206
56	.0163
57	.0127
58	.0098

Table 8.5. $\delta(s_{i0})$ when $\lambda_{i0} = 100, D_i = 1$.

s_{i0}	$\delta(s_{i0})$
100	.0398
101	.0351
102	.0308
103	.0268
104	.0233
105	.0200
106	.0172
107	.0146
108	.0124
109	.0104
110	.0087
115	.0032

Table 8.6. $\delta(s_{i0})$ when $\lambda_{i0} = 5, D_i = 10$.

s_{i0}	$\delta(s_{i0})$
49	.667
50	.563
51	.471
52	.389
53	.318
54	.258
55	.206
56	.163
57	.127
58	.098

Now let us consider another example. Suppose $r_{ij} = 0$, $A_{ij} = 5$, $\lambda_{ij} = .5$, $\lambda_{i0} = 5$ but $D_i = 10$. Thus $\lambda_{i0}D_i = 50$, as was the case in the previous example. To find $\delta(s_{i0})$ in this case we can again use the data in Table 8.4. The data in that table must be multiplied by the value of D_i to obtain the new values of $\delta(s_{i0})$, which are given in Table 8.6. As was the case in the first example, the average resupply time is dominated by the value of A_{ij}. Hence if a small amount of safety stock is carried at the depot, say 5 units so that $s_{i0} = 55$, then the average resupply time is 5.206 days. The expected number of units in resupply at the base is now 2.6030 units of which .103 units are attributable to depot delay.

Observe that as $\lambda_{i0}D_i$ remains constant but D_i increases, then $\delta(s_{i0})$ increases proportionately to the increase in D_i. For example, as D_i increased from 1 to 10, $\delta(s_{i0})$ increased by a factor of 10. If $\lambda_{ij} = .25$, $\lambda_{i0} = 2.5$, and $D_i = 20$, then $\delta(50) = 1.126$ days and $\delta(55) = .412$ days. Thus as D_i increases, while $\lambda_{i0}D_i$ remains constant, we see that $\delta(s_{i0})$ increases and can become a more significant portion of the average resupply time. Hence safety stock at the depot will become more important as D_i becomes larger while $\lambda_{i0}D_i$ remains constant.

In any case, the range of values that need to be evaluated explicitly in an optimization procedure is limited since $\delta(s_{i0})$ approaches small values quite quickly as $s_{i0} > \lfloor \lambda_i D_i \rfloor$. The range can usually be limited to two standard deviations of depot demand over the depot's resupply time. That is, the optimal value that s_{i0} assumes almost always is in the interval $\left[\lfloor \lambda_i D_i \rfloor, \lceil 2 \cdot (\lambda_i D_i)^{1/2} \rceil + \lfloor \lambda_i D_i \rfloor \right]$, assuming $\lfloor \lambda_i D_i \rfloor$ is the minimum depot stock level that is considered. Furthermore, when solving practical problems, the search for the optimal value of s_{i0} is limited to a subset of these values. Consider the values in Table 8.5. Observe that successive values do not differ by substantial amounts. Hence searches are limited to perhaps every second value in the interval given earlier. Thus when $\lambda_i D_i = 100$, the search for the optimal value of s_{i0} might be restricted to $100, 102, \ldots, 120$. As a consequence of these observations, a maximum of 10 to 15 possible values for s_{i0} are often explicitly considered in practice when employing the optimization methodology we will be discussing in subsequent sections.

8.1.2.1 Approximating the Stationary Probability Distribution for the Number of LRUs in Resupply

To calculate the expected number of base j backorders for LRU i, we must determine the stationary probability distribution for the number of units in resupply for LRU i at base j.

Let X_{ij} = random variable for the number of units in resupply for LRU i at base j in steady state. Clearly,

$$E[X_{ij}] = \lambda_{ij}T_{ij}$$
$$= r_{ij}\lambda_{ij}B_{ij} + (1 - r_{ij})\lambda_{ij}A_{ij} + (1 - r_{ij})\frac{\lambda_{ij}}{\lambda_{i0}}\mathcal{B}_D(s_{i0}).$$

To compute the variance of X_{ij} requires some additional analysis. We drop the LRU subscript in this analysis for ease of exposition.

Suppose there are N_D backorders at the depot for some LRU. Let N_j be the number of base j units backordered at the depot. The probability that $N_j = n_j$ when $N_D = n_D$, where $n_j \leq n_D$, is given by

$$P\{N_j = n_j | N_D = n_D\} = \binom{n_D}{n_j} \left(\frac{\hat{\lambda}_j}{\lambda_0} \right)^{n_j} \left(1 - \frac{\hat{\lambda}_j}{\lambda_0} \right)^{n_D - n_j},$$

where $\hat{\lambda}_j = (1 - r_j)\lambda_j$. Furthermore,

$$E[N_j|s_0] = E_{N_D}\left[E_{N_j}[N_j|N_D]\right] = E_{N_D}\left[\frac{\hat{\lambda}_j}{\lambda_0}N_D\right] = \frac{\hat{\lambda}_j}{\lambda_0}\mathcal{B}_D(s_0).$$

Thus the expected number of LRUs in backorder status corresponding to demands at base j is the fraction of total depot demand due to base j, on average, times the expected number of depot backorders given s_0.

The variance of N_j also depends on the depot stock level. We know that

$$\text{Var}(N_j|s_0) = E[N_j^2|s_0] - E[N_j|s_0]^2.$$

We also know that $E[N_j|s_0] = \frac{\hat{\lambda}_j}{\lambda_0}\mathcal{B}_D(s_0)$. Furthermore,

$$\begin{aligned}
E[N_j^2|s_0] &= E_{N_D}\left[E_{N_j}[N_j^2|N_D]\right] \\
&= E_{N_D}\left[\text{Var}(N_j|N_D) + E[N_j|N_D]^2\right] \\
&= E_{N_D}\left[N_D \cdot \frac{\hat{\lambda}_j}{\lambda_0}\cdot\left(1 - \frac{\hat{\lambda}_j}{\lambda_0}\right) + \left(\frac{\hat{\lambda}_j}{\lambda_0}N_D\right)^2\right] \\
&= \frac{\hat{\lambda}_j}{\lambda_0}\left(1 - \frac{\hat{\lambda}_j}{\lambda_0}\right)\mathcal{B}_D(s_0) + \left(\frac{\hat{\lambda}_j}{\lambda_0}\right)^2 \cdot E_{N_D}\left(N_D^2|s_0\right) \\
&= \frac{\hat{\lambda}_j}{\lambda_0}\left(1 - \frac{\hat{\lambda}_j}{\lambda_0}\right)\mathcal{B}_D(s_0) + \left(\frac{\hat{\lambda}_j}{\lambda_0}\right)^2\left[\text{Var}(N_D|s_0) + (E_{N_D}(N_D|s_0))^2\right] \\
&= \frac{\hat{\lambda}_j}{\lambda_0}\left(1 - \frac{\hat{\lambda}_j}{\lambda_0}\right)\mathcal{B}_D(s_0) + \left(\frac{\hat{\lambda}_j}{\lambda_0}\right)^2\left[\text{Var}(N_D|s_0) + (\mathcal{B}_D(s_0))^2\right].
\end{aligned}$$

Thus, by combining the above observations, we see that

$$\text{Var}(N_j|s_0) = \frac{\hat{\lambda}_j}{\lambda_0}\left(1 - \frac{\hat{\lambda}_j}{\lambda_0}\right)\mathcal{B}_D(s_0) + \left(\frac{\hat{\lambda}_j}{\lambda_0}\right)^2\text{Var}(N_D|s_0).$$

The $\text{Var}(N_D|s_0)$ can be computed as follows. Since

$$\text{Var}(N_D|s_0) = E[N_D^2|s_0] - (\mathcal{B}_D(s_0))^2,$$

we need to determine a method for calculating $E[N_D^2|s_0]$. Recall that N_D measures the number of depot backorders at a random point in time. Hence if $N_D = n_D$, then the number of units of the LRU in depot resupply must exceed the depot stock level by n_D. Thus if x measures the number of units of the LRU in depot resupply, there are n_D depot backorders for the LRU if and only if $x - s_0 = n_D$. Hence

$$E[N_D^2|s_0] = \sum_{x \geq s_0} (x - s_0)^2 p(x|\lambda_0 D),$$

where $p(x|\lambda_0 D)$ is the probability that x units are in depot resupply. From Palm's theorem, $p(x|\lambda_0 D) = e^{-\lambda_0 D} \frac{(\lambda_0 D)^x}{x!}$. Then

$$\begin{aligned}
E[N_D^2|s_0] &= \sum_{x \geq s_0} (x - (s_0 - 1) - 1)^2 p(x|\lambda_0 D) \\
&= \sum_{x \geq s_0} (x - (s_0 - 1))^2 p(x|\lambda_0 D) \\
&\quad - 2 \sum_{x \geq s_0} (x - s_0) p(x|\lambda_0 D) \\
&\quad - \sum_{x \geq s_0} p(x|\lambda_0 D) \\
&= \sum_{x \geq s_0} (x - (s_0 - 1))^2 p(x|\lambda_0 D) \\
&\quad - \mathcal{B}_D(s_0) - \sum_{x \geq s_0} (x - (s_0 - 1)) p(x|\lambda_0 D) \\
&= E[N_D^2|s_0 - 1] - \mathcal{B}_D(s_0) - \mathcal{B}_D(s_0 - 1).
\end{aligned}$$

We thus can determine $E[N_D^2|s_0]$ recursively, and therefore can determine $\mathrm{Var}(N_D|s_0)$ recursively. Observe that

$$\begin{aligned}
E[N_D^2|0] &= \sum_{x \geq 0} x^2 p(x|\lambda_0 D) \\
&= \mathrm{Var}[\text{Number of units in Depot resupply}] \\
&\quad + [\text{Expected number of units in depot resupply}]^2 \\
&= \lambda_0 D + (\lambda_0 D)^2
\end{aligned}$$

when base demand is described by a Poisson process.

Now that we have computed the mean and variance of N_D we are able to compute the variance of Y_j, where Y_j is the random variable that represents the demand over the order and ship time (A_j) plus the backordered demand due to the depot stock level s_0 at a random point in time at base j. Since the demands that cause depot backorders and those occurring during the transportation lead time to base j take place in non-overlapping intervals, and are therefore independent,

$$\text{Var}\,Y_j = \text{Var}(\text{Demand over } A_j)$$
$$+ \text{Var}(\text{Backordered demand due to depot stock } s_0),$$

$$\text{Var}\,Y_j = \hat{\lambda}_j A_j + \frac{\hat{\lambda}_j}{\lambda_0}\left(1 - \frac{\hat{\lambda}_j}{\lambda_0}\right)\mathcal{B}_D(s_0) + \left(\frac{\hat{\lambda}_j}{\lambda_0}\right)^2 \text{Var}(N_D|s_0),$$

where as before $\hat{\lambda}_j = (1 - r_j)\lambda_j$.

But to compute the expected number of backorders at base j for LRU i we must know the probability distribution of the number of units in resupply, X_{ij}. Rather than computing this distribution exactly, which is too computationally intensive to do, we approximate it, as Sherbrooke [317] did, with a negative binomial distribution with parameters

$$\mu_{ij} = r_{ij}\lambda_{ij}B_{ij} + (1 - r_{ij})\lambda_{ij}A_{ij} + (1 - r_{ij})\frac{\lambda_{ij}}{\lambda_{i0}}\mathcal{B}_D(s_{i0})$$

$$\sigma_{ij}^2 = r_{ij}\lambda_{ij}B_{ij} + (1 - r_{ij})\lambda_{ij}A_{ij}$$
$$+ \frac{(1 - r_{ij})\lambda_{ij}}{\lambda_{i0}}\left(1 - \frac{(1 - r_{ij})\lambda_{ij}}{\lambda_{i0}}\right)\mathcal{B}_D(s_{i0})$$
$$+ \frac{((1 - r_{ij})\lambda_{ij})^2}{\lambda_{i0}^2}\text{Var}(N_D|s_{i0}).$$

The appropriateness of the negative binomial approximation is discussed in Muckstadt [239]. We note that this approximation is quite accurate and hence is widely used in practice.

The negative binomial distribution has two parameters, call them p and r. These parameters must satisfy the two equations

$$\mu_{ij} = \frac{r(1 - p)}{p} \text{ and } \sigma_{ij}^2 = \frac{r(1 - p)}{p^2}.$$

Thus

$$1 > \frac{\mu_{ij}}{\sigma_{ij}^2} = p \text{ and } r = \frac{p\mu_{ij}}{1 - p}.$$

The probabilities can be computed recursively with

$$P\{X_{ij} = 0\} = p^r \text{ and } P\{X_{ij} = x\} = P\{X_{ij} = x - 1\}\frac{(r + x - 1)}{x}q,$$

where $q = 1 - p$.

The choice of the negative binomial distribution as an approximation to the probability distribution of X_{ij} was made for two key reasons. First, it is easily computed, which is essential in large scale applications. Second, it is an accurate approximation. A substantial amount of testing has been conducted to verify the validity of the approximations (Muckstadt [239]).

Additionally, in the Air Force and other implementations of the model we have just described, a negative binomial model was used for reasons other than those that have been discussed. A negative binomial model was used to represent the demand process as well in these approximations. That is, the negative binomial probability model replaces the Poisson model in all the demand probability model expressions found throughout this chapter. The negative binomial model was chosen in these applications to accommodate the high variance-to-mean ratios of the demand processes found in practice. A complete discussion of the variance-to-mean ratio of the demand process can be found in Crawford [78].

8.1.2.2   Finding Depot and Base LRU Stock Levels

We now turn to finding the best depot and base stock levels for each LRU, which are the stock levels that minimize the expected number of "holes" in aircraft, that is, that minimize the expected number of base LRU backorders at a random point in time subject to a constraint on investment in inventory:

$$\min \ \sum_i \sum_{j=1}^{m} \mathcal{B}_{ij}(s_{ij}|s_{i0})$$

subject to

$$\sum_i c_i \Big(s_{i0} + \sum_{j=1}^{m} s_{ij}\Big) \leq b, \tag{8.1}$$

$$s_{ij} = 0, 1, \ldots,$$

where c_i is the cost of one unit of LRU type i, b is the available budget to invest in the LRUs, and m is the number of bases.

Unfortunately Problem (8.1) is neither separable nor convex. It is not separable because the variables s_{ij} and s_{i0} interact when computing $\mathcal{B}_{ij}(s_{ij}|s_{i0}) = \sum_{x \geq s_{ij}} (x - s_{ij})p(x|\lambda_{ij}T_{ij}(s_{i0}))$, where $p(x|\lambda_{ij}T_{ij}(s_{i0}))$ is approximated by a negative binomial distribution with mean μ_{ij} and variance σ_{ij}^2, as defined previously. Both μ_{ij} and σ_{ij}^2 are functions of s_{i0}. We will discuss the convexity issue shortly.

There are many different ways to solve (8.1). One way is to use a marginal analysis approach and another is to employ a Lagrange multiplier method.

8.1.2.2.1   A Marginal Analysis Algorithm

We will first discuss the marginal analysis approach. Since the objective function is separable by LRU type, we first focus on a single LRU type. As before, we temporarily drop the LRU subscript to reduce the notation.

To determine a solution to (8.1) we analyze the relationships between total system LRU inventory and the expected number of total base backorders. We will construct this function first and will observe that this function need not be convex. Hence, we will construct its convex minorant and use this convex minorant as a basis for making budget allocations among different types of LRUs.

Next, define

$$\alpha(s, s_0) = \text{minimum total expected base backorders for}$$
$$\text{an LRU given that there are } s \text{ units of the}$$
$$\text{LRU in the system and the depot stock is } s_0.$$

Let S represent the set of values of s_0 that will be explicitly considered in the optimization process.

Let $s_0 \in S$. Then, given s_0, we can compute μ_j and σ_j^2 as well as $P\{X_j = x | s_0\} \equiv p_j(x | s_0)$. Recall that

$$\mathcal{B}_j(s_j | s_0) = \sum_{x > s_j} (x - s_j) p_j(x | s_0).$$

Also recall that $\mathcal{B}_j(s_j | s_0)$ is a convex function, given s_0, and satisfies

$$\mathcal{B}_j(s_j | s_0) = \mathcal{B}_j(s_j - 1 | s_0) - \left(1 - \sum_{x < s_j} p_j(x | s_0)\right).$$

Thus the reduction in backorders by adding one unit to the inventory at base j is $\left(1 - \sum_{x < s_j} p_j(x | s_0)\right)$. Note also that $\mathcal{B}_j(0 | s_0) = \mu_j$.

Now suppose that $s = s_0 + 1$. The question we must answer is to which base should this $(s_0 + 1)$st unit of stock be allocated. To begin the process, set $s_j = 0$ for all j. We want to place the unit at the base that has the largest reduction in expected base backorders resulting from the allocation; that is, assign the unit to the base having the largest value of $(1 - p_j(0 | s_0))$. Since the functions $\mathcal{B}_j(s_j | s_0)$ are convex given s_0 and are strictly decreasing as s_j increases, we can use a marginal analysis method to compute the best allocation of a stock of $s - s_0 = a$ units knowing the best allocation of $s - s_0 = a - 1$ units.

If $\tilde{s}_j(a - 1)$ represents the optimal stock level for base j given $a - 1$ units are available for allocation to the bases (given the value of s_0), then the base that is allocated the next unit of stock is the one that has the maximum value of

$$\left(1 - \sum_{x < \tilde{s}_j(a-1)+1} p_j(x|s_0)\right).$$

Observe that to calculate the marginal reduction in backorders for base j by adding an additional unit of stock to that base we need to determine

$$\sum_{x < \tilde{s}_j(a-1)+1} p_j(x|s_0) = \sum_{x < \tilde{s}_j(a-1)} p_j(x|s_0) + p_j(\tilde{s}_j(a-1)+1).$$

But $\sum_{x < \tilde{s}_j(a-1)} p_j(x|s_0)$ is known so that only $p_j(\tilde{s}_j(a-1)+1|s_0)$ must be computed. As observed earlier,

$$p_j(\tilde{s}_j(a-1)+1|s_0) = p(\tilde{s}_j(a-1)|s_0)\frac{(r+(\tilde{s}_j(a-1)+1)-1)}{\tilde{s}_j(a-1)+1}q.$$

Hence the function $\alpha(s,s_0)$ can be computed very quickly for a large range of s values. For example, we could choose to limit s by setting a minimum value for $\alpha(s,s_0)$. That is, when $\alpha(s,s_0) < \varepsilon$, then terminate the calculations.

The functions $\alpha(s,s_0)$ must be calculated for all $s_0 \in S$.

Next, define

$$\hat{\alpha}(s) = \min_{s_0} \alpha(s,s_0).$$

The function $\hat{\alpha}(s)$ is a strictly decreasing function of s; however, it need not be convex. As a consequence, we construct $\hat{\alpha}^c(s)$, where $\hat{\alpha}^c(s)$ is a piecewise linear convex minorant to the function $\hat{\alpha}(s)$. Let $\hat{S}_c$ be the values of s at which the slope of $\hat{\alpha}^c(s)$ changes.

For example, suppose $m = 10$ bases, each of which has a daily expected demand rate of .195 units. Further, assume the depot resupply time is 10 days and the depot-to-base transit time is 1 day for all bases. Figure 8.2 contains the graph of $\hat{\alpha}(s)$. Table 8.7 shows the values of $\hat{\alpha}(s)$ and s_0^*, the latter being the optimal depot stock levels corresponding to the various values of total system stock s. Note that the optimal depot stock level is not monotone non-decreasing as the total system stock increases. This illustrates why this type of optimization problem is not easily solved. Note also that $\hat{\alpha}(s)$ is not convex. Thus we construct its convex minorant, which is shown in Figure 8.3. In this case, $\hat{S}_c = \{35, 36, 41, 42, 43, 44, 45, 46, 48, 54, 55\}$.

Suppose we construct functions $\hat{\alpha}_i^c(s_i)$ for all LRU types i, where s_i is the total system stock for item type i. Rather than solving (8.1) for a specific target budget value b, we will construct a tradeoff curve of total base backorders over all LRU types as a

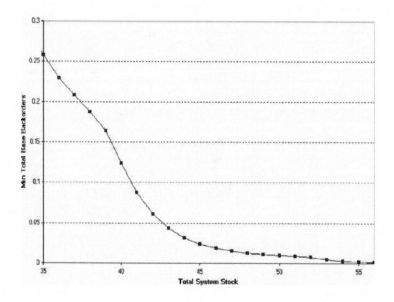

Fig. 8.2. Graph of $\hat{\alpha}(s)$.

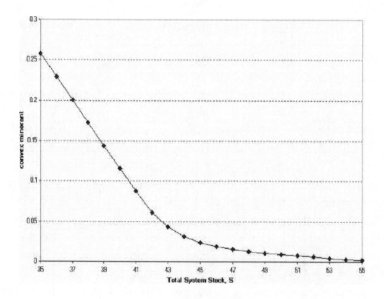

Fig. 8.3. Graph of $\hat{\alpha}^c(s)$.

Table 8.7. $\hat{\alpha}(s)$ and s_0^*.

s	$\hat{\alpha}(s)$	s_0^*	s	$\hat{\alpha}(s)$	s_0^*
35	.25798	25	46	.01867	26
36	.22889	26	47	.01565	27
37	.20787	26	48	.01234	28
38	.18685	26	49	.01117	28
39	.16421	25	50	.01000	28
40	.12383	20	51	.00882	28
41	.08727	21	52	.00765	28
42	.06120	22	53	.00465	23
43	.04324	23	54	.00285	24
44	.03128	24	55	.00183	25
45	.02355	25			

function of the total investment in all LRUs. This function is constructed using another marginal analysis algorithm.

Let $s_i^1, \ldots, s_i^{K_i}$ be the elements of $\hat{S}_c^i$, the set of total stock levels that will be considered for LRU type i based on the construction of the convex piecewise linear function $\hat{\alpha}_i^c(s_i)$. That is, the values s_i^k are the stock levels at which the slope of the total base backorder function $\hat{\alpha}_i^c(s_i)$ changes.

To begin the construction process, compute $\hat{\alpha}_i^c(s_i^1)$ for all items i. Let $\beta(1) = \sum_i \hat{\alpha}_i^c(s_i^1)$, and let $C(1) = \sum s_i^1 c_i$. Next, set $k_i = 1$ for all i and $\ell = 1$.

At the beginning of iteration $\ell \geq 1$, we have $C(\ell)$ and $\beta(\ell)$. For all i, compute

$$\Delta_i(s_i^{k_i}) = \frac{\hat{\alpha}_i^c(s_i^{k_i}) - \hat{\alpha}_i^c(s_i^{k_i+1})}{c_i(s_i^{k_i+1} - s_i^{k_i})},$$

the reduction in total expected base backorders by incrementing the stock level from $s_i^{k_i}$ to $s_i^{k_i+1}$ per dollar invested in LRU i.

Select i^* such that

$$\Delta_{i^*}(s_{i^*}^{k_{i^*}}) = \max_i \Delta_i(s_i^{k_i}).$$

Set

$$C(\ell+1) = C(\ell) + c_{i^*}(s_{i^*}^{k_{i^*}+1} - s_{i^*}^{k_{i^*}})$$

and

$$\beta(\ell+1) = \beta(\ell) - (\hat{\alpha}_{i^*}^c(s_{i^*}^{k_{i^*}}) - \hat{\alpha}_{i^*}^c(s_{i^*}^{k_{i^*}+1})).$$

Increment $k_{i^*} = k_{i^*} + 1$ and $\ell = \ell + 1$, and recompute the value of $\Delta_{i^*}(s_{i^*}^{k_{i^*}})$. The $\Delta_i(\cdot)$ values remain the same for all other values of i.

Repeat this process until there are no remaining values of k_i to consider, that is, $k_i = K_i$ for all i.

The process we have outlined can be used to compute the total expected base backorder function for a finite set of the investment levels in LRU system inventory, as depicted in Figure 8.4.

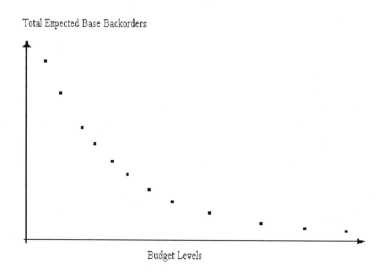

Fig. 8.4. Total expected base backorders for various budget levels generated by the algorithm.

8.1.2.2.2   A Lagrangian Method for Computing Depot and Base Stock Levels

Let us now describe an alternative method for computing the depot and base stock levels, that is, another method for solving (8.1). Roughly the algorithm works as follows. First construct a Lagrangian relaxation to (8.1) and then decompose it into a collection of single-item problems, which are solved one at a time. The constraint that is relaxed is the investment constraint. Thus there is only one Lagrangian multiplier in the relaxed formulation of our problem. For each item, given the multiplier value, determine the best depot and base stock levels using an enumeration method. This process is repeated over various multiplier values to determine the solution to (8.1).

This Lagrangian relaxation algorithm is based on an observation that the optimal solution to a relaxed problem yields an optimal solution to (8.1) for a certain value of b. By investigating an appropriate range of values for the Lagrangian multiplier we can provide good, if not necessarily optimal, solutions for (8.1).

Let us begin by constructing the Lagrangian relaxation:

$$\min_{s_{ij}} \sum_{i=1}^{n} \sum_{j=1}^{m} \mathcal{B}_{ij}(s_{ij}, s_{i0}) + \theta \sum_{i=1}^{n} c_i \sum_{j=0}^{m} s_{ij} - \theta b \qquad (8.2)$$

subject to $s_{ij} = 0, 1, \ldots$, where we continue to use the same notation as employed in the previous section. The parameter θ represents the multiplier corresponding to the constraint

$$\sum_{i=1}^{n} c_i \sum_{j=0}^{m} s_{ij} \leq b.$$

For this formulation, $\theta > 0$.

The Lagrangian relaxation may also be expressed as

$$\sum_{i=1}^{n} \left\{ \min \sum_{j=1}^{m} \mathcal{B}_{ij}(s_{ij}, s_{i0}) + \theta c_i \sum_{j=0}^{m} s_{ij} \right\} - \theta b$$

$$\text{subject to } s_{ij} = 0, 1, \ldots.$$

Hence the Lagrangian relaxation is separable by item, that is, we can solve

$$\min \sum_{j=1}^{m} \mathcal{B}_{ij}(s_{ij}, s_{i0}) + \theta c_i \sum_{j=0}^{m} s_{ij}, \qquad (8.3)$$

$$s_{ij} = 0, 1, \ldots,$$

for each item i independently of all other items.

In Chapter 7 we showed the expected backorder functions $\mathcal{B}_{ij}(s_{ij}, s_{i0})$ are convex functions of s_{ij} for a given value of s_{i0}. However, the function is not necessarily jointly convex in s_{ij} and s_{i0}, as we observed earlier. Thus to find an optimal solution to (8.1), given θ, we use an exhaustive search on the candidate depot stock levels.

Suppose θ assumes some value, say θ_ℓ, and s_{i0} is equal to ρ. Then we want to find the base stock levels that solve

$$W_{\rho\ell}^{i} = \min_{s_{ij}=0,1,\ldots} \left[\sum_{j=1}^{m} \left(\mathcal{B}_{ij}(s_{ij}, \rho) + \theta_\ell c_i s_{ij} \right) \right] + \theta_\ell c_i \rho. \qquad (8.4)$$

Notice that

$$\min_{s_{ij}=0,1,\dots} \left[\sum_{j=1}^{m} (\mathcal{B}_{ij}(s_{ij},\rho) + \theta_\ell c_i s_{ij}) \right]$$
$$= \sum_{j=1}^{m} \min \left(\mathcal{B}_{ij}(s_{ij},\rho) + \theta_\ell c_i s_{ij} \right), \tag{8.5}$$

that is, the relaxed optimization problem is also separable by base for each item, given θ_ℓ and ρ.

The solution to

$$\min \mathcal{B}_{ij}(s_{ij},\rho) + \theta_\ell c_i s_{ij}, \tag{8.6}$$
$$s_{ij} = 0, 1, \dots,$$

can be found quite easily. The objective function to Problem (8.6) is discretely convex in s_{ij}. Hence we can easily show, using a first difference argument, that the optimal value for s_{ij} is the smallest non-negative integer for which

$$\sum_{x \geq s_{ij}+1} p(x|\rho) \leq \theta_\ell c_i \text{ or } \sum_{x \leq s_{ij}} p(x|\rho) \geq 1 - \theta_\ell c_i,$$

where $p(x|\rho)$ is the probability that there are x units in resupply at base j for item i given $s_{i0} = \rho$. These probabilities are computed as described in Section 8.1.2.1.

Observe that whenever $\theta_\ell \geq 1/c_i$, the optimal value for s_{ij} is 0. Thus a necessary condition for $s_{ij} > 0$, for all i, is $\theta_\ell < \min_i 1/c_i = 1/\max_i c_i$.

We have shown how to find the optimal values of the base stock levels given θ_ℓ and ρ. To find an optimal set of depot and base stock levels for a given value of θ_ℓ, find the value of $s_{i0} = \rho$ for which

$$s_{i0} = \arg\min_\rho W^i_{\rho\ell} \tag{8.7}$$

and the corresponding base stock level values, s_{ij}. As we have discussed earlier, the range of depot stock levels that needs to be examined explicitly is limited because $\mathcal{B}_{i0}(s_{i0})/\lambda_{i0} \to 0$ rapidly as s_{i0} exceeds $\lambda_{i0}D$, the expected depot demand over the depot resupply time. In practical applications $\lfloor \lambda_{i0}D \rfloor$ is often a floor on s_{i0}.

Knowing the stock levels for the depot and bases, we also have the corresponding investment level for item i,

$$C_i(\theta_\ell) = c_i \sum_{j=0}^{m} s_{ij}. \tag{8.8}$$

The total investment level over all items, given $\theta = \theta_\ell$, is

$$\sum_{i=1}^{n} C_i(\theta_\ell). \tag{8.9}$$

Hence for every choice of θ there corresponds a required investment level in system stock.

For practical problems there are bounds on the value of θ, too. Recall that for s_{ij} to be positive, θ must be less than $1/c_i$. Hence a search for the optimal value is normally confined to the range $0 < \theta < 1/\max c_i$.

There are two very different ways that a Lagrangian-based algorithm can be employed in practice. In one approach a bisection method is used to find the optimal value of θ. Since a range of budget versus expected base level backorder values is often desired, a second approach is sometimes implemented. In this second approach, a set of M values for θ are prespecified, $0 < \theta_1 < \cdots < \theta_M < 1/\max c_i$. These values are set on the basis of experience in solving past problems. We now describe both approaches.

Algorithm 1

Step 1: Determine $\theta_{\min}$ and $\theta_{\max}$, the minimum and maximum values considered for θ. Set $\theta_1 = (\theta_{\min} + \theta_{\max})/2$ and $\ell = 1$

Step 2: Set $\theta = \theta_\ell$. For each item i, find the depot and base stock levels that yield $\min_\rho \{W_{\rho\ell}^i\}$.

Step 3: Calculate $C(\theta) = \sum_{i=1}^n c_i \sum_{j=0}^m s_{ij}(\theta)$. If $|C(\theta) - b| < \varepsilon$, stop; otherwise, if $C(\theta) > b$, set $\theta_{\min} = \theta_\ell$; otherwise set $\theta_{\max} = \theta_\ell$. Set $\theta_\ell = (\theta_{\max} + \theta_{\min})/2$, $\ell = \ell + 1$, and return to Step 2.

The number of iterations required to find a good solution clearly depends on the initial choices for $\theta_{\min}$ and $\theta_{\max}$ and ε. In real problems, when there is experience in choosing the initial range of θ values and ε is set to be $1/2\%$ of the budget, the number of iterations required normally does not exceed 10. Nonetheless, a very large number of calculations must be made repeatedly. The second algorithm requires fewer calculations, as we will see. The second algorithm is as follows.

Algorithm 2

Step 1: Select a set of M multiplier values

$$0 < \theta_1 < \theta_2 < \cdots < \theta_M < \frac{1}{\max c_i}$$

Step 2: For each item i, for each θ_ℓ, solve

$$\min_\rho W_{\rho\ell}^i$$

and obtain stock levels $s_{ij}(\theta_\ell)$ and $\sum_{j=1}^m \mathcal{B}_{ij}(s_{ij}(\theta_\ell), s_{i0}(\theta_\ell))$.

Step 3: Compute $C(\theta_\ell)$. Select the solution that has a budget closest to b.

As mentioned, it may appear that the two approaches require the same computational effort since the same calculations are performed in Step 2 in both cases. They are not as similar as one might initially believe, because of the manner in which the second algorithm is implemented.

For the second algorithm, let us examine how we compute $W^i_{\rho\ell}$. Recall that for a given value of depot stock for item i, $s_{i0} = \rho$, we find the optimal base stock levels by determining the smallest values of s_{ij} for which

$$\sum_{x \leq s_{ij}} p(x|\rho) \geq 1 - \theta_\ell c.$$

Observe that when $\ell_1 > \ell_2$,

$$s_{ij}(\theta_{\ell_1}) \leq s_{ij}(\theta_{\ell_2}).$$

Thus, when we implement Algorithm 2, for a given $\rho = s_{i0}$, we make a single pass through the base calculations and determine the optimal base stock levels for each value of θ. Observe that we do not need to repeat the calculations made when finding $s_{ij}(\theta_M)$ when we are finding $s_{ij}(\theta_{M-1})$, since $s_{ij}(\theta_M) \leq s_{ij}(\theta_{M-1})$, given that $\rho = s_{i0}$. In practice, fewer calculations are required to implement the second algorithm. The second algorithm is particularly useful when the goal is to understand the tradeoffs between investment in system stock versus expected base level backorders.

8.2   Waiting Time Analysis

The optimization problem we have formulated has as its objective the minimization of the total average number of outstanding LRU backorders at base level at a random point in time. While the average number may be low for a particular LRU at a base and the average waiting time may be low as well, there is a need to know how long a delay could be experienced in responding to a request for resupply. In other words, what is the probability distribution for the LRU resupply waiting time at either a base or the depot? At base level, this time corresponds to the time until an aircraft is again operational; at the depot, this time measures the time a resupply request is delayed prior to shipping a unit to the requesting base.

We will now derive the distribution of waiting times assuming demands for resupply occur according to a simple Poisson process and assuming resupply times are independent and identically distributed random variables with density $\psi(\cdot)$ with mean T. Our goal is to compute the probability that a failed LRU will wait longer than a time u given that a demand for resupply occurred at time t. We also assume a first-come, first-serve queue discipline is followed.

Let us begin by defining some notation. Let

$W(t)$ be the waiting time random variable for satisfying an LRU demand occurring at time t,

$I(t)$ be the net inventory random variable of the LRU just before time t,

$X(t)$ be the random variable for the number of LRU units in the resupply process just before time t, which does not include the demand occurring at time t,

$V_1(t,u)$ be a random variable measuring the number of LRUs completing repair in the interval $(t, t+u]$ that were in the resupply system at time t (i.e., failed during $(0,t)$ but were not repaired by time t),

$V_2(t,u)$ be a random variable measuring the number of LRUs completing repair in $(t, t+u]$ that entered the repair cycle during $(t, t+u]$,

$V(t,u) = V_1(t,u) + V_2(t,u)$, and

$F(x) = \int_0^x \psi(y)dy$, which is the probability that an LRU repair is completed in x or fewer time units.

The LRU resupply request at time t will remain unfilled for a period of time greater than u if and only if the number of units in resupply at the point in time t at which the resupply request is made is s units or greater (s is the stock level) and the number of units completing repair during $(t, t+u]$ is insufficient to satisfy that request by time $t+u$. Thus

$$W(t) > u \text{ if and only if } I(t) + V(t,u) + r(u) \leq 0,$$

where $r(u) = 1$ if the repair cycle time of the unit that failed at time t is less than or equal to u and is 0, otherwise.

Since $s = I(t) + X(t)$ for all t,

$$I(t) + V(t,u) + r(u) \leq 0 \text{ implies } s - X(t) + V(t,u) + r(u) \leq 0,$$

or $s \leq X(t) - V(t,u) - r(u) = X_1(t+u) - V_2(t,u) - r(u)$, where $X_1(t+u) = X(t) - V_1(t,u)$, that is, the number of LRUs in the resupply system at time $t+u$ that failed during $(0,t)$. Therefore

$$P\{W(t) > u\} = P\{X_1(t+u) - V_2(t,u) - r(u) \geq s\}.$$

Observe that the random variables $X_1(t+u)$, $V_2(t,u)$, and $r(u)$ are independent since LRUs requiring repair that arrive prior to t in no way influence arrivals requiring repair subsequent to time t in terms of timing, quantity, or repair times. The repair time of the failed LRU occurring at time t is also unaffected by those arrival times and the repair times of LRUs failing both prior to and subsequent to time t.

Let $p_1(t)$ be the probability that an arrival in $(0,t)$ remains in the resupply system at time $t+u$, that is,

$$p_1(t) = \frac{1}{t} \int_0^t (1 - F(t+u-v))dv$$
$$= \frac{1}{t} \int_u^{t+u} (1 - F(v))dv,$$

and p_0 be the probability that an LRU that failed during the interval $(t, t+u]$ completes its repair by time $t+u$, that is,

$$p_0 = \frac{1}{u} \int_0^u F(u-v)dv.$$

Then, following the reasoning presented in Chapter 7,

$$P\{X_1(t+u) = k\} = e^{-\lambda t p_1(t)} \frac{(\lambda t p_1(t))^k}{k!}$$
$$= e^{-\lambda \int_u^{t+u}(1-F(v))dv} \frac{(\lambda \int_u^{t+u}(1-F(v))dv)^k}{k!}$$

and

$$P\{V_2(t,u) = k\} = e^{-\lambda u p_0} \frac{(\lambda u p_0)^k}{k!}$$
$$= e^{-\lambda \int_0^u (1-F(v))dv} \frac{(\lambda \int_0^u F(u-v)dv)^k}{k!}.$$

To determine the probability that $W(t) > u$ we consider two cases. In the first case, $X_1(t+u) = s+k$, $k \geq 1$, and $V_2(t,u) < k$. That is, the number of LRU units remaining in the resupply system at time $t+u$ that correspond to LRU failures that occurred prior to time t is $s+k$, and additionally, less than k LRUs complete repair in $(t, t+u]$ of the units that failed in $(t, t+u]$. Since we assume demands are satisfied on a first-come, first-served basis, the LRU demand that occurred at time t cannot be satisfied by time $t+u$.

In the second case, $s+k$, $k \geq 0$, LRUs are in resupply at time $t+u$ that correspond to resupply requests that were placed prior to t. Furthermore, exactly k units completed repair during $(t, t+u]$ that correspond to LRUs entering the repair cycle during $(t, t+u]$. Thus, in this second case, the resupply request made at time t will remain unfilled if and only if the repair cycle time of the requesting unit exceeds u.

Combining these observations we see that

$$P\{W(t) > u\} = \sum_{k=1}^{\infty} \left[P\{X_1(t+u) = s+k\} \sum_{y=0}^{k-1} P\{V_2(t,u) = y\} \right]$$
$$+ \sum_{k=0}^{\infty} P\{X_1(t+u) = s+k\}\{1 - F(u)\} P\{V_2(t,u) = k\}.$$

Suppose $F(0) = 0$. Then the probability that the waiting time is 0 is given by

$$P\{W(t) = 0\} = \sum_{k=0}^{s-1} P\{X(t) = k\} = P\{I(t) > 0\}.$$

Observe that $P\{X_1(t+u) = k\}$ depends on t and u but $P\{V_2(t,u) = k\}$ is a function of u alone. Let

$$P\{X_1(u) = k\} = \lim_{t \to \infty} P\{X_1(t+u) = k\}$$
$$= e^{-\lambda \int_u^{\infty} (1 - F(v))dv} \frac{[\lambda \int_u^{\infty} (1 - F(v))dv]^k}{k!},$$

which represents the steady state probability that k LRUs that were in the resupply system at the time an LRU fails remain in the resupply system u time units later. Let W be a random variable that measures the steady state waiting time to satisfy a request for resupply. Then

$$P\{W > u\} = \sum_{k=1}^{\infty} \left[P\{X_1(u) = s+k\} \sum_{y=0}^{k-1} P\{V_2(u) = y\} \right]$$
$$+ \sum_{k=0}^{\infty} P\{X_1(u) = s+k\}\{1 - F(u)\} \cdot P\{V_2(u) = k\},$$

where $V_2(u)$ is the random variable measuring the number of LRUs that both arrive for repair and complete repair in a period of length u following the failure of an LRU.

This waiting time distribution could be used to establish minimum depot LRU stock levels. If there is a desire to ensure that replenishment of base inventories is not delayed by more than u days due to the lack of depot stock with probability α, then the minimum depot stock level can be set accordingly. Note that $P\{W(t) > u\}$ requires an explicit statement of the repair cycle time distribution $F(\cdot)$. Thus $F(\cdot)$ will have to be estimated to make the required calculations.

8.3   Exercises

8.1. Suppose we have a two-echelon base-depot system as discussed in Section 8.1.1. The demand process at each of 10 bases is a Poisson process, where each base has a demand rate of .5 units per day. Suppose the average depot-to-base order and ship time is either 1, 5, or 10 days. Further assume the average depot resupply time is either 15 or 30 days. All failed parts at the bases are repaired at the depot.

(a) Plot $\delta(s_{i0})$ for each base.
(b) Compute the average base resupply time as a function of s_{i0} for each depot-to-base order and ship time and each average depot resupply time combination. How important a factor is the depot stock level in each case?
(c) Suppose the demand at each base is described by a negative binomial distribution with an average of .5 units per day at each base. Suppose that the variance-to-mean ratio is identical at all bases, and is equal to either 1.01, 2, 5, or 10. Assuming demands are independent among the bases, what is the distribution of depot demand? Repeat the tasks indicated in (a) and (b) for these cases. How do the values compare with those computed initially?

8.2. Suppose we are managing the two-echelon system discussed in Section 8.1.1. Demands at each of 10 bases for an item occur according to a Poisson process with a rate of .5 units per day. The average depot-to-base resupply time, A_{ij}, for the item is 5 days for all 10 bases, and the average depot resupply time, D, is 30 days. Suppose the fraction of the failures that arise at each of the bases that are repaired at the bases is either .25, .5, or .75. Further assume the average base repair time for the item is 2 days.

Compute the mean and variance of the number of units in resupply for a base for each case for a range of values of the depot stock level. How does the variance-to-mean ratio change as the fraction of units repaired at the depot increases when the depot stock is equal to $\lfloor \lambda_0 D \rfloor$, where λ_0 is the average depot daily demand rate?

8.3. In Section 8.1.2.1 we derived expressions for the mean and variance for the number of units in resupply for an LRU at a base. The analysis was based on the assumption that the demand process is a Poisson process. Suppose that the demand process is negative binomially distributed rather than Poisson distributed. Construct new equations for the mean and variance of the number of units in resupply for LRU at a base. Carefully state your assumptions as you develop these expressions.

8.4. Assume there are three items whose stock levels need to be determined in a two-echelon system.

Removals of item i at base j occur according to a Poisson process at rate λ_{ij} units per day. When a removal occurs at a base, three events take place simultaneously and instantaneously. First, a unit of stock is withdrawn from base stock (if a unit is available); second, the failed unit is sent to the depot where it will be repaired; and third, the depot resupplies the base (if a unit is on hand). When units of stock are not available, backorders occur. If there is a backorder at a base, then an aircraft is grounded. If a backorder occurs at the depot, then there is a delay in resupplying the requesting base.

The depot repair cycle time for all bases for part type i is denoted by D_i, measured in days. The order and shipping time from the depot to each base is A_i days for item i. There are 10 bases in the system, numbered 1 through 10. The flying activity at bases 1 through 5 is the same and hence $\lambda_{i1} = \cdots = \lambda_{i5}$ for all i. Bases 6 through 8 have identical removal rates for all items, that is, $\lambda_{i6} = \lambda_{i7} = \lambda_{i8}$. Finally, bases 9 and 10 have identical removal rates for each of the three items, $\lambda_{i9} = \lambda_{i10}$. The data in the following table indicate the removal rates, depot repair cycle times, and unit costs for the three items. The value of A_i is two days for all three items for all bases.

Item	Base Removal Rate			Unit Costs (1000's of $)	Depot Repair Cycle Times
	Base 1–5	Base 6–8	Base 9–10		
1	1	0.5	0.25	3	10
2	1.5	0.75	0.375	4	8
3	0.5	0.25	0.125	5	10

Construct the convex function representing the relationship between minimum total expected base backorders and total system stock for the three items, that is, construct the functions $\hat{\alpha}_i^c(s_i)$.

Once these three functions have been created, construct the convex function that represents the relationships between minimum total base backorders (across items) and investment in stock for these three items.

What would the impact be if the values of D_i and A_i were reduced or increased by a factor of 50%?

9

Reorder Point, Lot Size Models: The Continuous Review Case

When there is a significant fixed cost incurred when placing an order with a supplier, it is no longer beneficial to order each time a demand occurs. In Chapter 2 we examined the impact of fixed costs on ordering quantities when the demand processes were known with certainty. Now we will study situations in which significant fixed ordering costs are incurred and the demand process is assumed to be a stationary stochastic process. Furthermore, we will assume a transaction reporting system exists. That is, we assume that we monitor the inventory levels continuously through time, and place an order for a quantity of stock, called the lot size, at a point in time so as to minimize the average annual cost of operations.

We will confine our attention to developing models for controlling inventories at a single location. To begin, we will also limit attention to managing a single item. Later, we will examine models in which we consider the interactions of many items on stocking decisions.

The type of model that we will discuss in greatest depth in this chapter is called a reorder point, lot size model. When employing this model, we assume a constant lot size of Q units is ordered whenever an order is placed. The timing of the placement of orders depends on the underlying demand process. In general, the best time to place an order can depend on two factors. First, we should know the amount of inventory on hand, on order, and, when appropriate, the number of outstanding backorders. Second, we may also have to know when the last demand occurred. In our analysis, we will ignore the second factor. We will only consider the inventory levels (on-hand, net inventory, or inventory position) when making a procurement decision. Thus when the appropriate inventory level reaches a particular value, called the reorder point, an order is placed. This reorder point is denoted by r. Implicitly, by ignoring the second factor, we are assuming that the demand process possesses the memory-less property. In fact, our analysis is largely based on the assumption that the demand process is a station-

J.A. Muckstadt and A. Sapra, *Principles of Inventory Management: When You Are Down to Four, Order More*, Springer Series in Operations Research and Financial Engineering, DOI 10.1007/978-0-387-68948-7_9, © Springer Science+Business Media, LLC 2010

ary Poisson process or some related process, such as a compound Poisson process. We will also construct approximations of the probability distribution for key random variables and of the objective function. These approximations are made only to simplify calculations; but, as we will discuss in detail, the results are just approximations.

The lot size–reorder point models are often denoted as (Q, r) models. Note that by their construction, such models are also based on the assumption that the reorder point is reached every time an order is to be placed. That is, there is no overshoot of the reorder point. When the demand process is a Poisson process, for example, demands occur for a single unit at a time. Hence if the reorder point is an integer, the reorder point will be reached each time an order is to be placed.

When the number of units demanded in a customer order is described by a second random variable, then there is no assurance that this reorder point will be hit in each order cycle. In these cases, policies other than a (Q, r) policy must be employed.

Most inventory control systems found in practice are not strictly continuous review or transaction reporting systems. While customer orders are often captured as they arise, orders are very frequently placed periodically, such as daily. For example, a car dealer records the removal of parts used to repair cars on a continuing basis, but normally places orders to replenish inventories only daily. Thus the (Q, r) models that we will construct are usually approximations. Nonetheless, these models are widely used in practice, and they most often provide nearly optimal values for the operating policy parameters.

Our primary goals in this chapter are to develop models based on the (Q, r) operating policy that can be used to find the policy parameter values that minimize average annual operating costs. We will also discuss how the optimal parameter values can be computed. The first model, which is an approximation, is widely discussed in texts and often used in practice. We will follow this discussion with the development of an exact model, and we will use it to demonstrate the effects of using the simple approximation model. The exact model is based on the assumption that the demand process is a Poisson process. We will also show how approximations to key performance measures can be constructed when a normal distribution is used to represent demand over a lead time. All of the aforementioned models and accompanying algorithms and heuristics used to compute values for Q and r are designed to optimize the performance for a single item managed at a single location. After completing these analyses, we will examine an approximation that was developed by the U.S. Air Force and currently deployed throughout the Department of Defense and elsewhere for managing large numbers of item types at a single location. These models consider system performance constraints (fill rates, backorder constraints, etc.) that capture the system effects of the reorder points and reorder quantities selected for each item. We will also briefly discuss models for cases in which demands are not always for single units. In these cases, an overshoot of the reorder point is possible and hence an alternative policy must be employed.

9.1 An Approximate Model When Backordering Is Permitted

The (Q, r) model we are about to discuss is an approximation, which is based on many assumptions. As we will see, this model is easy to understand and use; however, as is the case with all models, one must be aware of the impact that assumptions play on the resulting values of the policy parameters.

As we have mentioned, our goal is to construct a model, based on following a (Q, r) policy, for managing a single item at a single location. We will show how to find the optimal policy parameters, Q^* and r^*, derived from this model. These values will be found by minimizing the average annual costs that vary with the values of these policy parameters.

9.1.1 Assumptions

Let us begin by stating some of our assumptions. First, the costs considered in the model are (1) fixed order costs, denoted by K and measured in dollars per order; (2) backorder costs, denoted by π, which are charged for each unit that is backordered; and (3) carrying costs or holding costs, which are denoted by h, and are the cost of carrying a unit of stock on hand for a year.

Perhaps the key assumption in the development of the model is that there is never more than a single order outstanding at any point in time. This implies that whenever we reach the reorder point, there is no order outstanding. We will also assume that the reorder point is non-negative. While this is normally the case in practice, there are instances in which this assumption is not appropriate. As you read through this material, try to think of such cases. On the basis of our assumptions, the reorder point will be reached when the on-hand stock is non-negative. As we will see, normally the reorder point will be based on the system's inventory position. However, since in this case when the reorder point is reached there are no orders outstanding and there are also no backorders, then the inventory position equals the on-hand inventory at the time a procurement order is placed. The assumptions also imply that the demand over a lead time never exceeds Q. Otherwise another order would need to be placed and thus more than one order would be outstanding. Therefore, in practice, Q must be much larger than the expected demand over a lead time. For example, suppose a lead time is a week and the expected demand over a week is 100 units with a standard deviation of 10 units. Suppose further that Q is 400 units. Then it is very unlikely that demand over a lead time would exceed 400 units.

Another assumption is that backorders are not desirable. Let us define a cycle to be the time between the receipt of two successive orders. If backorders are incurred in a cycle, then they occur only at the end of a cycle. That is, the average number of out-standing backorders at a random point in time during a cycle is very small compared to the average amount of stock on hand. This assumption is illustrated in the following figure, in which we plot the dynamics of net inventory (on-hand minus backorders) and inventory position, where τ is the procurement lead time. Note that in the four cycles, backorders occurred in two of them. However, the magnitude of the cumulative back-orders over these four cycles is small so that the average number of units backordered at a random point in time in these cycles is quite small. Observe also that the length of a cycle varies from one cycle to the next owing to the random nature of the demand process.

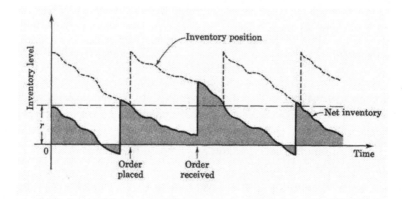

Fig. 9.1. Inventory dynamics.

Next, we assume that the demand process is a stationary continuous process, as illustrated in Figure 9.1. We let $f(\cdot)$ represent the density function for the demand over a lead time. We assume that the lead time can be a random variable; however,

$$\mu = \int_0^\infty x f(x)\,dx,$$

the expected lead time demand, is assumed to be much smaller than Q.

9.1.2 Constructing the Model

Let us now construct the objective function. Since all demand is met, that is, there are no lost sales, the expected variable procurement cost $C\lambda$ does not depend on the policy parameters and hence need not be included in the objective function. However, the fixed procurement cost does depend on the value of Q. Suppose λ is the average annual demand rate. Then the average number of orders placed per year is $\frac{\lambda}{Q}$. Since the fixed cost of placing an order is K, the average annual fixed ordering cost is $\frac{\lambda}{Q}K$.

To calculate the expected shortage cost per year, we determine the expected number of backorders per cycle, multiply this quantity by the expected number of cycles per year, $\frac{\lambda}{Q}$, and multiply the resultant amount by π.

Once the reorder point is reached in a cycle, the amount of demand that occurs from that point in time until the order arrives determines the number of backorders, if any, that will occur in that cycle. Let x represent the demand over the lead time. A backorder will occur if and only if $x > r$. If $x > r$, then $x - r$ measures the number of backordered units. Hence

$$\int_r^\infty (x - r) f(x) \, dx$$

measures the expected number of backorders per cycle. Consequently,

$$\pi \frac{\lambda}{Q} \int_r^\infty (x - r) f(x) \, dx$$

measures the expected annual backorder costs that will be incurred.

Finally, let us calculate the average annual cost of carrying inventory. Recall that

$$\text{net inventory} = \text{on-hand inventory} - \text{backorders}.$$

The time average of net inventory is

$$E[\text{net inventory}] = E[\text{on hand}] - E[\text{backorders}].$$

Recall from our earlier comments that we assume the costs are such that backorders are incurred in relatively small quantities and then only at the end of a cycle. As we discussed, this implies that the average number of outstanding backorders existing at a random point in time is negligible compared with the total inventory carried on hand over a cycle. That is,

$$E[\text{backorders}] \approx 0.$$

As a consequence,

$$E[\text{net inventory}] \approx E[\text{on hand}].$$

We will therefore approximate $E[\text{on hand}]$ by $E[\text{net inventory}]$.

Let s measure the safety stock. Recall that the safety stock measures the expected net inventory just before the arrival of an order and $s + Q$ is the expected on-hand inventory just after the arrival of an order. We assume that, in expectation, the demand is occurring at rate λ. Hence the average net inventory during a cycle is $\frac{1}{2}(Q+s)+\frac{1}{2}(s)=\frac{1}{2}Q+s$. But

$$\begin{aligned} s &= \int_0^\infty (r-x)f(x)\,dx \\ &= r-\mu. \end{aligned}$$

Thus we approximate the average on-hand inventory as

$$\frac{Q}{2}+r-\mu,$$

and the average annual cost of carrying inventory by

$$h\left[\frac{Q}{2}+r-\mu\right].$$

By combining the above expressions, we can state the average annual variable cost function, which we denote $C(Q,r)$, as follows:

$$C(Q,r) = \frac{\lambda}{Q}K+h\left[\frac{Q}{2}+r-\mu\right]+\frac{\pi\lambda}{Q}\int_r^\infty (x-r)f(x)\,dx. \qquad (9.1)$$

9.1.3 Finding Q^* and r^*

Next we will establish conditions that the optimal lot-size and reorder point values must satisfy. They are

$$\frac{\partial C}{\partial Q} = \frac{\partial C}{\partial r} = 0, \qquad (9.2)$$

assuming these conditions can be satisfied with $Q > 0$ and $r \geq 0$.

First, we see that

$$\frac{\partial C}{\partial Q} = -\frac{\lambda K}{Q^2}+\frac{h}{2}-\frac{\pi\lambda}{Q^2}\int_r^\infty (x-r)f(x)\,dx.$$

Letting

$$g(r) = \int_r^\infty (x-r)f(x)\,dx,$$

we can rewrite the above as

$$\frac{\partial C}{\partial Q} = -\frac{\lambda K}{Q^2}+\frac{h}{2}-\frac{\pi\lambda}{Q^2}g(r). \qquad (9.3)$$

To compute $\frac{\partial C}{\partial r}$ we must determine $\frac{dg(r)}{dr}$. Applying Leibnitz's rule, we find that

$$\frac{dg(r)}{dr} = -\int_r^\infty f(x)\, dx.$$

Let $\bar{F}(r) = \int_r^\infty f(x)\, dx$. Using these results, we next see that

$$\frac{\partial C}{\partial r} = h - \frac{\pi\lambda}{Q}\bar{F}(r). \tag{9.4}$$

Combining (9.2), (9.3), and (9.4) we obtain

$$Q^* = \sqrt{\frac{2\lambda(K + \pi g(r^*))}{h}} \tag{9.5}$$

and

$$\bar{F}(r^*) = \frac{Q^* h}{\pi\lambda}. \tag{9.6}$$

Observe that we now have two non-linear equations to solve to obtain Q^* and r^*. Thus a numerical procedure is required to find these values. Let us now present such an algorithm.

Observe from (9.5) that Q^* is always at least as large as the economic order quantity, $\sqrt{2\lambda K/h}$, which we denote by Q_E. In practice Q^* is normally greater than Q_E. This is an interesting observation. To avoid incurring backorders, and the corresponding backorder costs, purchase more each time an order is placed than you would if demand occurred at a known constant and continuous rate. Do this even though the expected carrying costs are higher than in the corresponding deterministic demand case.

Observe also from (9.6) that $\frac{Q^* h}{\pi\lambda}$ must be less than or equal to 1 for the relationship to make sense. This result occurs because we assumed that the expected number of backorders outstanding at a random point in time is negligible when constructing the approximate expression for the average on-hand inventory. That is, this model is based on the assumption that the expected number of backorders incurred per cycle, $g(r)$, is small relative to the average number of demands per cycle, Q.

The following procedure can be employed to find Q^* and r^*.

Algorithm for Finding Q^* and r^*.

Step 1: Set $Q_1 = Q_E, n = 1$.

Step 2: Find r_n that satisfies $\bar{F}(r_n) = \frac{Q_n h}{\pi\lambda}$.

Step 3: Using r_n, compute $Q_{n+1} = \sqrt{\frac{2\lambda(K + \pi g(r_n))}{h}}$.

Step 4: If $Q_{n+1} - Q_n \leq \varepsilon$, stop; otherwise, set $n = n + 1$ and return to Step 2. The desired tolerance, ε, satisfies $\varepsilon > 0$ and is assumed to be small relative to Q_E.

Observe from (9.5) and (9.6) that as Q increases, r decreases, or as r increases, Q decreases, which is intuitive, but also follows from the fact that

$$\frac{dQ}{dr} = -\frac{\pi\lambda}{h} \cdot f(r) \quad \text{and} \quad \frac{dr}{dQ} = -\frac{h}{\pi\lambda}\frac{1}{f(r)} < 0.$$

The algorithm is initiated with a known lower bound on Q^*, Q_E, which, in turn, yields an upper bound on r^*, r_1.

Convergence of this procedure depends on two critical factors. First, a solution to (9.5) and (9.6) must exist. This depends on the assumption that the average number of backorders outstanding at a random point in time is negligible and that the other assumptions on which the model was developed hold. Second, convergence of the proposed iterative procedure to an optimal solution can be guaranteed only if the objective function is convex. Unfortunately, the latter condition need not be true, although, as we are about to see, the objective function is convex in the region containing an optimal solution for commonly used density functions for demand over a lead time.

9.1.4 Convexity of the Objective Function

For $C(Q,r)$ to be convex, each term must be jointly convex in Q and r. Clearly, we see from (9.1) that the terms $\frac{\lambda K}{Q}$, $h\frac{Q}{2}$, and $h[r - \mu]$ are each convex functions of a single variable. The third term $\frac{\lambda}{Q}g(r)$ would need to be jointly convex in Q and r for $C(Q,r)$ to be convex. But $\frac{\lambda}{Q}g(r)$ need not be convex, even though for a given value of Q, it is convex in r and for a given value of r it is convex in Q. Let $E(Q,r) = \frac{\lambda}{Q}g(r)$. But for $E(Q,r)$ to be convex in any direction from any point in $Q > 0$ and $r \geq 0$, the Hessian matrix

$$\begin{bmatrix} \frac{\partial^2 E}{\partial Q^2} & \frac{\partial^2 E}{\partial Q \partial r} \\ \frac{\partial^2 E}{\partial Q \partial r} & \frac{\partial^2 E}{\partial r^2} \end{bmatrix}$$

must be positive semi-definite. This implies that the diagonal elements of this matrix are non-negative and that its determinant is also non-negative. The following theorem establishes when $E(Q,r)$ is convex.

Theorem 9.1. *Suppose there exists a real number a such that $f(x)$ is differentiable and non-increasing for $x \geq a$. Then $E(Q,r)$ is convex in the region $Q > 0$, $a \leq r < \infty$.*

Proof. Suppose $f(x)$ is the density for lead time demand and

$$g(r) = \int_r^\infty (x-r)f(x)\,dx.$$

But, by integrating by parts, we see that

$$\int_r^\infty \bar{F}(x)\,dx = x\bar{F}(x)\big|_r^\infty + \int_r^\infty xf(x)\,dx = -r\int_r^\infty f(x)\,dx + \int_r^\infty xf(x)\,dx$$
$$= \int_r^\infty (x-r)f(x)\,dx,$$

assuming $\lim_{x\to\infty} x\bar{F}(x) = 0$, which is the case for most commonly used distributions. Thus $E(Q,r) = \frac{\lambda}{Q}\int_r^\infty \bar{F}(x)\,dx$.

The Hessian matrix is

$$\begin{bmatrix} \frac{2\lambda}{Q^3}\int_r^\infty \bar{F}(x)\,dx & \frac{\lambda}{Q^2}\bar{F}(r) \\ \frac{\lambda}{Q^2}\bar{F}(r) & \frac{\lambda}{Q}f(r) \end{bmatrix}.$$

Clearly the diagonal elements are non-negative. If $E(Q,r)$ is convex in the region $Q > 0$ and $a \le r < \infty$, then the determinant of the Hessian matrix must be non-negative. The determinant is

$$\frac{2f(r)\lambda^2}{Q^4}\int_r^\infty \bar{F}(x)\,dx - \frac{\lambda^2}{Q^4}(\bar{F}(r))^2$$
$$= \frac{\lambda^2}{Q^4}\left\{ 2f(r)\int_r^\infty \bar{F}(x)\,dx - (\bar{F}(r))^2 \right\}.$$

Let the term within the brackets be denoted by $H(r)$. Observe that

$$\frac{dH}{dr} = -2f(r)\bar{F}(r) + 2f'(r)\int_r^\infty \bar{F}(x)\,dx + 2f(r)\bar{F}(r)$$
$$= 2f'(r)\int_r^\infty \bar{F}(x)\,dx.$$

Since $f'(r) \le 0$, $\frac{dH}{dr} \le 0$, too. Furthermore, $\lim_{r\to\infty} H(r) = 0$, which implies that $H(r) \ge 0$ for $r \ge a$. $\square$

Observe that if $f(x)$ has an exponential distribution, then $a = 0$; if $f(x)$ has a normal distribution, then a equals the expected demand over a lead time, μ. If backorders are to be avoided, then it is likely that the optimal safety stock will be non-negative, and therefore $r^* \ge \mu$. Because the algorithm begins with $r_1 \ge r^*$ and r decreases at each iteration of the procedure, the search will be confined to the region in which the objective function is convex. This is a particularly useful observation since, in practice, lead time demand is very often approximated by a normal distribution. Let us now examine this special case more closely.

9.1.5 Lead Time Demand Is Normally Distributed

As mentioned, a special case arises when $f(x)$ is a normal density used to approximate the lead time demand with expected lead time demand denoted by μ and standard deviation denoted by σ. Recall from (9.6) that r is chosen to satisfy $\bar{F}(r) = \frac{Qh}{\pi\lambda}$. Thus it is imperative to note that the only portion of the distribution function used to represent demand over a lead time that is of importance is its upper tail. Since

$$f(x) = \frac{1}{\sqrt{2\pi}\sigma} \, e^{-\frac{1}{2}(\frac{x-\mu}{\sigma})^2}$$

is the normal density, this upper tail is relatively "light." That is, the exponent increases as the square of x, which means that the area in the tail becomes small quickly.

Compare this with the case when we approximate the lead time density with an exponential distribution with mean μ. Then $\bar{F}(x) = e^{-x/\mu}$, which has a heavier upper tail since the exponent is proportional to x rather than x^2. Selecting a model to represent lead time demand requires only selecting the probability model whose upper tail matches the data pertaining to forecast errors in an accurate manner. By choosing a normal distribution as this model, we are implicitly saying that the upper tail behaves in a way that is represented by that type of density function.

Let us now see how to compute $g(r)$ when $f(x)$ is a normal density. Recall that

$$g(r) = \int_r^\infty (x-r)f(x)\,dx = \int_r^\infty xf(x)\,dx - r\bar{F}(r).$$

But $\int_r^\infty xf(x)\,dx = \int_r^\infty \frac{x}{\sigma}\phi\left(\frac{x-\mu}{\sigma}\right)\,dx$, where $\phi(\cdot)$ denotes the standard normal density. Letting $y = \frac{x-\mu}{\sigma}$,

$$\int_r^\infty \frac{x}{\sigma}\phi\left(\frac{x-\mu}{\sigma}\right)\,dx = \sigma\int_{\frac{r-\mu}{\sigma}}^\infty y\phi(y)\,dy + \mu\int_{\frac{r-\mu}{\sigma}}^\infty \phi(y)\,dy.$$

Further observe that

$$\int_z^\infty y\phi(y)\,dy = \frac{1}{\sqrt{2\pi}}\int_z^\infty ye^{-\frac{1}{2}y^2}\,dy$$

and can be easily evaluated by letting $w = \frac{1}{2}y^2$ so that $dw = y\,dy$. Then

$$\frac{1}{\sqrt{2\pi}} \int_z^\infty y e^{-\frac{1}{2}y^2} dy = \frac{1}{\sqrt{2\pi}} \int_{\frac{1}{2}z^2}^\infty e^{-w} dw = \frac{e^{-\frac{1}{2}z^2}}{\sqrt{2\pi}} = \phi(z).$$

Therefore

$$g(r) = \sigma\phi\left(\frac{r-\mu}{\sigma}\right) + \mu\bar{\Phi}\left(\frac{r-\mu}{\sigma}\right) - r\bar{\Phi}\left(\frac{r-\mu}{\sigma}\right)$$

$$= (\mu - r)\bar{\Phi}\left(\frac{r-\mu}{\sigma}\right) + \sigma\phi\left(\frac{r-\mu}{\sigma}\right) \tag{9.7}$$

$$= \sigma\left[\phi\left(\frac{r-\mu}{\sigma}\right) - \left(\frac{r-\mu}{\sigma}\right)\bar{\Phi}\left(\frac{r-\mu}{\sigma}\right)\right], \tag{9.8}$$

$3[0.2154586 - 1.11 \times 0.13350] = 3 \times (0.06727) = 0.20181$

where $\bar{\Phi}(z) = \int_z^\infty \phi(u)\, du$, the complementary cumulative standard normal distribution function.

Then, when lead time demand is normally distributed,

$$C(Q,r) = \frac{\lambda}{Q}K + h\left[\frac{Q}{2} + r - \mu\right] + \frac{\pi\lambda}{Q}\left[(\mu - r)\bar{\Phi}\left(\frac{r-\mu}{\sigma}\right) + \sigma\phi\left(\frac{r-\mu}{\sigma}\right)\right].$$

Let us now illustrate the mechanics of the algorithm on a few example problems. In the first example, suppose $\lambda = 100$, $K = 200$, $h = 4$, and $\pi = 30$, and that lead time demand is normally distributed with $\mu = 10$ and $\sigma = 3$. To find Q^* and r^*, we will make use of some tables to compute $g(r)$. Let $k = \frac{r-\mu}{\sigma}$, or $r = \mu + k\sigma$, and

$$G(k) = \phi(k) - k\bar{\Phi}(k).$$

Then, using (9.8), we see that $g(r) = \sigma G(k)$. The function $G(k)$, the standard normal loss function, is available in Microsoft's Excel library or in tabular form in some texts. Note that $g(r)$ is a linear function of σ. Double σ for a given value of r (or k) and $g(r)$ doubles as well. This is an important property and is one reason why the normal distribution is used in commercial applications.

Returning to the example problem, we see that the algorithm proceeds as follows.

Iteration 1

Step 1: $Q_E = \sqrt{\dfrac{2 \cdot 100 \cdot 200}{4}} = 100 = Q_1, n = 1.$

Step 2: $\bar{\Phi}(r_1) = \dfrac{Q_1 h}{\pi\lambda} = \dfrac{Q_1}{750} = .1333$ or

$$r_1 = 10 + (1.11 \times 3) = 13.33 \qquad (k = 1.11),$$
$$g(r_1) = 3 \cdot (.06727) = .20181.$$

$3\left[0.07182 - 1.11 \times 0.133\right]$

$3[0.$

Step 3: $Q_2 = \sqrt{10000 + \dfrac{200 \cdot (.20181) \cdot 30}{4}} = 101.5.$

Step 4: If $\varepsilon = .1, Q_2 - Q_1 > \varepsilon$. Set $n = 2$ and return to Step 2 to initiate the second iteration.

Iteration 2

Step 2:

$$\bar{\Phi}(r_2) = .1353 \text{ and}$$
$$r_2 = (1.10) \cdot (3) + 10 = 13.3,$$
$$(k = 1.10)$$
$$g(r_2) = 3 \cdot (.06862) = .20586.$$

Step 3:

$$Q_3 = \sqrt{10000 + 1500 \cdot (.20586)}$$
$$= 101.5.$$

Step 4: Since $Q_3 - Q_2 < \varepsilon$, stop. $Q^* \cong 101.5$ and $r^* = 13.3$ $(k = 1.10)$.

Now suppose that again $\lambda = 100$, $K = 200$, and $h = 4$. But now suppose $\mu = 50$ and $\sigma = 25$. Table 9.1 shows how Q^* and r^* vary as a function of π. As expected, the desire to avoid stockouts during a cycle increases as the stockout cost increases. The value of r^* increases substantially as π increases, and r^* is a concave increasing function of π. In contrast, Q^* is relatively insensitive to the value of π. Once $\bar{\Phi}(r_1)$, as a function of π, assumes a value less than .15, Q^* is almost constant. Note that Q^* is a decreasing convex function of π.

Table 9.1. Q^* and r^* as a function of π

π	Q^*	r^*
15	117.33	62.25
30	113.8	75.75
60	111.88	86.00
120	110.43	94.75

9.1.6 Alternative Heuristics for Computing Lot Sizes and Reorder Points

There are many approaches to computing lot sizes and reorder points in practice. A common method for computing safety stock is to set its value at some number of weeks of supply. Some companies choose to set safety stock equal to a given number of estimated standard deviations of lead time demand. Such practices can lead to very poor and costly results.

Other simple rules are employed to compute Q and r, too. Since stockout costs are inherently difficult to determine exactly, companies sometimes choose to set performance targets rather than to establish the value of π.

Two performance constraints are commonly employed. In fact, we have observed that practitioners often do not know the difference between the two. The first performance constraint stipulates the maximum probability of running out of stock during a cycle. The second is a fill rate target, that is, the fraction of customers whose demands are satisfied on time. The latter is often expressed as

$$1 - \frac{\text{expected shortages per year}}{\text{expected demand per year}}.$$

Furthermore, the expected number of shortages per year is stated as the expected number of cycles per year times the expected number of shortages per cycle, where the expected number of shortages per cycle is calculated as $g(r)$. Thus the fill rate is approximated as

$$1 - \frac{\frac{\lambda}{Q}g(r)}{\lambda} = 1 - \frac{g(r)}{Q}.$$

But remember, $\frac{g(r)}{Q}$ is just an approximation. It is accurate only when there is only a maximum of one order outstanding at a point in time. As we will see in the following section, $\frac{1}{Q}\{g(r) - g(r+Q)\}$ provides an exact method for measuring the fraction of customer demands that are not satisfied immediately. The term $g(r+Q)$ measures the expected demand in excess of $r+Q$ over a lead time. If we assume demand over a lead time cannot exceed Q, then $g(r+Q) = 0$.

Suppose, rather than providing a value for π, we assume that Q and r are determined as follows. Let a be the maximum probability of a stockout during a lead time. Then $\bar{F}(r) \leq a$. To find the values for Q and r, solve the following optimization problem:

$$\min_{Q,r\geq 0} \frac{\lambda K}{Q} + h\left[\frac{Q}{2} + r - \mu\right] \tag{9.9}$$

subject to

$$\bar{F}(r) \leq a. \tag{9.10}$$

Notice that Q and r can be determined separately. The optimizing value of Q is Q_E. Since $h \cdot r$ is minimized by making r as small as possible, $\bar{F}(r) = a$ in the optimal solution, and hence $r = \bar{F}^{-1}(a)$. But is this a formulation that is reasonable? We ask you to discuss this model as an exercise.

Although the model described in (9.9) and (9.10) is usually not stated formally, the solution derived from this model is often found in commercial software.

Another possible approach, which avoids some of the pitfalls present in (9.9) and (9.10) is as follows. Suppose we solve the problem

$$\min_{Q,r \geq 0} C(Q,r) \tag{9.11}$$

subject to

$$\bar{F}(r) = a, \tag{9.12}$$

while avoiding specifying the value of π. Since (9.12) holds, we know that $r = \bar{F}^{-1}(a)$. Recall also from our earlier model (Equation (9.6)) that

$$\bar{F}(r) = \frac{Qh}{\pi\lambda} \quad \text{or} \quad \pi = \frac{Qh}{\bar{F}(r) \cdot \lambda}. \tag{9.13}$$

One of the Karush–Kuhn–Tucker conditions for Problem (9.11) and (9.12) implies that

$$Q = \sqrt{\frac{2\lambda(K + \pi g(r))}{h}}, \tag{9.14}$$

which is the same expression as (9.5). Suppose we use (9.13) to estimate the value of π and substitute this value for π in (9.14). Then we have the approximate value

$$Q = \frac{g(r)}{a} + \sqrt{\left[\frac{g(r)}{a}\right]^2 + \frac{2\lambda K}{h}}. \tag{9.15}$$

Observe that this proposed approximation avoids a major problem that arises when employing the first approximation.

To illustrate this approach, suppose $\lambda = 100$, $K = 200$, $h = 4$, and demand is normally distributed with $\mu = 50$ and $\sigma = 25$. Suppose $a = .15$.

First, $r = \bar{F}^{-1}(.15) = 75.75$ or $k = 1.03$, approximately. Then from (9.15)

$$Q = \frac{1.967}{.15} + \sqrt{\left[\frac{1.967}{.15}\right]^2 + 10000}$$
$$= 13.11 + 100.85 = 113.96.$$

Note that if $\pi = \frac{(113.96) \cdot 4}{(.15)(100)} = 30.38 \approx 30$, we obtain approximately the same answer as presented in Table 9.1.

Let us analyze one additional approximate model in which a fill rate constraint is imposed, where

$$1 - \frac{g(r)}{Q}$$

is used as an approximation to the fill rate. Suppose the fill rate target is set so that $\frac{g(r)}{Q} = a$. Then the average annual variable cost expression is

$$\frac{\lambda K}{Q} + h\left[\frac{Q}{2} + r - \mu\right] + \pi\lambda a. \tag{9.16}$$

Since the last term in (9.16) is a constant, the goal is then to find Q and r that

$$\min_{Q,r \geq 0} \frac{\lambda K}{Q} + h\left[\frac{Q}{2} + r - \mu\right] \tag{9.17}$$

subject to

$$\frac{g(r)}{Q} = a. \tag{9.18}$$

The Karush–Kuhn–Tucker conditions for this problem are as follows, where θ is the Lagrange multiplier corresponding to constraint (9.18).

$$-\frac{\lambda K}{Q^2} + \frac{h}{2} - \frac{\theta g(r)}{Q^2} = 0, \tag{9.19}$$

$$h - \frac{\theta \bar{F}(r)}{Q} = 0, \tag{9.20}$$

$$\frac{g(r)}{Q} = a. \tag{9.21}$$

Note that from (9.20) we get

$$\theta = \frac{Qh}{\bar{F}(r)},$$

which implies that θ is an imputed value for $\pi\lambda$. Additionally, we get

$$Q = \frac{g(r)}{\bar{F}(r)} + \sqrt{\left[\frac{g(r)}{\bar{F}(r)}\right]^2 + \frac{2\lambda K}{h}}, \tag{9.22}$$

which is the same value obtained when

$$\pi = \frac{Qh}{\bar{F}(r)\lambda}$$

is substituted into (9.5) in the original approximation model.

We now illustrate how (9.19) through (9.22) can be used to construct a heuristic for solving Problem (9.17) and (9.18).

Heuristic for Solving Problem (9.17), (9.18)

Step 1: Set $Q_1 = Q_E$ and $n = 1$.

Step 2: Find r_n such that $g(r_n) = aQ_n$.

Step 3: Using (9.22), calculate Q_{n+1}.

Step 4: If $Q_{n+1} - Q_n < \varepsilon$, stop; otherwise, set $n = n + 1$ and return to Step 2.

Let us show how this heuristic applies to finding Q^* and r^* for the following parameter settings. Suppose $\lambda = 100, K = 200, h = 4$, demand is normally distributed with mean demand over a lead time of 10 and a standard deviation of 3 units. Suppose $a = .002$. Let $\varepsilon = .1$.

Then the heuristic proceeds as follows.

Iteration 1

Step 1: $Q_1 = Q_{EOQ} = 100$.

Step 2: $g(r_1) = (.002)(100) = .2 = 3 \cdot G(k_1)$ or $k_1 = 1.12$ and $r_1 = 13.36$.

Step 3:

$$Q_2 = \frac{.2}{.1314} + \sqrt{\left(\frac{.2}{.1314}\right)^2 + 10000}$$
$$= 101.53.$$

Iteration 2

Step 2:

$$g(r_2) = (.002)(101.53) = .203$$
$$= 3 \cdot G(k_2).$$

or $k_2 = 1.11$ and $r_2 = 13.33$.

Step 3: $Q_3 = 101.53$.

Step 4: Since $|Q_3 - Q_2| < \varepsilon$, stop, with $r = 13.33$ and $Q = 101.53$.

Observe that this solution closely matches the one found in the first example in Section 9.1.5, where $\frac{g(r^*)}{Q^*} = .00203$.

9.1.7   Final Comments on the Approximate Model

The models discussed in Section 9.1 are all based on the key assumption that no more than one order should be outstanding at any point in time. Furthermore, the models all employed an approximation of the average quantity of on-hand inventory. This approximation is based on the assumption that the average number of outstanding backorders at a random point in time is negligible. These assumptions hold only when Q is so much greater than μ so that the likelihood of more than one order will be outstanding at any point in time is negligible, and when the backorder cost is sufficiently large so that the average number of outstanding backorders at a random point in time is very small compared with the average amount of on-hand inventory.

In Section 9.2 we will develop exact models that are based on different, and, in many ways, less restrictive assumptions. In particular, these models will not require us to make either of the key assumptions discussed in the previous paragraph.

9.2   An Exact Model

Let us now develop a model that is based on an exact representation of the stationary probability distribution of net inventory. Using this distribution, we will be able to compute the exact values of the expectations found in the model's objective function. This exact representation will be developed on the basis of the assumption that the underlying demand process is a stationary Poisson process. This implies that units are demanded one by one and that the time between arriving demands is exponentially distributed. We let λ represent the annual demand rate. We further assume that the order lead time is a constant, which we denote by τ and is measured in years. Finally, since the demand process is described by a discrete random variable, both Q and r will be integer-valued.

Let us make an important observation. Since we assume the demand process is a Poisson process, it is possible for demand over a lead time to be large enough so that more than one order can be outstanding at a point in time. Hence we will no longer assume that the maximum number of outstanding orders at any point in time is one. Furthermore, this implies that we cannot use the on-hand inventory level as a basis for placing orders. The reorder point will now be based on the system's inventory position.

Recall that the inventory position random variable measures the amount of on-hand plus on-order minus backorders at any point in time. Note that if demand over a lead time is large and Q is small, then several orders may be outstanding at a point in time. Once Q units are demanded following the placement of the preceding order, a new procurement order is placed. Thus in this system, when a demand arises that reduces the inventory position to r units, an order is placed immediately for Q units, which raises the inventory position to $r+Q$ units. Consequently, the inventory position random variable assumes only one of the values $r+1, \ldots, r+Q$. Since a customer demand occurring when the inventory position random variable equals $r+1$ automatically triggers an order to be placed, the inventory position random variable does not assume the value of r for a positive amount of time.

Our analysis will proceed as follows. We will first derive the stationary probability distribution of the inventory position random variable. Using this distribution, we will determine the stationary probability distribution for net inventory, which, in turn, will permit us to compute a variety of performance measures.

9.2.1 Determining the Stationary Distribution of the Inventory Position Random Variable

Recall that the inventory position random variable assumes only values in the set $\{r+1, \ldots, r+Q\}$ for a positive amount of time. Let I represent the inventory position random variable.

The graph in Figure 9.2 indicates the transitions the random variable I makes through time. The length of time in any state $r+j$, $j=1, \ldots, Q$, is exponentially distributed with mean $1/\lambda$. The time until the next demand is independent of the state the system is in, that is, the value of $r+j$. A cycle in this system corresponds to the length of time between the placement of successive procurement orders, that is, the length of time between entering state

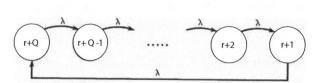

Fig. 9.2. Transition graph of the inventory position random variable.

$r+Q$, the initiation of a cycle, and the termination of a cycle, which occurs when the Qth subsequent demand occurs in the cycle. The expected cycle length is just Q/λ. Note that the process we have described is a renewal process. As a consequence,

$$P[I = r+j] = \frac{E[\text{time the system is in state } r+j \text{ during a cycle}]}{E[\text{cycle length}]}$$

$$= \frac{\frac{1}{\lambda}}{\frac{Q}{\lambda}} = \frac{1}{Q}. \tag{9.23}$$

Thus the inventory position random variable possesses a uniform distribution. Note that as long as the arrival process is a renewal process, I will have a uniform distribution.

9.2.2 *Determining the Stationary Distribution of the Net Inventory Random Variable*

Now that we have determined that I has a uniform distribution, we can use this result to establish the stationary probability distribution for the net inventory random variable, which we denote by N.

Suppose we step into the system's operation at a random point in time t. At time $t - \tau$, I must assume one of the values $r+1,\ldots,r+Q$. Suppose at time $t - \tau$, $I = r+j$.

Then, at time t, N can assume only one of the values $r+j, r+j-1, r+j-2, \ldots$. N cannot exceed $r+j$ since any unit ordered subsequent to time $t - \tau$ will not arrive prior to time t. Furthermore, every unit ordered by time $t - \tau$ will have arrived by time t. Hence $r+j$ units are available by time t to meet the demands that arise in the interval of time $(t - \tau, t]$. Thus the net inventory will be n units at time t if and only if exactly $r+j-n$ units were demanded in the period of length τ preceding t.

Let

$$p(y; \lambda\tau) = e^{-\lambda\tau} \frac{(\lambda\tau)^y}{y!}, \tag{9.24}$$

$P[N = n]$ be the probability that the net inventory random variable equals n, and

$$\mathcal{P}(y; \lambda\tau) = \sum_{u=y}^{\infty} p(u; \lambda\tau). \tag{9.25}$$

Suppose first that $n \leq r$. The inventory position at time $t - \tau$ can assume any value in $\{r+1,\ldots,r+Q\}$ and have $N = n$ at time t. The probability that $N = n$ at time t given that $I = r+j$ at time $t - \tau$ is $p(r+j-n; \lambda\tau)$, which is independent of t. Furthermore, $P[I = r+j] = \frac{1}{Q}$ and therefore

$$
\begin{aligned}
P[N = n] &= \sum_{j=1}^{Q} p(r+j-n; \lambda\tau) P[I = r+j] \\
&= \frac{1}{Q} \sum_{j=1}^{Q} p(r+j-n; \lambda\tau) \\
&= \frac{1}{Q}\{\mathcal{P}(r+1-n; \lambda\tau) - \mathcal{P}(r+Q+1-n; \lambda\tau)\} \tag{9.26}
\end{aligned}
$$

when $n \le r$.

Suppose next that $n > r$. Then the inventory position cannot be less than n at time $t - \tau$. In fact, the inventory position at time $t - \tau$ must be an element of the set $\{n, n + 1, \ldots, r + Q\}$. Consequently,

$$
\begin{aligned}
P[N = n] &= \frac{1}{Q} \sum_{j=n-r}^{Q} p(r + j - n; \lambda \tau) \\
&= \frac{1}{Q} \sum_{u=0}^{r+Q-n} p(u; \lambda \tau) \\
&= \frac{1}{Q} \left\{ 1 - \sum_{u=r+Q-n+1}^{\infty} p(u; \lambda \tau) \right\} \\
&= \frac{1}{Q} \{ 1 - \mathcal{P}(r + Q - n + 1; \lambda \tau) \}
\end{aligned}
\tag{9.27}
$$

when $r + 1 \le n \le r + Q$.

Now that we have shown how to compute $P[N = n]$, let us compute several performance measures of interest.

9.2.3 Computing Performance Measures

There are several performance measures that are of importance, the probability that an arriving customer will not find stock on the shelf (which is the probability that net inventory is not positive), the average number of backorder incidents per year, the average number of backorders outstanding at a random point in time, and the average amount of on-hand inventory.

Let the probability that the system is out of stock at a random point in time be denoted by P_{out}. By the PASTA (Poisson arrivals see time averages) principle, this is also the probability that an arriving customer will find no stock on the shelf. When $N = -n$, $n = 0, 1, \ldots$, there is no stock on the shelf and therefore

$$
P_{\text{out}} = \sum_{n=0}^{\infty} P[N = -n].
\tag{9.28}
$$

From (9.26), again assuming $r \ge 0$,

$$P_{\text{out}} = \frac{1}{Q} \sum_{n=0}^{\infty} \{\mathcal{P}(n+r+1;\lambda\tau) - \mathcal{P}(n+r+Q+1;\lambda\tau)\}$$

$$= \frac{1}{Q} \left\{ \sum_{u=r+1}^{\infty} \mathcal{P}(u;\lambda\tau) - \sum_{u=r+Q+1}^{\infty} \mathcal{P}(u;\lambda\tau) \right\}$$

$$= \frac{1}{Q} \{g(r) - g(r+Q)\}, \tag{9.29}$$

where $g(j) = \sum_{u=j+1}^{\infty} \mathcal{P}(u;\lambda\tau)$. Hence $g(r)$ measures the expected demand in excess of r over a period of length τ. Likewise, $g(r+Q)$ equals the expected demand in excess of $r+Q$ over a period of length τ. Recall that in our approximate model we assumed that $g(r+Q) = 0$.

The fill rate is given by $1 - P_{\text{out}}$. Note that by ignoring the term $g(r+Q)$, we overstate the fill rate. The extent of this overstatement depends of course on the magnitude of $g(r+Q)$.

The expected number of backorder incidents per year equals the expected number of demands per year, λ, times the probability that an arriving order will not be satisfied immediately. Let $E(Q,r)$ denote this quantity. Then

$$E(Q,r) = \lambda P_{\text{out}}. \tag{9.30}$$

In the approximate model developed in Section 9.1, we assumed

$$E(Q,r) = \frac{\lambda}{Q} g(r),$$

which overstates the true expected number of backorders per year.

Next, let us compute the average number of backorders outstanding at a random point in time, which we denote by $B(Q,r)$. Since backorders exist when N is negative,

$$B(Q,r) = \sum_{n=1}^{\infty} nP[N = -n].$$

From (9.26),

$$B(Q,r) = \frac{1}{Q} \sum_{n=1}^{\infty} n\{\mathcal{P}(n+r+1;\lambda\tau) - \mathcal{P}(n+r+Q+1;\lambda\tau)\}$$

$$= \frac{1}{Q} \left\{ \sum_{u=r+1}^{\infty} (u-(r+1))\mathcal{P}(u;\lambda\tau) - \sum_{u=r+Q+1}^{\infty} (u-(r+Q+1))\mathcal{P}(u;\lambda\tau) \right\}$$

$$= \frac{1}{Q} \{G(r) - G(r+Q)\}, \tag{9.31}$$

where $G(j) = \sum_{u=j+1}^{\infty}(u-(j+1))\mathcal{P}(u;\lambda\tau)$. Note that

$$G(j) = \frac{(\lambda\tau)^2}{2}\mathcal{P}(j-1;\lambda\tau) - \lambda\tau j \mathcal{P}(j;\lambda\tau) + \frac{(j+1)j}{2}\mathcal{P}(j+1;\lambda\tau),$$

which we leave as an exercise to establish. Also note that $G(j)$ measures the expected time-weighted demand in excess of j over a period of length τ, that is, the time-weighted backorders.

The final performance measure we will compute is the expected on-hand inventory at a random point in time. Let $OH(Q,r)$ denote this quantity.

Recall that

$$E[I] = OH(Q,r) + E[\text{on order}] - B(Q,r).$$

Since, from Little's law, $E[\text{on order}] = \lambda\tau = \mu$ and $E[I] = \frac{Q+1}{2} + r$, we have

$$OH(Q,r) = \frac{Q+1}{2} + r - \mu + B(Q,r). \tag{9.32}$$

9.2.4   Average Annual Cost Expression

Now that we have shown how the various performance measures can be calculated exactly, we are in a position to state the corresponding exact expression for the average annual variable costs. We will use the nomenclature introduced in Section 9.1, with one addition. Let $\bar{\pi}$ represent the cost incurred per unit backordered for a year. Then

$$C(Q,r) = \frac{\lambda K}{Q} + h\left[\frac{Q+1}{2} + r - \mu\right] + \pi E(Q,r) + (h+\bar{\pi})B(Q,r). \tag{9.33}$$

9.2.5   Waiting Time Analysis

Let us now explore how the probability distribution for the number of backorders outstanding at a random point in time can be used to ascertain the expectation and variance of the length of time customers will wait for service.

Suppose we let $\bar{N}$ be a random variable for the number of outstanding backorders in equilibrium, and W be the random variable for the length of time an arriving customer waits to be served in equilibrium. Clearly these two random variables are related. Again, suppose $r \geq 0$. Then, intuitively, the number of backorders observed at a random point in time matches the demand that arises while customers wait to receive their goods.

That is, $P[\bar{N} = n] = P[\text{demand over the random waiting time} = n]$. This result can be proven rigorously, although we will not do so. As a consequence,

$$E[\bar{N}] = \lambda E[W],$$

which also follows from Little's law and from the following analysis.

Suppose we know that a waiting time is of length $W = w$. Then $P[\bar{N} = n|W = w] = e^{-\lambda w}(\lambda w)^n/n!$. Suppose $h(w)$ is the density function for the length of a waiting time. We do not know what functional form $h(w)$ takes, but, as we will see, we do not need to know. The unconditional probability for the number of units backordered at a random point in time is

$$P[\bar{N} = n] = \int_0^\infty P[\bar{N} = n|W = w]h(w)\,dw$$

$$= \int_0^\infty (e^{-\lambda w}(\lambda w)^n/n!)\,h(w)\,dw. \tag{9.34}$$

The probability generating function for the random variable $\bar{N}$ is

$$E[z^{\bar{N}}] = \sum_{n=0}^\infty z^n P[\bar{N} = n]. \tag{9.35}$$

Recall that we can find the mean and variance, as well as other moments, of $\bar{N}$ as follows. Compute

$$\frac{dE[z^{\bar{N}}]}{dz} = \sum_{n=0}^\infty nz^{n-1}P[\bar{N} = n], \tag{9.36}$$

$$\frac{d^2E[z^{\bar{N}}]}{dz^2} = \sum_{n=0}^\infty n(n-1)z^{n-2}P[\bar{N} = n] \tag{9.37}$$

and evaluate (9.36) and (9.37) when $z = 1$. Equation (9.36) evaluated at $z = 1$ provides $E[\bar{N}]$ and (9.37) evaluated at $z = 1$ yields $E[\bar{N}(\bar{N} - 1)] = E[\bar{N}^2] - E[\bar{N}]$. Hence the variance of $\bar{N}$,

$$\text{Var}\,\bar{N} = E[\bar{N}^2] - E[\bar{N}]^2,$$

can be found as well using (9.36) and (9.37).

Observe that

$$E(z^{\bar{N}}) = \sum_{n=0}^{\infty} z^n \int_0^{\infty} e^{-\lambda w} \frac{(\lambda w)^n}{n!} h(w) \, dw$$

$$= \int_0^{\infty} e^{-\lambda w} \left[\sum_{n=0}^{\infty} \frac{(\lambda w z)^n}{n!} \right] h(w) \, dw$$

$$= \int_0^{\infty} e^{-(1-z)\lambda w} h(w) \, dw. \tag{9.38}$$

Thus $E(z^{\bar{N}})$ is equal to the Laplace transform of the waiting time random variable with the parameter of the transform taking on the value $(1-z)\lambda$. Taking the derivative of the right-hand side of (9.38), and evaluating at $z = 1$, again establishes

$$E[\bar{N}] = \lambda E[W]. \tag{9.39}$$

By taking the second derivative of (9.38) with respect to z and evaluating at $z = 1$, we are able ultimately to show that

$$\operatorname{Var} \bar{N} = \lambda E[W] + \lambda^2 \operatorname{Var} W. \tag{9.40}$$

From (9.31) we showed that

$$E[\bar{N}] = B(Q,r) = \frac{1}{Q} \{ G(r) - G(r+Q) \}.$$

Furthermore, we can also compute $E[\bar{N}^2]$ using the probability distribution for the net inventory random variable, N. Hence we can establish the variance of the random variable W as

$$\operatorname{Var} W = \frac{1}{\lambda^2} [\operatorname{Var} \bar{N} - B(Q,r)]. \tag{9.41}$$

Knowing the mean and variance of W can provide valuable information pertaining to the level of service that will be provided to customers. In practice, the backorder costs are often adjusted in the cost model so that waiting times meet marketing objectives.

9.2.6 Continuous Approximations: The General Case

The exact analysis presented in the previous subsections is often used as the basis for other approximations. Rather than assuming the demand process is a Poisson process, let us assume that demand is represented by a continuous random variable with $f(x)$ and $\bar{F}(x)$ representing the density and complementary cumulative distribution functions for lead time demand, respectively. Let μ and σ^2 be the mean and variance of the lead time demand random variable.

As was the case in the development of the exact model, we will first establish how the probability distribution for the inventory position random variable is estimated. Then we will use this distribution to find the distribution of net inventory. Next we will state the various performance measures that are used in our cost model.

Recall that the inventory position random variable is uniformly distributed over the integers $r+1,\ldots,r+Q$ when the demand process is a Poisson process. In general, the probability distribution for I is not uniform, as we will illustrate subsequently. However, in our approximation, we will assume that I has a continuous distribution and that I is uniformly distributed over the interval $[r,r+Q]$. A somewhat better approximation results when the interval $[r+1/2,r+1/2+Q]$ is used as the range of the random variable I. To apply this more accurate approximation, substitute $r+1/2$ for r in what follows. We have chosen the range $[r,r+Q]$ solely to simplify the notation.

Next, let us develop an approximation for our net inventory random variable, N. As in the exact case, we will consider two situations. In the first, let $x \leq r$.

Letting $n(x)$ represent the density function for net inventory, we see that

$$n(x) = \frac{1}{Q}\int_r^{r+Q} f(y-x)\,dy$$
$$= \frac{1}{Q}\{\bar{F}(r-x) - \bar{F}(r+Q-x)\}, \; x \leq r, \tag{9.42}$$

where, as before, we assume $r \geq 0$.

In the second case, $r < x \leq r+Q$. Then

$$n(x) = \frac{1}{Q}\int_x^{r+Q} f(y-x)\,dy$$
$$= \frac{1}{Q}\{1 - \bar{F}(r+Q-x)\}. \tag{9.43}$$

The probability that the system is out of stock at a random point in time, P_{out}, is approximated by

$$P_{\text{out}} = \int_0^\infty n(-x)\,dx = \frac{1}{Q}\int_0^\infty [\bar{F}(r+x) - \bar{F}(r+Q+x)]\,dx \tag{9.44}$$

and the expected number of stockouts is approximated by

$$E(Q,r) = \lambda P_{\text{out}} = \frac{\lambda}{Q}\int_0^\infty [\bar{F}(r+x) - \bar{F}(r+Q+x)]\,dx. \tag{9.45}$$

The expected number of backorders outstanding at a random point in time is given by

$$B(Q,r) = \int_0^\infty xn(-x)\,dx = \frac{1}{Q}\int_0^\infty x[\bar{F}(r+x) - \bar{F}(r+Q+x)]\,dx. \qquad (9.46)$$

As before, let $\bar{N}$ be the random variable measuring the number of backorders outstanding at a random point in time and W be the random variable for customer waiting time. Then

$$B(Q,r) = E[\bar{N}] = \lambda E[W]$$

holds exactly. An approximation between the variances of $\bar{N}$ and W is as follows

$$\mathrm{Var}\,\bar{N} = aE[\bar{N}] + \lambda^2\,\mathrm{Var}\,W, \qquad (9.47)$$

where a is the variance-to-mean ratio of the demand process. In the Poisson case, $a = 1$.

Interestingly, the function $B(Q,r)$ is jointly convex in Q and r. This result can be proven by constructing the corresponding Hessian matrix and observing that its diagonal elements are positive and its determinant is non-negative. We leave the proof of this result as an exercise. Unfortunately, $E(Q,r)$ is not a convex function of Q and r in general.

The final performance measure is the average on-hand inventory at a random point in time, which we denote by $OH(Q,r)$. This expectation is

$$OH(Q,r) = \frac{Q}{2} + r - \mu + B(Q,r), \qquad (9.48)$$

which is almost identical to the expression in the exact case.

Let us now examine two special cases.

9.2.7   A Continuous Approximation: Normal Distribution

An important special case occurs when the demand over a lead time is approximated by a normal distribution. That is,

$$f(x) = \frac{1}{\sqrt{2\pi}\sigma}e^{-\frac{1}{2}\left(\frac{x-\mu}{\sigma}\right)^2}.$$

The density function for net inventory, $n(x)$, is expressed in this case as

$$n(x) = \begin{cases} \dfrac{1}{Q}\left[\bar{\Phi}\left(\dfrac{r-x-\mu}{\sigma}\right) - \bar{\Phi}\left(\dfrac{r+Q-x-\mu}{\sigma}\right)\right], & x \leq r \\[3mm] \dfrac{1}{Q}\left[1 - \bar{\Phi}\left(\dfrac{r+Q-x-\mu}{\sigma}\right)\right], & r < x \leq r+Q, \end{cases} \tag{9.49}$$

where $\bar{\Phi}(\cdot)$ is the complementary cumulative standard normal distribution function.

The probability of the system being out of stock at a random point in time, P_{out}, is given by

$$P_{\text{out}} = \int_0^\infty n(-x)\, dx = \frac{1}{Q}\int_0^\infty \left[\bar{\Phi}\left(\frac{r+x-\mu}{\sigma}\right) - \bar{\Phi}\left(\frac{r+Q-\mu+x}{\sigma}\right)\right] dx. \tag{9.50}$$

The integral $\int_0^\infty \bar{\Phi}\left(\frac{u+x-\mu}{\sigma}\right) dx = \int_u^\infty \bar{\Phi}\left(\frac{x-\mu}{\sigma}\right) dx$, which is the expected demand in excess of u over a lead time, and can be written as

$$g(u) = \sigma\phi\left(\frac{u-\mu}{\sigma}\right) - (u-\mu)\bar{\Phi}\left(\frac{u-\mu}{\sigma}\right), \tag{9.51}$$

where $\phi(\cdot)$ is the standard normal density function. We developed this expression earlier in this chapter in (9.7). Hence

$$P_{\text{out}} = \frac{1}{Q}[g(r) - g(r+Q)]. \tag{9.52}$$

The expected number of backorders incurred per year, denoted by $E(Q,r)$, is

$$E(Q,r) = \lambda P_{\text{out}}. \tag{9.53}$$

The average number of backorders existing at a random point in time, $B(Q,r)$, is expressed as

$$B(Q,r) = \int_0^\infty xn(-x)\, dx = \frac{1}{Q}\int_0^\infty x\left[\bar{\Phi}\left(\frac{r+x-\mu}{\sigma}\right) - \bar{\Phi}\left(\frac{r+Q+x-\mu}{\sigma}\right)\right] dx. \tag{9.54}$$

Note that

$$\int_0^\infty x\bar{\Phi}\left(\frac{y+x-\mu}{\sigma}\right) dx = \sigma\int_{\frac{y-\mu}{\sigma}}^\infty (\sigma z + \mu - y)\bar{\Phi}(z)\, dz$$

$$= \sigma^2\int_{\frac{y-\mu}{\sigma}}^\infty z\bar{\Phi}(z)\, dz + (\mu - y)\sigma\int_{\frac{y-\mu}{\sigma}}^\infty \bar{\Phi}(z)\, dz.$$

Furthermore,

$$\int_{\frac{y-\mu}{\sigma}}^{\infty} \bar{\Phi}(z)\,dz = -\left(\frac{y-\mu}{\sigma}\right)\bar{\Phi}\left(\frac{y-\mu}{\sigma}\right) + \int_{\frac{y-\mu}{\sigma}}^{\infty} z\phi(z)\,dz$$

$$= -\left(\frac{y-\mu}{\sigma}\right)\bar{\Phi}\left(\frac{y-\mu}{\sigma}\right) + \phi\left(\frac{y-\mu}{\sigma}\right)$$

and

$$\int_{\frac{y-\mu}{\sigma}}^{\infty} z\bar{\Phi}(z)\,dz = -\frac{1}{2}\left(\frac{y-\mu}{\sigma}\right)^2\bar{\Phi}\left(\frac{y-\mu}{\sigma}\right) + \frac{1}{2}\int_{\frac{y-\mu}{\sigma}}^{\infty} z^2\phi(z)\,dz.$$

Also,

$$\int_{\frac{y-\mu}{\sigma}}^{\infty} z^2\phi(z)\,dz = \frac{1}{\sqrt{2\pi}}\int_{\frac{y-\mu}{\sigma}}^{\infty} z(ze^{-z^2/2})\,dz$$

$$= \left(\frac{y-\mu}{\sigma}\right)\phi\left(\frac{y-\mu}{\sigma}\right) + \bar{\Phi}\left(\frac{y-\mu}{\sigma}\right)$$

so that, after a small amount of algebra, we find that

$$\int_0^{\infty} x\bar{\Phi}\left(\frac{y+x-\mu}{\sigma}\right)dx = \frac{1}{2}\left[\sigma^2 + (y-\mu)^2\right]\bar{\Phi}\left(\frac{y-\mu}{\sigma}\right)$$

$$-\frac{\sigma}{2}(y-\mu)\phi\left(\frac{y-\mu}{\sigma}\right). \tag{9.55}$$

Let

$$G(y) = \frac{1}{2}[\sigma^2 + (y-\mu)^2]\bar{\Phi}\left(\frac{y-\mu}{\sigma}\right) - \frac{\sigma}{2}(y-\mu)\phi\left(\frac{y-\mu}{\sigma}\right), \tag{9.56}$$

the expected time-weighted demand in excess of y over a lead time. Thus

$$B(Q,r) = \frac{1}{Q}[G(r) - G(r+Q)]. \tag{9.57}$$

Finally, the expected on-hand inventory at a random point in time, $OH(Q,r)$, is

$$OH(Q,r) = \frac{Q}{2} + r - \mu + B(Q,r). \tag{9.58}$$

9.2.8 Another Continuous Approximation: Laplace Distribution

Earlier we pointed out that only the upper tail portion of the lead time demand distribution is of importance in determining Q and r, that the normal distribution has a relatively light tail, and that an exponential distribution has an upper tail that is heavier than the normal distribution's upper tail.

An alternative model for the lead time demand distribution is the Laplace distribution, which has an exponential upper tail. In particular, the density function for the Laplace distribution is

$$f(x) = \frac{\sqrt{2}}{2\sigma} e^{-\sqrt{2}\left|\frac{x-\mu}{\sigma}\right|}, \quad -\infty < x < \infty. \tag{9.59}$$

Note that this is a two-parameter distribution and possesses a symmetric density about its mean, μ, as does the normal distribution. But, unlike the normal distribution, it is an exponential function in both directions from the mean. By symmetric we imply that

$$\bar{F}(\mu + k\sigma) = 1 - \bar{F}(\mu - k\sigma) = \frac{1}{2} e^{-\sqrt{2}k},$$

where $k \geq 0$ represents the number of standard deviations of demand. Thus the Laplace distribution has a standard distribution form as does the normal distribution. Because it does have a standard distribution form, it has desirable mathematical properties which make the model particularly attractive to use in practical applications. We will subsequently discuss how this distribution has been deployed to manage millions of items in the Department of Defense.

Let us now see how the form of this distribution affects the probability distribution for net inventory and subsequently the mathematical representation of the performance measures of interest.

Recall that we can express the reorder point in terms of the expected lead time demand, μ, and the safety stock, $k\sigma$, that is, $r = \mu + k\sigma$. In what follows we will think of k as the decision variable rather than r. Furthermore, for practical reasons and for simplicity of presentation, let us assume that $k \geq 0$. Thus we assume that safety stock will not be negative. By making this assumption, we will act as if the density function for lead time demand can be expressed as

$$f(x) = \frac{\sqrt{2}}{2\sigma} e^{-\sqrt{2}\left(\frac{x-\mu}{\sigma}\right)}. \tag{9.60}$$

We have removed the absolute value requirement in the exponent. This will not have any impact on the model as long as $k \geq 0$.

The basis for our analysis is the assumption made earlier that the inventory position random variable is uniformly distributed over the range $[r, r+Q]$. As a consequence, the density function for net inventory can be stated as

$$n(x) = \frac{1}{Q} \int_r^{r+Q} \frac{\sqrt{2}}{2\sigma} e^{-\sqrt{2}\frac{(y-\mu-x)}{\sigma}} \, dy$$

$$= \frac{1}{Q} \int_{\frac{r-\mu-x}{\sigma}}^{\frac{r+Q-\mu-x}{\sigma}} \frac{\sqrt{2}}{2} e^{-\sqrt{2}y} \, dy$$

$$= \frac{1}{2Q} e^{-\sqrt{2}\left(\frac{k\sigma-x}{\sigma}\right)} (1 - e^{-\sqrt{2}\frac{Q}{\sigma}}), \qquad (9.61)$$

when $x < r$ with $r = \mu + k\sigma$.

Then the probability that the system is out of stock is approximated as

$$P_{\text{out}} = \int_0^\infty n(-x) \, dx = \int_0^\infty \frac{1}{2Q} e^{-\sqrt{2}\left(\frac{k\sigma+x}{\sigma}\right)} (1 - e^{-\sqrt{2}\frac{Q}{\sigma}}) \, dx$$

$$= \frac{1}{2\sqrt{2}} \frac{\sigma}{Q} (1 - e^{-\sqrt{2}\frac{Q}{\sigma}}) e^{-\sqrt{2}k}. \qquad (9.62)$$

Furthermore,

$$E(Q,k) = \lambda P_{\text{out}} = \frac{\lambda}{2\sqrt{2}} \frac{\sigma}{Q} (1 - e^{-\sqrt{2}\frac{Q}{\sigma}}) e^{-\sqrt{2}k} \qquad (9.63)$$

$$B(Q,k) = \int_0^\infty x n(-x) \, dx = \frac{(1 - e^{-\sqrt{2}\frac{Q}{\sigma}}) e^{-\sqrt{2}k}}{2Q} \int_0^\infty x e^{-\frac{\sqrt{2}x}{\sigma}} \, dx$$

$$= \frac{\sigma^2}{4Q} (1 - e^{-\sqrt{2}\frac{Q}{\sigma}}) e^{-\sqrt{2}k}, \qquad (9.64)$$

and

$$OH(Q,k) = \frac{Q}{2} + r - \mu + B(Q,k) = \frac{Q}{2} + k\sigma + B(Q,k). \qquad (9.65)$$

Observe that all the performance measures are trivial to compute. This is one reason that this model is useful in practice. However, the main reason it has found use in practice is that the Laplace distribution represents forecast errors more accurately than does the normal distribution in many applications. We will return to this model in Section 9.2.9.2 .

9.2.9 Optimization

Using the expressions that we have developed, let us find Q^* and r^* that minimize the general average annual cost expression that we developed earlier,

$$C(Q,r) = \frac{\lambda}{Q}K + h\left[\frac{Q}{2} + r - \mu\right] + \pi E[Q,r] + [\bar{\pi} + h]B(Q,r).$$

Assume first that demand is approximated with a continuous distribution. Later we will study the discrete demand case.

9.2.9.1 Normal Demand Model

Necessary conditions for Q and r to be an optimal lot size and reorder point, respectively, require

$$\frac{\partial C}{\partial Q} = \frac{\partial C}{\partial r} = 0.$$

Let us focus on a special case. Let us assume that $\bar{\pi} = 0$, that is, there is only a backorder cost charged in proportion to the average number of backorder incidents per year. In addition, let us assume the fixed ordering costs are large and that $Q \geq \mu + 2\sigma$. Suppose $Q = \mu + 2\sigma$. Furthermore, as a conservative estimate, assume that there is no safety stock, that is, $r = \mu$. Also, assume that the lead time demand distribution has a relatively high coefficient of variation, where $\frac{\sigma}{\mu}$ is the coefficient of variation. The normal distribution model is useful in practical settings when $\frac{\sigma}{\mu} < \frac{1}{2}$. When $\frac{\sigma}{\mu}$ assumes larger values, the tails of the normal distribution may not represent the data well in practice. Under the normality assumptions let us evaluate $g(r+Q)$ and $G(r+Q)$. Now

$$g(r+Q) = \sigma\phi\left(\frac{r+Q-\mu}{\sigma}\right) - (r+Q-\mu)\bar{\Phi}\left(\frac{r+Q-\mu}{\sigma}\right)$$

$$= \frac{\mu}{2}(.000007145)$$

and, using (9.56),

$$G(r+Q) = (2.125)\cdot\mu^2\cdot\bar{\Phi}(4) - \frac{\mu^2}{2}\phi(4)$$

$$= \frac{\mu^2}{2}[4.25\bar{\Phi}(4) - \phi(4)]$$

$$= \mu^2[.0000173].$$

Hence, unless μ substantially exceeds 10^5 these quantities are negligible.

Assuming $g(r+Q) = G(r+Q) = 0$ and $\bar{\pi} = 0$, and setting $\frac{\partial C}{\partial Q} = \frac{\partial C}{\partial r} = 0$, we obtain

$$Q = \sqrt{\frac{2\lambda(K + \pi g(r)) + 2hG(r)}{h}} \qquad (9.66)$$

and

$$\left[1 - \frac{h}{\pi\lambda}(r-\mu)\right] \bar{\Phi}\left(\frac{r-\mu}{\sigma}\right) + \frac{h}{\pi\lambda}\sigma\phi\left(\frac{r-\mu}{\sigma}\right) = \frac{Qh}{\pi\lambda}. \tag{9.67}$$

Observe that when $\frac{h}{\pi\lambda}$ is small and $2hG(r)$ is small compared to $2\lambda(K+\pi g(r))$, then

$$Q \approx \sqrt{\frac{2\lambda(K+\pi g(r))}{h}}$$

and

$$\bar{\Phi}\left(\frac{r-\mu}{\sigma}\right) \approx \frac{Qh}{\pi\lambda}; \tag{9.68}$$

but, these are the same as (9.5) and (9.6) which we developed in Section 9.1.3.

Let us consider the example discussed in Section 9.1.5. In that example, $\lambda = 100$, $K = 200$, $h = 4$, $\pi = 30$, and $\mu = 10$ with $\sigma = 3$. Then $\frac{h}{\pi\lambda} = .00133$ and, even if the safety stock is zero, $2 \cdot G(r) = 4.5$. But $\frac{2\lambda K}{h} = 10000$, and thus the effect on the choice of Q would be minimal if the term involving $G(r)$ is ignored in (9.66).

Similarly, the effect of finding r using (9.67) rather than (9.68) would be very small in this and in similar situations.

Let us consider another situation. Suppose that $\pi = 0$, so that our cost model is

$$C(Q,r) = \frac{\lambda}{Q}K + h\left[\frac{Q}{2} + r - \mu\right] + [\bar{\pi}+h]B(Q,r). \tag{9.69}$$

Recall that $B(Q,r)$ is jointly convex in Q and r, and therefore a local optimal solution is also a global optimal solution.

The first-order conditions, which are necessary and sufficient conditions, are (assuming $r \geq -1$)

$$\begin{aligned}
0 = \frac{\partial C(Q,r)}{\partial r} &= h + [\bar{\pi}+h]\frac{\partial B(Q,r)}{\partial r} \\
&= h - \frac{[\bar{\pi}+h]}{Q}\left\{\sigma\left[\phi\left(\frac{r-\mu}{\sigma}\right) - \phi\left(\frac{r+Q-\mu}{\sigma}\right)\right]\right. \\
&\quad \left. + (\mu-r)\left[\bar{\Phi}\left(\frac{r-\mu}{\sigma}\right) - \bar{\Phi}\left(\frac{r+Q-\mu}{\sigma}\right)\right]\right\}
\end{aligned} \tag{9.70}$$

and

$$0 = \frac{\partial C(Q,r)}{\partial Q} = -\frac{\lambda K}{Q^2} + \frac{h}{2} + [\bar\pi + h]\frac{\partial B(Q,r)}{\partial Q}$$

$$= -\frac{\lambda K}{Q^2} + \frac{h}{2} - \frac{[\bar\pi + h]}{Q}\left\{\sigma\phi\left(\frac{r+Q-\mu}{\sigma}\right)\right.$$

$$+ (\mu - (r+Q))\bar\Phi\left(\frac{r+Q-\mu}{\sigma}\right)\right\}$$

$$-\frac{(\bar\pi + h)}{Q^2}\{G(r) - G(r+Q)\}, \tag{9.71}$$

where $G(y) = \frac{1}{2}[\sigma^2 + (y-\mu)^2]\bar\Phi\left(\frac{y-\mu}{\sigma}\right) - \frac{\sigma}{2}(y-\mu)\phi\left(\frac{y-\mu}{\sigma}\right)$, as in equation (9.56).

Suppose that Q is sufficiently large because K is large, so that all the terms in these expressions involving $Q+r$, that is, $\phi\left(\frac{Q+r-\mu}{\sigma}\right)$, $\bar\Phi\left(\frac{Q+r-\mu}{\sigma}\right)$, and $G(r+Q)$, are negligible in value relative to the other terms. Then (9.70) becomes

$$0 = h - \frac{[\bar\pi + h]}{Q}g(r)$$

and (9.71) becomes

$$0 = -\frac{\lambda K}{Q^2} + \frac{h}{2} - \frac{\bar\pi + h}{Q^2}G(r)$$

or

$$g(r) = \frac{Qh}{\bar\pi + h} \tag{9.72}$$

and

$$Q = \sqrt{\frac{2(\lambda K + [\bar\pi + h]G(r))}{h}}. \tag{9.73}$$

Then (9.72) and (9.73) can be used to find approximate values of Q^* and r^* by employing the following procedure.

Algorithm for Finding Q^* and r^*

Step 1: Set $Q_1 = Q_E, n = 1$.

Step 2: Using Q_n, find r_n, where $g(r_n) = \frac{Q_n h}{\bar\pi + h}$.

Step 3: Using r_n, find Q_{n+1}, where

$$Q_{n+1} = \sqrt{\frac{2(\lambda K + (\bar\pi + h)G(r_n))}{h}}$$

Step 4: When $|Q_{n+1} - Q_n| < \varepsilon$, stop; otherwise, set $n = n+1$ and return to Step 2.

While this algorithm is simple, finding the value of r_n that satisfies (9.72) for a given value of Q requires a numerical search. Although this search is simple to accomplish, it is more computationally intensive than the heuristic developed in Section 9.1.5.

9.2.9.2 Laplace Demand Model

Suppose that the probability distribution of demand over a lead time is a Laplace distribution. As mentioned earlier, one reason the Laplace distribution is used in practice is because of the mathematical simplicity of the resulting cost function and the relative ease of finding the values of Q^* and r^*. Let us compare the results of using the Laplace model rather than the normal model to describe lead time demand. To make the comparison, assume that $\pi = 0$. Then our average annual variable cost function is

$$C(Q,r) = \frac{\lambda K}{Q} + h \left[\frac{Q}{2} + r - \mu \right] + [\bar{\pi} + h] B(Q,r). \tag{9.74}$$

When $r = \mu + k\sigma$, we can express this function in terms of Q and k as

$$C(Q,k) = \frac{\lambda K}{Q} + h \left[\frac{Q}{2} + k\sigma \right] + [\bar{\pi} + h] \frac{\sigma^2}{4Q}(1 - e^{-\sqrt{2}\frac{Q}{\sigma}}) e^{-\sqrt{2}k}, \tag{9.75}$$

where we use (9.64) to determine $B(Q,r)$.

Observe that the backorder expression contains the term $(1 - e^{-\sqrt{2}\frac{Q}{\sigma}})$. When Q is large compared with σ, then $e^{-\sqrt{2}\frac{Q}{\sigma}} \approx 0$. For one of our earlier examples where Q was about 100 and $\sigma = 25$, $e^{-4\sqrt{2}} = .0035$. In situations like this one, where Q_E/σ is sufficiently large so that $e^{-\sqrt{2}\frac{Q}{\sigma}}$ is negligible, we can approximate (9.75) with

$$C(Q,k) = \frac{\lambda K}{Q} + h \left[\frac{Q}{2} + k\sigma \right] + [\bar{\pi} + h] \frac{\sigma^2}{4Q} e^{-\sqrt{2}k}. \tag{9.76}$$

The optimality conditions when (9.76) is used as the objective function are

$$\frac{\partial C(Q,k)}{\partial Q} = -\frac{\lambda K}{Q^2} + \frac{h}{2} - [\bar{\pi} + h] \frac{\sigma^2}{4Q^2} e^{-\sqrt{2}k} = 0 \tag{9.77}$$

and

$$\frac{\partial C(Q,k)}{\partial k} = h\sigma - [\bar{\pi} + h] \frac{\sigma^2}{4Q} \sqrt{2} e^{-\sqrt{2}k} = 0. \tag{9.78}$$

Then, from (9.77),

$$Q = \sqrt{\frac{2(\lambda K + (\bar{\pi}+h)\frac{\sigma^2}{4}e^{-\sqrt{2}k})}{h}} \tag{9.79}$$

and, from (9.78),

$$k = -\frac{1}{\sqrt{2}}\ln\left\{\frac{4Qh}{(\bar{\pi}+h)\sigma\sqrt{2}}\right\}. \tag{9.80}$$

Now suppose we substitute (9.80) into (9.79). Then

$$Q = \sqrt{\frac{2(\lambda K + (h\sigma Q)/\sqrt{2})}{h}}. \tag{9.81}$$

Squaring both sides and solving for Q results in

$$Q = \frac{\sigma}{\sqrt{2}} + \sqrt{\frac{2\lambda K}{h} + \left(\frac{\sigma}{\sqrt{2}}\right)^2}. \tag{9.82}$$

Thus, no search is required to find Q^* and k^*. Simply solve for Q^* using (9.82) and substitute the result into (9.80) to find k^* and $r^* = \mu + k^*\sigma$. Thus Q is only a function of σ and not of k in this case. When $\sigma \to 0, Q$ becomes Q_E. As mentioned, the simplicity of this result has great appeal. Thus when the distribution of forecast errors can be accurately approximated by a Laplace distribution, there is a compelling reason to use the Laplace distribution to represent lead time demand and to establish Q^* and r^*.

We also note that the value of Q obtained using (9.82) provides an upper bound on Q when $e^{-\sqrt{2}\frac{Q}{\sigma}}$ cannot be neglected in the model. Thus, in general, $Q_E \leq Q^* \leq \frac{\sigma}{\sqrt{2}} + \sqrt{\frac{2\lambda K}{h} + \left(\frac{\sigma}{\sqrt{2}}\right)^2}$ when the Laplace distribution is used to represent the distribution of lead time demand.

Similar results can be obtained if we assume $\bar{\pi} = 0$ and $\pi > 0$. We leave this as an exercise.

9.2.9.3   Exact Poisson Model

Let us return to the exact model that we developed in Section 9.2.1–9.2.4 and discuss how we can find optimal values for Q and r. Recall that the demand is a discrete process, a Poisson process. We will develop our optimization approach for the case where $\pi = 0$. In this case, our objective function can be written as

$$C(Q,r) = \frac{\lambda K}{Q} + h\left[\frac{Q+1}{2} + r - \mu\right] + [\bar{\pi}+h]B(Q,r). \tag{9.83}$$

Recall that

$$B(Q,r) = \frac{1}{Q}\{G(r) - G(r+Q)\},$$

where

$$G(y) = \frac{(\lambda\tau)^2}{2}\mathcal{P}(y-1;\lambda\tau) - \lambda\tau y\mathcal{P}(y;\lambda\tau) + \frac{y(y+1)}{2}\mathcal{P}(y+1;\lambda\tau) \qquad (9.84)$$

and $\mathcal{P}(y;\lambda\tau) = \sum_{u=y}^{\infty}e^{-\lambda\tau}\frac{(\lambda\tau)^u}{u!}$.

To find the optimal values for Q and r we construct the first difference functions

$$\Delta C_Q(Q,r) = C(Q+1,r) - C(Q,r) \qquad (9.85)$$

and

$$\Delta C_r(Q,r) = C(Q,r+1) - C(Q,r). \qquad (9.86)$$

We seek the pair Q^* and r^* for which Q^* is the smallest value of $Q, Q \geq 1$ and integer for which

$$\Delta C_Q(Q^*,r^*) \geq 0$$

and for which r^* is the smallest integer value of $r, r \geq -1$, for which

$$\Delta C_r(Q^*,r^*) \geq 0.$$

Then

$$\Delta C_Q(Q,r) = [\lambda K + (\bar{\pi}+h)G(r)]\left[\frac{1}{Q+1} - \frac{1}{Q}\right]$$
$$-(\bar{\pi}+h)\left[\frac{G(r+Q+1)}{Q+1} - \frac{G(r+Q)}{Q}\right] + \frac{h}{2} \qquad (9.87)$$

and

$$\Delta C_r(Q,r) = h + \frac{1}{Q}[\bar{\pi}+h][(G(r+1) - G(r+1+Q)) - (G(r) - G(r+Q))]. \qquad (9.88)$$

Then (9.87) and (9.88), along with (9.84), can be used to calculate Q^* and r^*. Begin with $Q_1 = \max[1, \lfloor Q_E \rfloor]$, where $\lfloor x \rfloor$ is the integer portion of x, and use (9.88) to find the smallest r for which $\Delta C_r(Q,r) \geq 0$. Then use the resulting value of r in (9.87) to find the smallest value of Q for $\Delta C_Q(Q,r) \geq 0$. Repeat this process until convergence occurs.

For example, suppose $\lambda\tau = 2$, $K = 1$, $\lambda = 8$, $h = 4$ and $\bar{\pi} = 36$. Then $Q_1 = 2$. Using (9.88), with $Q_1 = 2$, $r_1 = 2$. Then from (9.87), with $r_1 = 2$, we find $Q_2 = 3$. With $Q_2 = 3$, $r_2 = 2$, again. Thus $Q^* = 3$ and $r^* = 2$.

When Q is large relative to $\lambda\tau$, $G(r+Q) \approx 0$. In these cases, which occur frequently, we can ignore the terms $G(r+Q)$ and $G(r+1+Q)$ in expressions (9.87) and (9.88). Then, in these cases, condition (9.87) reduces to finding the smallest Q for which

$$Q(Q+1) \geq \frac{2\lambda}{h}\left[K + \frac{\bar{\pi}+h}{\lambda}G(r)\right] \qquad (9.89)$$

and condition (9.88) reduces to finding the smallest value of r for which

$$(\lambda\tau - (r+1))\mathcal{P}(r+1;\lambda\tau) + (r+1)p(r+1;\lambda\tau) \leq \frac{Qh}{\bar{\pi}+h}. \qquad (9.90)$$

Again assume that $\lambda\tau = 2$, $\lambda = 8$, $h = 4$ and $\bar{\pi} = 36$; however, now let $K = 16$. Let us find Q^* and r^* for this case.

Clearly Q^* must satisfy

$$Q(Q+1) \geq \frac{2\lambda K}{h}. \qquad (9.91)$$

Hence let $Q_1 = 8$, which is the smallest integer satisfying expression (9.91). Then $r_1 = 1$ from (9.90). When $r_1 = 1$, $G(1) = .86465$. Then we find the smallest value of Q for which $Q(Q+1) \geq 81.293$, or $Q_2 = 9$. Next find r_2, where the right-hand side of (9.90) is now .9. The result is $r_2 = 1$, and the process terminates with $Q^* = 9$ and $r^* = 1$.

9.2.10 Additional Observations: Compound Poisson Demand Process, Uncertain Lead Times

We have developed models in this chapter based on the assumption that customers always order one unit. But this is not always the case. For example, bookstores order various quantities of a particular book title when they place orders on publishers. Customer orders may arise according a Poisson process; however, the amount required per customer order is uncertain and is assumed to be described by another random variable. The timing of orders and order quantities are assumed to be independent of each other. When the customer order process is a Poisson process and the order quantity is also a random variable, then the lead time demand process is a compound Poisson demand process. Note that if the distribution of customer order quantities has a high variance-to-mean ratio, then the demand process can be very lumpy, which implies lead time demand can be highly variable.

Let us assume for simplicity that the order size random variable is integer-valued. We will also assume that the probability that an arriving customer orders a single unit is positive. This latter assumption will guarantee that we can have an ergodic Markov chain describe the transitions of the inventory position random variable.

Suppose a customer arrives and asks for a quantity y of the item but only $n < y$ units are on the shelf. We assume the customer takes the n units and backorders the remaining $y - n$ units. Thus backorders will be counted in units and not in customer orders that remain to be filled.

There are two types of policies that we will examine for managing inventories when lead time demand is a compound Poisson process. The first is called an (nQ, r) policy. This type of policy operates as follows. Suppose a customer places an order at time t for an amount that lowers the inventory position to r or lower. Let m be this value. At that instant, a procurement order is placed that is an integer n multiple of the basic lot size Q. Then n multiples of Q are immediately ordered so that $r < m + nQ \leq r + Q$. Note that the result of placing the procurement order does not necessarily increase the inventory position to its maximum possible value, $r + Q$. When $Q = 1$ the inventory position will always be raised to its maximum level. This type of policy is, in general, not optimal.

The second type of policy is normally called an (s, S) policy in the inventory literature. When a customer order arrives at time t that lowers the inventory position below s $(= r + 1)$, an order is immediately placed that raises the inventory position to S. Thus if after a customer order arrives the inventory position drops to $m < s = r + 1$, a procurement order is placed for an amount of $S - m$. This type of policy is optimal for the cost model that we will consider. A proof of the optimality of this type of policy is presented in the following chapter.

Let us study each of these two policy types in greater detail. Specifically, we will show how the distribution for the random variable I, the inventory position, can be computed in each case. Because the random variables for the arrival of customer orders and the order quantities are independent, the probability distribution for the net inventory random variable N can be computed as

$$P[N = n] = \sum_{i=r+1}^{S} P[\text{demand over the lead time} = i - n] \cdot P[I = i], n \leq s,$$

and

$$P[N = n] = \sum_{i=n}^{S} P[\text{demand over the lead time} = i - n] \cdot P[I = i], n > s,$$

where $P[\text{demand over a lead time} = y]$ has a compound Poisson distribution. Again we assume the lead time is a known constant τ. Also, let Z be a random variable describing the customer order size. Since the lead time demand distribution is compound Poisson, the lead time demand has mean $\lambda \tau E(Z)$ and variance $(\lambda \tau)^2 E[Z^2]$. Then we can easily compute the performance measures of interest and can construct the average annual cost functions using the concepts developed in Section 9.2.2 through 9.2.4.

9.2.10.1   Finding the Stationary Distribution of the Inventory Position Random Variable When an (nQ, r) Policy Is Followed

Recall that when following an (nQ, r) policy, a procurement order is placed at a time when a customer order is received that reduces the inventory position, call it m, to less than or equal to r. Then an order is placed for nQ units, where n satisfies $r < m + nQ \leq r + Q$. Thus after the procurement order is placed, the inventory position will be in one of the Q possible states $r + 1, \ldots, r + Q$.

Let I represent the inventory position random variable. Our goal is to find $P[I = r + i]$, $i = 1, \ldots, Q$. Note that transitions that I experiences occur when a customer order is received. Suppose $I = r + i$ just prior to the order's arrival, and $I = r + j$ just after the order's receipt. The transitions can be represented by a Markov chain. Let v_{ij} represent the transition probabilities corresponding to this chain, that is, v_{ij} represents the probability the inventory position moves from $r + i$ to $r + j$ as a result of the customer's order. We now show how the values of v_{ij} can be computed and how they can be used to find $P[I = r + i]$.

First, suppose $j \leq i$. Then $I = r + j$ after a customer order is received when $I = r + i$ prior to its receipt if and only if z, the size of the customer demand, satisfies $z \in \{i - j, i - j + Q, i - j + 2Q, \ldots, i - j + nQ, \ldots\}$. Let $P[Z = i - j + nQ]$ represent the probability that the demand placed by an arriving customer is equal to $i - j + nQ$. Then

$$v_{ij} = \sum_{n=0}^{\infty} P[Z = i - j + nQ], \quad \text{when } j \leq i. \tag{9.92}$$

Second, suppose $j > i$. In this case the inventory position cannot be in state $r + j$ unless the customer's demand was at least $i - j + Q$. Therefore

$$v_{ij} = \sum_{n=1}^{\infty} P[Z = i - j + nQ], \quad \text{when } j > i. \tag{9.93}$$

Since the transitions of I are governed by a Markov chain,

$$P[I = r + j] = \sum_{i=1}^{Q} P[I = r + i] v_{ij}, \quad \text{where } j = 1, \ldots, Q. \tag{9.94}$$

Let us now evaluate $\sum_{i=1}^{Q} v_{ij}$. Note that

$$
\begin{aligned}
\sum_{i=1}^{Q} v_{ij} &= \sum_{i=j}^{Q} v_{ij} + \sum_{i=1}^{j-1} v_{ij} \\
&= \sum_{i=j}^{Q} \left\{ \sum_{n=0}^{\infty} P[Z = i - j + nQ] \right\} + \sum_{i=1}^{j-1} \left\{ \sum_{n=1}^{\infty} P[Z = i - j + nQ] \right\} \\
&= \sum_{n=0}^{\infty} \left\{ \sum_{i=j}^{Q} P[Z = i - j + nQ] \right\} + \sum_{n=1}^{\infty} \left\{ \sum_{i=1}^{j-1} P[Z = i - j + nQ] \right\} \\
&= \sum_{n=0}^{\infty} \left[\sum_{u=0}^{Q-j} P[Z = u + nQ] + \sum_{u=1}^{j-1} P[Z = Q + nQ - u] \right] = 1, \quad (9.95)
\end{aligned}
$$

that is, $\sum_{i=1}^{Q} v_{ij} = 1$. Hence $P[I = r + i] = P[I = r + j]$ for all $i, j = 1, \ldots, Q$ satisfies (9.94). Since $\sum_{i=1}^{Q} P[I = r + i] = 1$ as well,

$$
P[I = r + i] = \frac{1}{Q}, \ i = 1, \ldots, Q. \quad (9.96)
$$

Remember we assumed that $P[Z = 1] > 0$. To see why this is necessary, consider the following example. Suppose $Q = 2$ and $P[Z = 2] = 1$. Then it is impossible for the inventory position to equal $r + 1$ assuming the inventory position begins in the state $r + 2 = r + Q$. That is, $P[I = r + Q] = 1$ and not $\frac{1}{2}$, which it would be if the inventory position random variable were uniformly distributed.

9.2.10.2   Establishing the Probability Distribution of the Inventory Position Random Variable When an (s, S) Policy Is Employed

When an (s, S) policy is followed, the inventory position random variable is usually not uniformly distributed. Every time a procurement order is placed, the inventory position assumes the value of S. If $P[Z = 1] < 1$, then the inventory position need not be in the state $S - 1$ in each cycle. Therefore, $P[I = S] \neq P[I = S - 1]$. Hence the distribution of I is more complicated to calculate.

We will now show two ways to find the steady state probability distribution of I. Remember that every time a customer demand occurs, the value of the inventory position changes. This change process can be represented by a transition matrix for given values of s and S. The first method that we will present is based on constructing this transition matrix.

Recall that when following an (s, S) policy, an order is placed whenever a customer would lower the inventory position below s. When this occurs, an order is placed that immediately raises the inventory position to S units. Let $Q = S - (s-1)$. For example, suppose $S = 7$ and $s = 5$, so that $Q = 3$. In this case, the inventory position random variable can only assume any of the values 5, 6, or 7. The transition matrix, which we call P, for changes to the inventory position in this example is as follows:

$$
P \equiv \begin{array}{c} s \\ s-1 \\ s-2 \end{array}
\begin{pmatrix}
\overset{s}{p(Z \geq 3)} & \overset{s-1}{p(Z=1)} & \overset{s-2}{p(Z=2)} \\
p(Z \geq 2) & 0 & p(Z=1) \\
p(Z \geq 1) & 0 & 0
\end{pmatrix},
$$

where Z is the random variable associated with the customer's order size and $p(\cdot)$ is the probability of the indicated event. Observe that P depends only on Q, not on the particular values of S and s. As long as $S - (s-1)$ remains constant, P represents the transition matrix.

Suppose $p(Z = 1) = .2$, $p(Z = 2) = .4$, and $p(Z \geq 3) = .4$. Then

$$
P \equiv \begin{array}{c} s \\ s-1 \\ s-2 \end{array}
\begin{pmatrix}
\overset{s}{0.4} & \overset{s-1}{0.2} & \overset{s-2}{0.4} \\
0.8 & 0 & 0.2 \\
1 & 0 & 0
\end{pmatrix}.
$$

The steady state probability distribution for I can be found by solving the equations

$$
\pi P = \pi,
$$
$$
\sum \pi_i = 1,
$$

where $\pi = (\pi_S, \pi_{S-1}, \pi_{S-2})$ is the row vector of respective probabilities. We can ignore one of the equations since there are four equations and three unknowns. Then we have

$$
\pi_S \cdot p(Z = 1) = \pi_{S-1},
$$
$$
\pi_S \cdot p(Z = 2) + \pi_{S-1} \cdot p(Z = 1) = \pi_{S-2},
$$
$$
\pi_S + \pi_{S-1} + \pi_{S-2} = 1.
$$

Thus

$$
\pi_{S-2} = \pi_S \cdot p(Z = 2) + \pi_S \cdot (p(Z = 1))^2
$$

and

$$
\pi_S + \pi_S \cdot p(Z = 2) + \pi_S \cdot (p(Z = 2) + (p(Z = 1))^2) = 1,
$$

so then

$$\pi_S = \frac{1}{1 + p(Z=1) + p(Z=2) + (p(Z=1))^2}.$$

For our example, $\pi_S = \frac{1}{1.64} = 0.610$, $\pi_{S-1} = \frac{0.2}{1.64} = 0.122$, and $\pi_{S-2} = \frac{0.44}{1.64} = 0.268$.

As this example illustrates, computing the values of the stationary distribution is easily accomplished owing to the structure of the transition matrix P. Intuitively, the value of π_i indicates the long-run fraction of time that the inventory position is in state i.

An alternative way to compute this probability distribution is to construct a renewal equation. Let us first show how to construct this equation and then illustrate how to compute the probabilities using it.

The renewal equation is expressed as

$$M_j = 1 + \sum_{i=1}^{j} p(Z=i) \cdot M_{j-i}, \quad 1 \leq j \leq Q, \tag{9.97}$$

where $M_0 = 0$. For $0 \leq j \leq Q$,

$$\pi_{S-j} = p(I = S - j) = \frac{M_{j+1} - M_j}{M_Q}. \tag{9.98}$$

Let us return to our example. Using (9.97) we see that

$$M_1 = 1,$$
$$M_2 = 1 + p(Z=1) \cdot M_1,$$
$$M_3 = 1 + P(Z=1) \cdot M_2 + p(Z=2) \cdot M_1,$$

and therefore $M_1 = 1$, $M_2 = 1.2$, and $M_3 = 1.64$. From (9.98) we again see that $\pi_S = 0.610$, $\pi_{S-1} = 0.122$, and $\pi_{S-2} = 0.268$. The number M_3 measures the expected number of customer orders that are received between replenishment orders.

9.2.10.3   Constructing an Objective Function

As in Section 9.2.2, we can now compute the distribution function for the net inventory random variable N. As we stated earlier, since the demand process is a compound Poisson process, the inventory position random variable and the demand during the lead time random variable are independent. Therefore we can easily calculate

$$P[N = n] = \sum_{i=r+1}^{r+Q} P[Y = i - n] \cdot P[I = i], \, n \leq r, \tag{9.99}$$

and

$$P[N = n] = \sum_{i=n}^{r+Q} P[Y = i - n] \cdot P[I = i], \, r + Q \geq n \geq r + 1, \tag{9.100}$$

where Y is the random variable for lead time demand, which has a compound Poisson distribution.

Then we can calculate the performance measures as follows. For the (nQ, r) policy we can use (9.29), (9.30), (9.31), and (9.32). Functions $g(\cdot)$ and $G(\cdot)$ are calculated using the compound Poisson lead time demand model rather than the Poisson model. Also, replace $\lambda \tau$ with $\lambda \cdot E[Z] \cdot \tau$ which is the expected lead time demand.

For the (s, S) policy, with $s = r + 1$,

$$\begin{aligned} P_{\text{out}} &= P[N \leq 0] \\ &= \sum_{i=r+1}^{S} P[I = i] \cdot P[Y \geq i], \end{aligned} \tag{9.101}$$

$$\begin{aligned} B(s, S) &= \sum_{n=0}^{\infty} nP[N = -n] \\ &= \sum_{i=r+1}^{S} P[I = i] \cdot E[(Y - i)^{+}], \end{aligned} \tag{9.102}$$

and

$$\begin{aligned} OH(s, S) &= E[I] - E[Y] + B(s, S) \\ &= \sum_{i=1}^{Q} \frac{M_i}{M_Q} + s - 1 - \lambda \cdot E[Z] \cdot \tau + B(s, S), \end{aligned} \tag{9.103}$$

where $E[(Y - i)^{+}] = \sum_{y>i}(y - i)P[Y = y]$ and M_j is computed using (9.97).

The fixed ordering cost term in the objective function can be computed as follows when an (nQ, r) policy is used to control inventory. Let us assume that the fixed cost K is incurred independently of the size of the procurement order.

A procurement order is placed when a customer order arrives with probability

$$\sum_{i=r+1}^{r+Q} P[I = i] \cdot P[Z \geq i - r] = \frac{1}{Q} \sum_{i=r+1}^{r+Q} P[Z \geq i - r].$$

Since customer orders arrive at the rate λ, then the expected number of orders placed per year is

$$\frac{\lambda}{Q} \sum_{i=r+1}^{r+Q} P[Z \geq i-r] = \frac{\lambda}{Q} \sum_{u=1}^{Q} P[Z \geq u]. \tag{9.104}$$

Hence the expected fixed order cost incurred per year when following this policy is

$$\frac{\lambda K}{Q} \sum_{u=1}^{Q} P[Z \geq u]. \tag{9.105}$$

When an (s, S) policy is followed, $\frac{1}{M_Q}$ is the probability that an arriving customer order results in a procurement order. Thus in this case the expected fixed order cost incurred per year is $\frac{\lambda K}{M_Q}$.

9.2.11 Stochastic Lead Times

Suppose we return to the situation where demands are of unit size and the demand process is a Poisson process. But now let us assume that the lead time is no longer a constant but is a random variable with a gamma distribution. In practice, a gamma distribution is often used to model lead times. Let us compute the distribution function for lead time demand in this case.

The gamma density function for the lead time is given by

$$f(\tau) = \frac{e^{-\tau/b} \tau^{a-1}}{b^a \Gamma(a)}, \tag{9.106}$$

where $E[\tau] = ab$ and $\mathrm{Var}\,\tau = ab^2$. Then

$$
\begin{aligned}
P[Y = y] &= \int_0^\infty P[Y = y|\tau] f(\tau)\, d\tau \\
&= \int_0^\infty e^{-\lambda \tau} \frac{(\lambda \tau)^y}{y!} \frac{e^{-\tau/b} \tau^{a-1}}{b^a \Gamma(a)}\, d\tau \\
&= \frac{\lambda^y}{b^a \Gamma(a) y!} \int_0^\infty e^{-(\lambda + 1/b)\tau} \tau^{a+y-1}\, d\tau \\
&= \frac{\lambda^y}{b^a \Gamma(a) y!} \cdot \left[\frac{b}{1 + \lambda b}\right]^{a+y} [\Gamma(a+y)] \\
&= \frac{\Gamma(a+y)}{\Gamma(a) y!} \cdot \left[\frac{1}{1 + \lambda b}\right]^a \left[\frac{\lambda b}{1 + \lambda b}\right]^y, \tag{9.107}
\end{aligned}
$$

or Y has a negative binomial distribution with $p = \frac{1}{\lambda b + 1}$. Then

$$E[Y] = a\frac{(1-p)}{p} \ \text{ and } \ \text{Var}\,Y = \frac{a(1-p)}{p^2}. \tag{9.108}$$

Suppose, in a more general way, that the demand process is a compound Poisson process. Let X represent the number of customer orders over a lead time. Then

$$E[X] = \lambda E[\tau] \tag{9.109}$$

and

$$\text{Var}\,X = \lambda E[\tau] + \lambda^2 \text{Var}\,\tau. \tag{9.110}$$

As earlier, let Y and Z represent the random variables for lead time demand and order size, respectively. Then

$$E[Y] = E[X] \cdot E[Z] \tag{9.111}$$

and

$$\text{Var}\,Y = E[X] \cdot \text{Var}\,Z + \text{Var}\,X \cdot [E(Z)]^2. \tag{9.112}$$

Thus

$$E[Y] = \lambda \cdot E[\tau] \cdot E[Z] \tag{9.113}$$

and

$$\text{Var}\,Y = \lambda \cdot E[Z^2] \cdot E[\tau] + (\lambda E[Z])^2 \text{Var}\,\tau. \tag{9.114}$$

In practice these expressions for $E[Y]$ and $\text{Var}\,Y$ are often used to estimate parameters of a negative binomial distribution when a discrete distribution is desired. When a continuous two-parameter distribution, such as a normal or gamma distribution, is used to approximate the distribution of lead time demand, expressions (9.113) and (9.114) are used to estimate the distribution's mean and variance.

In yet more general cases, where the demand process is neither a Poisson nor a compound Poisson process, the following expressions are used to estimate the mean and variance of lead time demand:

$$E[Y] = \lambda E[Z] \cdot E[\tau] \tag{9.115}$$

and

$$\begin{aligned} \text{Var}\,Y = {}& \lambda \cdot [\text{Var}\,Z + a \cdot [E[Z]]^2] \cdot E[\tau] \\ & + [\lambda \cdot E[Z]]^2 \cdot \text{Var}\,\tau, \end{aligned} \tag{9.116}$$

where a is the variance-to-mean ratio of the underlying demand process.

9.3 A Multi-Item Model

Inventory systems contain many item types. Hence it is important to develop inventory plans that consider the interactions among the items. In some cases, the interactions occur because of space limitations in warehouses. In others, there may be a maximum investment in inventory that is desired. In yet other situations, there may be customer service performance targets that are to be achieved across items rather than for individual items. In the exercises, we will ask you to consider some of these problem types.

In this section we will briefly discuss one multi-item model that has been successfully used by the Department of Defense for over three decades. The model is used to establish stock levels for consumable items, that is, those that are not subject to repair. Jet engines are repairable when they fail, but transistors are not repairable. Jet engines are expensive, while transistors are not. The model is thus used to manage the less expensive and non-repairable items.

The model, which is used to manage hundreds of thousands of items each stocked at a single location, is a continuous review model. It is called the Presutti–Trepp model [271], and named after its creators who worked at the Headquarters, Air Force Logistics Command. The goal is to set reorder quantities and reorder points for each item so as to minimize fixed ordering costs and inventory carrying costs while insuring that a supply service goal is achieved.

Obviously, demand for items in this environment is highly uncertain. After considerable experimentation and analysis conducted jointly by the Air Force and the Army, the Laplace distribution was selected to model the demand process over a lead time. This distribution, which we discussed in Section 9.2.9.2, is used in the Presutti–Trepp model to find Q and r for each item type.

Presutti and Trepp developed four possible optimization models that could be used to determine the best reorder points and quantities. These differed in the manner in which holding costs were charged and the type of performance constraint employed. Inventory holding costs were charged on the basis of either the average amount of inventory held or the average inventory position. Recall that the inventory position is calculated to include the amount of on-order inventory. The argument for using the inventory position to charge holding costs was based on two factors. First, the money was allocated for purchased goods and sometimes items were paid for when the order is placed. Since the inventory carrying cost factor used by the Department of Defense is largely based on the cost of capital, it made sense to use the inventory position. Second, as we will see, computing stock levels is much easier when using the average inventory position rather than average on-hand inventory as the basis for calculating holding costs.

The performance constraints are also of two possible types. The first limits the total number of expected backordered units per year. The second measure constrains the total average number of outstanding backorders at a random point in time. The first performance measure is an incident-oriented measure, while the second is a time-weighted backorder measure.

Let

λ_i = demand rate for item i,
K_i = fixed order cost for item i,
h_i = holding cost rate for item i,
σ_i = standard deviation of lead time demand for item i,
Q_i, r_i be the reorder quantity and reorder point for item i, and
z_i be an essentiality factor for item i.

This essentiality factor is used to reflect the relative importance of stockouts. Some items are more essential than others in the conduct of military operations, and hence it is less desirable to have backorders for these item types.

Recall that, on the basis of the assumptions and the analysis presented in Section 9.2.9.2,

$$E[I] = E[\text{inventory position}] = \frac{Q}{2} + r, \tag{9.117}$$

the average on-hand inventory is

$$OH(Q,r) = \frac{Q}{2} + r - \mu + B(Q,r) = \frac{Q}{2} + k\sigma + B(Q,r), \tag{9.118}$$

the average number of outstanding backorders at a random point in time is

$$B(Q,k) = \frac{1}{4}\frac{\sigma^2}{Q}(1 - e^{-\sqrt{2}\frac{Q}{\sigma}})e^{-\sqrt{2}k}, \tag{9.119}$$

and the expected number of backorder incidents per year is

$$E(Q,k) = \frac{\lambda}{2\sqrt{2}}\frac{\sigma}{Q}(1 - e^{-\sqrt{2}\frac{Q}{\sigma}})e^{-\sqrt{2}k}. \tag{9.120}$$

9.3.1 Model 1

In this first model, holding costs are charged in proportion to the average amount of on-hand inventory and the performance constraint limits the total number of essentiality-weighted backorder incidents per year. Then the goal is to

$$\underset{Q_i \geq 0}{\text{minimize}}\ C_1(\bar{Q},\bar{k}) = \sum_i \left\{ \frac{\lambda_i K_i}{Q_i} + h_i \left[\frac{Q_i}{2} + k_i \sigma_i + \frac{1}{4} \frac{\sigma_i^2}{Q_i} \left(1 - e^{-\sqrt{2}\frac{Q_i}{\sigma_i}} \right) e^{-\sqrt{2}k_i} \right] \right\}$$

$$(9.121)$$

subject to

$$A_1(\bar{Q},\bar{k}) = \sum_i \frac{1}{2\sqrt{2}} \frac{z_i \lambda_i \sigma_i}{Q_i} \left[1 - e^{-\sqrt{2}\frac{Q_i}{\sigma_i}} \right] e^{-\sqrt{2}k_i} \leq a_1, \qquad (9.122)$$

where a_1 is the maximum average number of backorder incidents permitted per year and $\bar{Q}$ and $\bar{k}$ are vectors of the order quantities and safety stock levels, respectively.

To obtain the solution to this problem, we construct the Lagrangian function

$$L_1(\bar{Q},\bar{r}) = C_1(\bar{Q},\bar{k}) + \theta A_1(\bar{Q},\bar{k}), \qquad (9.123)$$

where θ is the Lagrange multiplier associated with expression (9.122). Calculating $\frac{\partial L_1}{\partial k_i}$ and setting $\frac{\partial L_1}{\partial k_i} = 0$, we get

$$k_i = -\frac{1}{\sqrt{2}} \ln \left[\frac{Q_i h_i}{\frac{1}{2} \left(1 - e^{-\sqrt{2}\frac{Q_i}{\sigma_i}} \right) \left(\frac{h_i \sigma_i}{\sqrt{2}} + \theta z_i \lambda_i \right)} \right], \qquad (9.124)$$

which implies that the constraint can be expressed as

$$\sum_i \left\{ \frac{z_i \lambda_i \sigma_i h_i}{\frac{h_i \sigma_i}{\sqrt{2}} + \theta z_i \lambda_i} \right\} = \sqrt{2} a_1, \qquad (9.125)$$

where we assume the constraint is active.

By setting $\frac{\partial L_1}{\partial Q_i} = 0$ and substituting (9.124) for k_i, we get

$$-\frac{K_i \lambda_i}{Q_i^2} + \frac{h_i}{2} - \frac{h_i \sigma_i}{\sqrt{2} \left(1 - e^{-\sqrt{2}\frac{Q_i}{\sigma_i}} \right)} \left[\frac{1 - e^{-\sqrt{2}\frac{Q_i}{\sigma_i}}}{Q_i} - \frac{\sqrt{2}}{\sigma_i} e^{-\sqrt{2}\frac{Q_i}{\sigma_i}} \right] = 0. \qquad (9.126)$$

We will postpone our discussion of how we find the value of Q_i.

Observe that we need to know this value of Q_i and also θ to find k_i. But θ must satisfy condition (9.125). Observe that as θ increases the left-hand side of expression (9.125) decreases, and vice versa. Hence a simple line search method can be used to find a good approximation to θ relatively quickly.

9.3.2 Model 2

The holding costs in the second model are charged in proportion to the average inventory position. The performance constraint is the same as in the first model. The goal is to find Q_i^* and k_i^* that

$$
\underset{Q_i \geq 0}{\text{minimize}} \; C_2(\bar{Q}, \bar{k}) = \sum_i \left\{ \frac{\lambda_i K_i}{Q_i} + h_i \left(\frac{Q_i}{2} + k_i \sigma_i \right) \right\} \tag{9.127}
$$

subject to constraint (9.122).

Again construct the Lagrangian function $L_2(\bar{Q}, \bar{k})$ corresponding to this problem. By taking the partial derivative of this function with respect to k_i, setting the resulting quantity to zero, and solving for k_i, we get

$$
k_i = -\frac{1}{\sqrt{2}} \ln \left[\frac{Q_i h_i}{\frac{1}{2} \theta z_i \lambda_i \left(1 - e^{-\sqrt{2}\frac{Q_i}{\sigma_i}}\right)} \right]. \tag{9.128}
$$

Substituting this result into expression (9.122) yields

$$
\theta = \frac{1}{\sqrt{2}a_1} \sum_i \sigma_i h_i. \tag{9.129}
$$

Furthermore, computing $\frac{\partial L_2}{\partial Q_i} = 0$ we again obtain (9.126), after substituting for k_i. Hence Q_i has the same value in both models.

Note that in this case we can evaluate θ directly without a search, which makes this model desirable to use in practice.

9.3.3 Model 3

Now suppose that we charge holding costs proportional to on-hand inventory. Further, let us constrain the essentiality-weighted number of outstanding backorders at a random point in time. Then our goal is to find Q_i^* and k_i that

$$
\underset{Q_i \geq 0}{\text{minimize}} \; C_3(\bar{Q}, \bar{k}) = \sum_i \left\{ \frac{\lambda_i K_i}{Q_i} + h_i \left[\frac{Q_i}{2} + k_i \sigma_i + \frac{1}{4}\frac{\sigma_i^2}{Q_i}\left(1 - e^{-\sqrt{2}\frac{Q_i}{\sigma_i}}\right) e^{-\sqrt{2}k_i} \right] \right\} \tag{9.130}
$$

subject to

$$\sum_i \frac{1}{4} \frac{z_i \sigma_i^2}{Q_i} (1 - e^{-\sqrt{2}\frac{Q_i}{\sigma_i}}) e^{-\sqrt{2}k_i} \leq a_2, \tag{9.131}$$

where a_2 represents the limit on the total average number of outstanding essentiality-weighted backorders permitted at a random point in time. Let $L_3(\bar{Q}, \bar{k})$ be the Lagrangian corresponding to this problem. The optimal values for Q_i and k_i satisfy $\frac{\partial L_3}{\partial k_i} = \frac{\partial L_3}{\partial Q_i} = 0$. The equation $\frac{\partial L_3}{\partial k_i} = 0$ results in

$$k_i = -\frac{1}{\sqrt{2}} \ln \left[\frac{\sqrt{2} Q_i h_i}{\frac{1}{2}\sigma_i (1 - e^{-\sqrt{2}\frac{Q_i}{\theta_i}})(h_i + \theta z_i)} \right]. \tag{9.132}$$

By substituting this value of k_i into the constraint we obtain

$$\sum_i \frac{z_i h_i \sigma_i}{(h_i + \theta z_i)} = \sqrt{2} a_2. \tag{9.133}$$

Observe that there is no explicit expression for θ. A search must be used to find θ.
    To find Q^*, we use $\frac{\partial L_3}{\partial Q_i} = 0$. This equation again reduces to (9.126).

9.3.4 Model 4

In our final model, we charge holding costs proportional to the average inventory position. The performance constraint again limits the total average number of essentiality-weighted backorders outstanding at a random point in time. The goal in this model is to

$$\underset{Q_i \geq 0}{\text{minimize}}\, C_4(\bar{Q}, \bar{k}) = \sum_i \left\{ \frac{\lambda_i K_i}{Q_i} + h_i \left(\frac{Q_i}{2} + k_i \sigma_i \right) \right\} \tag{9.134}$$

subject to constraint (9.131).
    Let $L_4(\bar{Q}, \bar{k})$ be the corresponding Lagrangian. Then $\frac{\partial L_4}{\partial k_i} = 0$ implies

$$k_i = -\frac{1}{\sqrt{2}} \ln \left[\frac{\sqrt{2} Q_i h_i}{\frac{1}{2}\theta z_i \sigma_i (1 - e^{-\sqrt{2}\frac{Q_i}{\sigma_i}})} \right], \tag{9.135}$$

which, when combined with (9.131), implies that

$$\theta = \sum_i \frac{\sigma_i h_i}{\sqrt{2} a_2}. \tag{9.136}$$

As in the other three models, $\frac{\partial L_4}{\partial Q_i} = 0$ implies (9.126) must hold.

Hence Q_i assumes the same value in all the models.

9.3.5 Finding Q_i

In general we can use (9.126) to find Q_i. To do so requires evaluating this expression using a search method. However, as discussed earlier in Section 9.2.9.2,

$$Q_i = \frac{\sigma_i}{\sqrt{2}} + \sqrt{\frac{2\lambda_i K_i}{h_i} + \left(\frac{\sigma_i}{\sqrt{2}}\right)^2} \qquad (9.137)$$

provides an upper bound on Q_i^*.

Thus $1 \leq Q \leq \frac{\sigma}{\sqrt{2}} + \sqrt{\frac{2\lambda K}{h} + \left(\frac{\sigma}{\sqrt{2}}\right)^2}$, which provides the upper and lower bounds for a search for the optimal value. Furthermore, recall that when $Q/\sigma > 3$, $e^{-\sqrt{2}\frac{Q}{\sigma}} < .0144$ so that (9.137) can be safely used to approximate Q^* accurately.

9.4 Exercises

9.1. Recall the approximation model developed in Section 9.1.6 for setting Q and r:

$$\min \frac{\lambda K}{Q} + h\left[\frac{Q}{2} + r - \mu\right]$$

subject to $\bar{F}(r) \leq a$. Discuss the effect of using this model to find the optimal values for Q and r.

9.2. Suppose customer demand is generated by a Poisson process with rate λ. Assume the procurement lead time is a constant τ. Then show that

$$G(j) = \sum_{u=j+1}^{\infty} (u - (j+1))\mathcal{P}(u; \lambda\tau)$$

$$= \frac{(\lambda\tau)^2}{2}\mathcal{P}(j-1; \lambda\tau) - (\lambda\tau)j\mathcal{P}(j; \lambda\tau)$$

$$+ \frac{(j+1)}{2}j\mathcal{P}(j+1; \lambda\tau),$$

where $\mathcal{P}(j;\lambda\tau) = \sum_{u=j}^{\infty} e^{-\lambda\tau} \frac{(\lambda\tau)^u}{u!}$.

9.3. Show that

$$\sum_{j=r}^{\infty} \mathcal{P}(j;\lambda\tau) = (\lambda\tau)\mathcal{P}(r-1;\lambda\tau) + (1-r)\mathcal{P}(r;\lambda\tau).$$

9.4. Prove that the Hessian matrix corresponding to the function $B(Q,r)$ is positive semi-definite.

9.5. Suppose the Laplace distribution is used to represent the demand over a lead time. Further, assume that $\pi > 0$ and $\bar{\pi} = 0$. Develop a method for finding Q^* and r^* in this case. Review the discussion in Section 9.2.9.2 as a guide for developing your procedure.

9.6. Verify expressions (9.87) and (9.88).

9.7. Assume that n items are stocked at a particular warehouse. The warehouse manager is concerned about the number of shipments from suppliers received each year and wants the inventory policy to limit the expected number of such receipts. Suppose λ_i, h_i, τ_i, $\bar{\pi}_i$, and σ_i represent, respectively, the demand rate for item i per year, the holding cost incurred by holding a unit of item i for a year, τ_i the lead time length for item i, the cost of carrying a backorder for a year for item i, and the standard deviation of lead time demand for item i.

Suppose the goal is to minimize the average annual costs of carrying inventory and time-weighted backorders while limiting the expected number of orders placed per year.

First, suppose the demand process for each item type is a Poisson process. Develop the corresponding optimization model and an algorithm for computing the reorder point and reorder quantities for each item.

Second, suppose the demand over a lead time is represented by a Laplace distribution for each item. Again develop the corresponding optimization model and construct an algorithm for finding the values for Q_i and r_i.

9.8. In Section 9.1 we developed an approximation model based on the assumption that demand in excess of supply is backordered. Suppose now that demand is lost when inventory is not available to meet demand. The goal now is to find the values of Q and r that minimize the average annual fixed ordering costs, holding costs and lost sales costs. Using the notation found in Section 9.1, develop a model and a solution procedure for finding Q^* and r^*. In the model π now represents the cost of a lost sale. As you develop the model you will have to determine the expected cycle length and the expected number of orders placed per year. Observe that this latter quantity is no longer $\frac{\lambda}{Q}$, since some demand is lost. Thus during a cycle, demand equals Q units plus possibly some lost sales. Hence the expected demand during a cycle is equal to Q plus the expected lost sales. You will have to make some assumption as to how to deal with this observation as you construct your model and solution methodology. Remember, as you develop your solution methodology that you are assuming that lost sales will be a small fraction of total sales and that Q is sufficiently large so that no more than one order is outstanding at any point in time.

9.9. Suppose $\phi(\cdot)$ and $\bar{\Phi}(\cdot)$ represent the standard normal density and complementary cumulative distribution functions, respectively. Show that

$$\int_r^\infty x\phi(x)\,dx = \phi(r),$$

$$\int_r^\infty x^2\phi(x)\,dx = \bar{\Phi}(r) + r\phi(r),$$

$$\int_r^\infty \bar{\Phi}(x)\,dx = \phi(r) - r\bar{\Phi}(r), \text{ and}$$

$$\int_r^\infty x\bar{\Phi}(x)\,dx = -\frac{1}{2}r^2\bar{\Phi}(r) + \frac{1}{2}\left[\bar{\Phi}(r) + r\phi(r)\right]$$

$$= \frac{1}{2}\left[(1-r^2)\bar{\Phi}(r) + r\phi(r)\right].$$

9.10. Suppose $p(x;\lambda t)$ represents the probability that x units are demanded through a period of length t. Let λt be the expected demand over this period, where λ is the demand rate. Furthermore, let $\sigma_x^2 \cdot t$ represent the variance of this demand process over this period of time. Suppose the procurement lead time is a random variable with mean μ_t and variance σ_t^2. Let Y be a random variable representing demand over a procurement lead time. Show that

$$E[Y] = \lambda\mu_t \text{ and}$$

$$\text{Var}[Y] = \mu_t\sigma_x^2 + \lambda^2\sigma_t^2.$$

9.11. Suppose demands for an item are generated by a compound Poisson process. The customer arrival rate is λ units per unit of time. Furthermore, the number of units ordered per arriving customer has a geometric distribution. Construct the probability distribution for the number of units required to meet demand over a time interval of length t. Show how to find the mean and variance of this distribution by constructing and using the corresponding generating function. This type of compound Poisson process is called a "stuttering" Poisson process. Illustrate your results assuming $\lambda = 2$, $t = 1$, and the mean of the compounding geometric distribution is 2.

9.12. Llenroc Electronics (LE) stocks circuit boards found in radar sets used by commercial airlines. The inventory of these boards and many other electronic items used by airlines are held in LE's national warehouse. One such board has a normal purchase price of $100. Since this board is widely used, the airlines purchase an average of 2000 units per year from LE. LE in turn purchases these boards from a manufacturer. The estimated fixed cost to place an order by LE is $800. LE uses a holding cost rate of .20 on an annual basis. Airlines do not carry much inventory and hence rely on their suppliers to respond immediately to requests for parts. Since demand varies over time, occasionally LE is out of stock for this board when an airline places an order. When this occurs, LE places an emergency order on the manufacturer. The cost of acquiring and shipping the unit to the airline in these emergency situations is $2000. After some analysis, LE planners have decided to approximate the distribution of demand over a lead time using a normal distribution with a mean of 200 units and a standard deviation of 20 units. Determine the order quantity and reorder point that minimize the expected average annual variable cost.

9.13. Suppose demands on a system occur according to a Poisson process. Furthermore, suppose the procurement lead time is a known constant, τ. When demands arise and the system is out of stock they are lost. Let us assume that a (Q, r) policy is used to manage the system's stock. As in the backorder case, the inventory position random variable assumes values in the set $\{r+1, \ldots, r+Q\}$. Is the inventory position random variable uniformly distributed over these values? Why? In the backorder case, it is possible to have any non-negative integer number of outstanding orders. But in the lost sales case, there can never be more than $\lfloor (Q+r)/Q \rfloor$ orders outstanding. Hence when $r/Q < 1$, no more than one order will ever be outstanding. Let us suppose that this is the case.

Derive the exact cost model when there is never more than a single order outstanding. Do this by constructing an expression for the expected cost per cycle and multiplying the result by the expected number of cycles per year.

Show that the expected amount of time the system is out of stock per cycle is given by

$$T = \int_0^\tau \lambda \cdot (\tau - t) \cdot e^{-\lambda t} \frac{(\lambda t)^{r-1}}{(r-1)!} \, dt,$$

where λ is the rate at which demands occur. Using this result, show that the expected number of cycles per year is $\lambda/(Q + \lambda T)$.

Then compute the expected number of lost sales per cycle. The lost sales cost per cycle is π times the expected number of lost sales per cycle. Lastly, compute the expected number of unit years of inventory held per cycle. To do this, first calculate the expected number of unit years of stock held until an order is placed, and then the expected amount held over the lead time. To make these calculations, compute the probability distribution of the on-hand stock just after the arrival of a procurement order. Note that the on-hand stock at this time ranges from Q to $Q + r$.

9.14. An example of the heuristic approximation algorithm for finding Q and r was given in Section 9.1.5. In the first example in that section we assumed that demand is normally distributed. Suppose demand over a lead time is approximated by a Laplace distribution with the same parameter values. Find Q^* and r^* in this case. How do the results compare with those found in Section 9.1.5?

9.15. Suppose the demand over a lead time follows an exponential distribution. Develop equations to find Q^* and r^* for the model presented in Section 9.1.3.

9.16. Solve the first example problem in Section 9.1.5 assuming lead time demand is exponentially distributed. Let the mean of this distribution equal the mean of the normal distribution in that example. All other parameter values remain the same.

9.17. Assume demand arrives according to a Poisson process. Further assume that lead times are gamma distributed and that orders do not cross. That is, procurement orders arrive in the sequence in which they are placed. Develop the cost model for this case.

9.18. Llenroc Electronics stocks parts used by airlines. One such part has a demand rate of 12 units per year, and demands arrive according to a Poisson process. For every order that is placed, Llenroc incurs a fixed order cost of $40. The cost to carry one unit of stock of this item for a year is $100, that is, $h = \$100$. Demands that occur when the system is out of stock are backordered. The cost of a shortage lasting one year is estimated to be $5000, that is, $\bar{\pi} = \$5000$. The lead time is one month. Find Q^* and r^*.

9.19. In Problem 9.18, suppose there is no backorder cost, but there is a constraint on the fill rate. Suppose the goal is to minimize the average annual cost of ordering and holding inventory subject to the constraint that the fill rate must be at least .99, that is, $P_{out} \leq .01$. Find Q^* and r^* in this case.

9.20. Suppose demand over a lead time for a part is normally distributed with an expected demand of 20 units and a standard deviation of 5 units. The fixed ordering cost is $100, the annual demand rate is 500 units, the cost to hold a unit in stock for a year is $10, and the cost per backorder incident is $100. Assuming excess demand is backordered, find Q^* and r^* using the algorithm presented in Section 9.1.3.

9.21. Suppose demand over a lead time is Laplace distributed. Using the data in Problem 9.20, find Q^* and r^*. Compare the results with those obtained from Problem 9.20.

9.22. The Llenroc hospital orders a certain pharmaceutical. The supplier of this pharmaceutical sells them to the Llenroc hospital in cartons containing 12 doses. Demand for this pharmaceutical occurs according to a Poisson process at a rate of 4 doses per day. Assume the clinic in which it is used operates for 250 days per year. The procurement lead time from the supplier is one day. The cost of a carton of the pharmaceutical is $480. The fixed ordering cost is $5. The backorder cost $\bar{\pi}$ is $500, and the annual holding cost factor is such that $h = 16$. The hospital manages its inventories using an (nQ,r) policy. In this case $Q = 12$. Find n^* and r^* so as to minimize the average annual fixed ordering, holding, and backorder costs.

9.23. In Section 9.2 we developed an exact model for the (Q,r) policy when demand was a stationary Poisson process and when backorders are allowed. But suppose all demand occurring when the system is out of stock is lost. Further suppose $\tau = 0$.

In this case, modify the results obtained in Section 9.2. In particular, find the probability distribution of the inventory position and net inventory random variables. What do you observe? Next, develop a cost model and procedure for finding the optimal values for Q and r. Can $r > 0$? Is it optimal for r to be negative?

9.24. Verify (9.124) through (9.126).

10

Lot Sizing Models: The Periodic Review Case

We now turn our attention to managing inventory in a periodic review setting in which there are fixed costs incurred when placing orders. Thus we are examining the same type of problem presented in the previous chapter, but doing so now in a periodic review rather than a continuous review operational environment. As was the case in our study of continuous review systems, owing to the presence of these fixed ordering costs, it may no longer be economical to place an order in each period of the planning horizon. The fixed cost requires that the order be large enough to justify incurring this cost. This leads to a policy in which there are two critical numbers: the reorder point and the order-up-to level. This is the (s, S) policy introduced in the preceding chapter. When this policy is followed, an order is placed if and only if the inventory position at the beginning of a period is less than or equal to the reorder point. If an order is placed, then the size of the order will be such that the inventory position after placing the order will be equal to the order-up-to level. This means that the size of the order has to be at least the difference between the order-up-to level and the reorder point.

This problem can have several variations. The major variations are due to lead time, length of the horizon, type of solution, and restrictions on the order quantity. The lead time could be zero, positive but a known integer number of periods, or random. The length of the horizon could be finite or infinite. We will discuss many of these cases in this chapter. The type of solution derived could be optimal or just approximate. The allure of an approximate solution comes from the computational requirements associated with computing an optimal solution. We will discuss an algorithm to compute an approximate solution for the finite-horizon case, an algorithm to compute the optimal solution for an infinite-horizon problem, and lastly, a heuristic that is effective in certain circumstances.

J.A. Muckstadt and A. Sapra, *Principles of Inventory Management: When You Are Down to Four, Order More*, Springer Series in Operations Research and Financial Engineering, DOI 10.1007/978-0-387-68948-7_10, © Springer Science+Business Media, LLC 2010

10.1 Notation

To begin, suppose we want to manage our system over a finite horizon in which there are T periods. To simplify our discussion, let us assume that the order lead time is zero. As a consequence, the inventory position and net inventory random variables are the same. Let us now describe the manner in which the system operates. At the beginning of period t, the net inventory, which we denote by x_t, is observed, and a decision is made as to whether or not to place an order. If an order is placed, then a fixed cost K and a cost proportional to the quantity purchased are incurred. By assumption, the order is delivered instantaneously. The inventory after the order placement decision is denoted by y_t. Consequently, $y_t > x_t$ if an order is placed; otherwise $y_t = x_t$. Thus, the total ordering cost is equal to

$$K\delta_t + c(y_t - x_t),$$

where $\delta_t = 1$ if $y_t > x_t$, that is, when an order is placed, but $\delta_t = 0$ when no order is placed, and c is the per-unit purchase cost.

Through the rest of the period, demand is realized. If demand is less than y_t, then there is inventory left over at the end of the period, and a holding cost is charged in proportion to this amount at the rate of h dollars per unit. Let D_t be the random variable for demand in period t. The total expected holding cost at the end of period t is $hE[\max(y_t - D_t, 0)]$. On the other hand, if the demand is more than the net inventory y_t, then the full demand cannot be satisfied on time, and the excess demand is assumed to be backordered. Let each unit of backordered demand incur a penalty at the rate of b dollars per unit. Once again, if we take the expectation over demand, the expected backorder cost is $bE[\max(D_t - y_t, 0)]$ at the end of period t. Let us denote the total cost of order placement and inventory management for one period by $V_t^1(x_t, y_t)$.

Since $(y_t - D_t)^+ = \max(y_t - D_t, 0)$ and $(D_t - y_t)^+ = \max(D_t - y_t, 0)$, the expression for one period's expected cost is equal to

$$V_t^1(x_t, y_t) = K\delta_t + c(y_t - x_t) + hE[(y_t - D_t)^+] + bE[(D_t - y_t)^+].$$

But this is not the end of it. The decisions made in period t will have an impact on period $t+1$ through the inventory or backorders that exist at the end of period t. For example, if we order too much in period t, it may take several periods before this inventory can be used, and we will have to pay holding costs on this inventory in the future. Similarly, if we order too little, then we may have to place a bigger order next period which will increase the cost for that period. Consequently, we add one more term which indicates the future impact of today's decisions. This term is a recursive term and is equal to the optimal cost from the next period through the end of the planning horizon given that the net inventory at the beginning of the next period before the order placement decision will be $y_t - D_t$. We denote this term as $E[v_{t+1}(y_t - D_t)]$. We will define $v_t(\cdot)$ shortly. The cost from period t through the end of the horizon if an order of size $y_t - x_t$ is placed in period t is

$$V_t(x_t, y_t) = K\delta_t + c(y_t - x_t) + hE[(y_t - D_t)^+]$$
$$+ bE[(D_t - y_t)^+] + E[v_{t+1}(y_t - D_t)].$$

Suppose we now minimize the right-hand side over y_t, our decision variable. Then we will get the optimal cost from period t through the end of the horizon as a function of x_t. Since we denoted the optimal cost in period $t+1$ onwards by $v_{t+1}(\cdot)$, we will denote the optimal cost for period t onwards by $v_t(x_t)$. Therefore,

$$v_t(x_t) = \min_{y_t \geq x_t} K\delta_t + c(y_t - x_t) + hE[(y_t - D_t)^+]$$
$$+ bE[(D_t - y_t)^+] + E[v_{t+1}(y_t - D_t)]. \tag{10.1}$$

The above equation is valid for any period $t = 1, 2, \ldots, T$. Since the planning horizon terminates at the end of period T, what is the meaning of v_{T+1}? Traditionally, it is assumed that any leftover inventory at the end of the horizon is salvaged at some value which we denote by w per unit. The salvage value is assumed to be less than or equal to the cost of the item c. On the other hand, if there is a backlog at the end of period T, this backlog is fulfilled by purchasing the requisite quantity at the rate of c per unit. Therefore,

$$v_{T+1}(x) = -wx, \quad x \geq 0,$$
$$= -cx, \quad x < 0.$$

Theorem 10.1. *For the formulation in (10.1), there exists an optimal policy involving two critical numbers s_t and S_t such that if $x_t \leq s_t$, an order is placed to raise the inventory level to S_t. Otherwise, if $x_t > s_t$, then no order is placed.*

The form of the policy as characterized in the above theorem is not surprising. It is the same as the one discussed in the previous chapter. Because of the fixed cost of placing an order, it makes sense to wait until the inventory has been depleted sufficiently and the requirement has become large enough to justify placing the order. We will prove the above theorem in Section 10.4 for a special case in which the salvage value is equal to the purchasing cost. That proof assumes that the procurement lead times are zero in length. However, the proof also holds for the cases where lead times are positive and non-crossing. In such situations, the expected cost functions are computed on the basis of demand over a lead time in addition to the demand in a period. The proof also holds when the discounting of future costs is included in the model.

10.2 An Approximation Algorithm

We now present an approximation algorithm for the system studied in the previous section. An approximation algorithm provides approximate solutions to problems for which an exact solution is difficult to compute. It is different from a heuristic in the sense that we can establish a bound analytically on the worst-case performance for an approximation algorithm. Recall that we have stated that the optimal policy is of the (s_t, S_t) type. Thus, our goal is to develop a procedure for computing approximate values of s_t and S_t.

Let us alter our description of the operating environment slightly so that we can obtain a bound on the cost relative to the optimal cost. Now we will assume that the demand in period t is known at the beginning of period t. The demands for periods $t+1, t+2, \ldots, T$ remain unknown. When the ordering decision is made for period t, all the past information is assumed to be known, including demands and orders placed, if any, in periods $1, 2, \ldots, t$.

We assume that any units carried to the end of the horizon become worthless. That is, $w = 0$. This assumption is not necessary as such, but imposing it makes the analysis somewhat simpler. We provide steps of the algorithm in the following subsection.

In presenting the algorithm, we use D_t or d_t to indicate whether demand is unknown or known. As before, D_t is a random variable, whereas d_t is the value assumed by random variable D_t.

10.2.1 Algorithm

The algorithm has the following steps:

Step 1: Set $t^* = 0$ and $t = 1$. t^* is the last period in which an order was placed and t is the current period. $S_0 = 0$.

Step 2: If $t = T$, place an order equal to d_T plus any backlog carried over from period $T - 1$ if the entering inventory is insufficient to meet these requirements. Otherwise, if $t < T$, compute the total backordering cost in periods $\{t^* + 1, t^* + 2, \ldots, t\}$. The backordering cost is computed easily since demand for periods $t^* + 1, t^* + 2, \ldots, t$ and the order-up-to level in period t^*, S_{t^*}, are all known. The net inventory at the beginning of period $t^* + 1$ is $S_{t^*} - d_{t^*}$. We know, by assumption, that no order is placed in period $t^* + 1$. After a demand equal to d_{t^*+1} is realized, the net inventory at the end of period $t^* + 1$ is $S_{t^*} - d_{t^*} - d_{t^*+1}$. A backordering cost is incurred only if this quantity is negative and is equal to

$$b\max(d_{t^*} + d_{t^*+1} - S_{t^*}, 0) = b(d_{t^*} + d_{t^*+1} - S_{t^*})^+.$$

Note that we do not have to compute the expectation of the backordered quantity because we know it completely. In a similar way, the net inventory at the end of period $t^* + 2$ is $S_{t^*} - d_{t^*} - d_{t^*+1} - d_{t^*+2}$ since no order was placed in period $t^* + 2$ either. Once again, a backordering cost is incurred only if this quantity is negative which is equal to

$$b\max(d_{t^*} + d_{t^*+1} + d_{t^*+2} - S_{t^*}, 0) = b(d_{t^*} + d_{t^*+1} + d_{t^*+2} - S_{t^*})^+.$$

We can write similar expressions for each of the periods $\{t^* + 1, t^* + 2, \ldots, t\}$. With these terms combined, the total backordering cost for periods $\{t^* + 1, t^* + 2, \ldots, t\}$ is

$$b(d_{t^*} + d_{t^*+1} - S_{t^*})^+ + b(d_{t^*} + d_{t^*+1} + d_{t^*+2} - S_{t^*})^+$$
$$+ \cdots + b(d_{t^*} + d_{t^*+1} + d_{t^*+2} + \cdots + d_t - S_{t^*})^+.$$

Step 3: If the cost in Step 2 is more than K, place an order in period t and set $t^* = t$; otherwise, go to Step 4. If an order is placed, the order quantity is equal to the maximum possible quantity such that the projected holding cost over future periods in the planning horizon is less than or equal to K.

Unlike the backordering cost computation in Step 2, we now have to compute an expectation because future demands are unknown. Our goal is to find the order quantity Q, which has to be at least the demand in period t plus any backlog carried into period t. To see this, suppose Q is less than d_t plus the incoming backlog to period t. Then there will be backlog at the end of period t and the holding cost will be zero. Since the net inventory at the beginning of period $t + 1$ is negative, the holding cost through the rest of the planning horizon is zero. But we assumed that we picked an order quantity such that the expected holding cost from period t through the end of the planning horizon was equal to $K > 0$. Clearly, there is a contradiction and so our assertion that Q could be less than the sum of d_t and any backlog carried into period t from period $t - 1$ is not true.

Note that we would place an order only when on-hand inventory (if any) at the beginning of period t is not sufficient to satisfy the demand in period t. The quantity that is on hand at the end of period t is the same as the net inventory at the beginning of period $t + 1$, x_{t+1}. If demand in period $t + 1$ is less than x_{t+1}, then on-hand inventory at the end of period $t+1$ is equal to $x_{t+1} - D_{t+1}$; otherwise, the on-hand inventory is 0. Therefore, the expected holding cost is equal to

$$hE \max(x_{t+1} - D_{t+1}, 0) = hE(x_{t+1} - D_{t+1})^+.$$

For period $t+2$, the net inventory at the beginning of the period is $x_{t+1} - D_{t+1}$. The holding cost is incurred only if the net inventory at the end of period $t+2$, $x_{t+1} - D_{t+1} - D_{t+2}$, is positive. Thus, the expected holding cost in period $t+2$ is equal to

$$hE \max(x_{t+1} - D_{t+1} - D_{t+2}, 0) = hE(x_{t+1} - D_{t+1} - D_{t+2})^+.$$

In a similar way, we can compute the expected holding costs for periods $t+3, \dots, T$. For example, for period T, the expected holding cost is equal to

$$hE \max(x_{t+1} - D_{t+1} - D_{t+2} - \cdots - D_T, 0) = hE(x_{t+1} - D_{t+1} - D_{t+2} - \cdots - D_T)^+.$$

Thus, the total expected holding cost from period t through the end of the horizon due to the order placed in period t is equal to

$$H_t(Q) = hE(x_{t+1} - D_{t+1})^+ + hE(x_{t+1} - D_{t+1} - D_{t+2})^+ \\ + \cdots + hE(x_{t+1} - D_{t+1} - D_{t+2} - \cdots - D_T)^+,$$

where $x_{t+1} = x_t + Q - d_t$.

Note that the above expression is equal to the marginal holding cost *only* due to the order placed in period t. Does it matter when the next order after period t is placed? The answer is no. The holding cost in a future period t' after t is incurred only if the on-hand inventory at the end of period t' is positive. If that happens to be the case, then no order could have been placed in period t', as we have noted. Thus, incurring holding costs in a future period is equivalent to not placing an order in that period for these units.

The order quantity is the maximum possible value of Q such that $H_t(Q) \leq K$. Set $t = t + 1$ and go back to Step 2.

Step 4: If the cost in step 2 is less than K, do not place an order in period t. Set $t = t + 1$ and go to Step 2.

The expected cost of following this algorithm is at most 3 times the expected cost of following the optimal algorithm. Note that this bound is only for costs in expectation; this does not mean that in every possible scenario the cost of this algorithm is at most three times the optimal cost. The proof, which can be found in Levi [217], uses cost balancing among the three types of costs: the holding cost, the backorder cost, and the fixed ordering cost. Consequently, this algorithm is called the *triple balancing algorithm* (TBA).

Observe that the above algorithm is implemented on a concurrent basis. At the beginning of each period, we assess the past information and make a decision accordingly. Therefore, the selected policy in a given period is a function of what has transpired until then. The following examples illustrate Steps 2, 3, and 4 in the above algorithm.

Suppose the current period is 3 and a decision has to be made on whether or not to place an order in this period, and, if so, what the order quantity should be. Suppose the following information is known at this point. The last order was placed in period 1. The inventory after the delivery of the order, but before satisfying demand in period 1, was 26. The demands in periods 1 and 2 were 10 and 20, respectively. The demand in period 3 is going to be 10. Further, demands in periods 4 and 5 can equal either 10 or 20 units with probabilities $1/3$ and $2/3$, respectively, and independently. In addition, assume $K = 50$, $h = 1$, and $b = 10$.

The algorithm in this case works as follows. In accordance with Step 2, we first determine whether or not to place an order in period 3. This requires computing the backorder costs for periods 1, 2, and 3. The net inventory at the end of period 1 is $26 - 10 = 16$ and so the backorder cost is zero. The net inventory at the end of period 2 (and so at the beginning of period 3) is $26 - 10 - 20 = -4$, leading to a backlog cost of $(10)(4) = 40$. If no order is placed in period 3, then the backorder quantity at the end of period 3 equals the backorder quantity at the end of period 2 plus demand in period 3, or $4 + 10 = 14$ units. Thus, the backorder cost would be $(10)(14) = 140$. Therefore, the total backorder cost over periods 1 through 3 is $140 + 40 = 180$, which is greater than the fixed ordering cost. Consequently, an order must be placed.

We move to Step 3 to determine the order quantity (Q). The order quantity will be at least as large as the demand in period 3 plus the backlog from period 2. Thus, $Q \geq 14$. Now, we set the order quantity so that the expected holding cost due to the order quantity until the end of the horizon balances with the fixed cost. The holding cost in period 3 is equal to $(1)(Q - 14)$. The net inventory at the beginning of period 4 is $x_4 = Q - 14$. The expected holding cost in period 4 is equal to

$$(1)\left(\frac{1}{3}\right)\max(x_4 - 10, 0) + (1)\left(\frac{2}{3}\right)\max(x_4 - 20, 0).$$

To compute the expected holding cost for period 5, we have to know the distribution of the sum of demands for periods 4 and 5. The net inventory at the end of period 5 will be positive only if this sum is less than the net inventory at the beginning of period 4, x_4. By assumption, the sum of demands of periods 4 and 5 is equal to 20 with probability $1/9$, 30 with probability $4/9$, and 40 with probability $4/9$. Therefore, the expected holding cost for period 5 is equal to

$$(1)\left(\frac{1}{9}\right)\max(x_4-20,0)+(1)\left(\frac{4}{9}\right)\max(x_4-30,0)+(1)\left(\frac{4}{9}\right)\max(x_4-40,0).$$

Therefore, the total holding cost from period 3 until the end of the planning horizon is

$$H_3(Q) = Q-14+\frac{1}{3}\max(x_4-10,0)+\frac{2}{3}\max(x_4-20,0)$$

$$+\frac{1}{9}\max(x_4-20,0)+\frac{4}{9}\max(x_4-30,0)+\frac{4}{9}\max(x_4-40,0)$$

$$= Q-14+\frac{1}{3}\max(Q-14-10,0)+\frac{2}{3}\max(Q-14-20,0)$$

$$+\frac{1}{9}\max(Q-14-20,0)+\frac{4}{9}\max(Q-14-30,0)+\frac{4}{9}\max(Q-14-40,0),$$

where we obtain the second equation by substituting $x_4 = Q-14$ in the first equation.

To find the value of $Q = Q^*$ such that $H_3(Q^*) = K$, first let us hypothesize that $Q^* \geq 54$. Then, $H_3(Q) = 3Q-92$. Setting it equal to $K = 50$, we get $Q^* = 47.33$, which violates the hypothesis that the order quantity is greater than or equal to 54. Next, hypothesize that Q^* lies between 44 and 54. Then $H_3(Q) = (23/9)Q-68$, which when set equal to $K = 50$ yields $Q^* = 46.17$, which satisfies our hypothesis that Q^* lies between 44 and 54. Hence, the optimal order quantity is 46.17.

Let us now look at a related example. Suppose the demand is normally distributed in periods 4 and 5 with a mean of 15 and a variance of 100. Assume the other data remain unchanged. Let us now determine the order quantity in period 3.

Clearly, the analysis for making the ordering decision in period 3 remains the same as in the last example. To compute the order quantity, we want to compute the expected holding cost for period 4 and period 5; for period 3 the holding cost remains equal to $(1)(Q-14)$. The net inventory at the beginning of period 4, x_4, is equal to $Q-14$. For period 4, the expected holding cost is equal to

$$hE(x_4-D_4)^+ = (1)\int_0^{x_4}(x_4-z)\frac{1}{\sqrt{2\pi}\sigma_4}e^{-\frac{(z-\mu_4)^2}{2\sigma_4^2}}\,dz,$$

where we assume $\mu_4 = 15$ and $\sigma_4 = 10$. As before, to compute the expected holding cost for period 5, we have to know the distribution of the sum of demands for periods 4 and 5. Assuming independence of demand from period to period, the cumulative demand for periods 4 and 5 is normally distributed as $N(30,200)$. The expected holding cost for period 5 is equal to

$$hE\left(x_4 - D_4 - D_5\right)^+ = (1)\int_0^{x_4}(x_4-z)\frac{1}{\sqrt{2\pi}\sigma_5}e^{-\frac{(x_4-\mu_5)^2}{2\sigma_5^2}}\,dx,$$

with $\mu_5 = 30$ and $\sigma_5 = \sqrt{200}$. Therefore, the total holding cost from period 3 through the end of the horizon is equal to

$$H_3(Q) = Q - 14 + \int_0^{x_4}(x_4-z)\frac{1}{\sqrt{2\pi}\sigma_4}e^{-\frac{(z-\mu_4)^2}{2\sigma_4^2}}\,dz + \int_0^{x_4}(x_4-z)\frac{1}{\sqrt{2\pi}\sigma_5}e^{-\frac{(z-\mu_5)^2}{2\sigma_5^2}}\,dz$$

$$= Q - 14 + \int_0^{Q-14}(Q-14-z)\frac{1}{\sqrt{2\pi}\sigma_4}e^{-\frac{(z-\mu_4)^2}{2\sigma_4^2}}\,dz$$

$$+ \int_0^{Q-14}(Q-14-z)\frac{1}{\sqrt{2\pi}\sigma_5}e^{-\frac{(z-\mu_5)^2}{2\sigma_5^2}}\,dz,$$

where the second equation is obtained by substituting $x_4 = Q - 14$. To obtain the optimal order quantity, we set $H_3(Q) = K$ and conduct a simple numerical search to solve for Q. Doing so, we find that $Q^* = 44.83$.

Note that the holding cost for periods 3, 4 and 5 in the example are all increasing in Q, which implies that $H_3(Q)$ is increasing in Q. An increasing function takes any value only once, which means that there is only one value of Q for which $H_3(Q) = K$.

10.3   Algorithm for Computing a Stationary Policy

In this section, we will present an algorithm for computing the *optimal* (s,S) policy for an infinite-horizon problem. The optimal policy minimizes the long-run average cost of managing the inventory. Note that the same criterion was used for the EOQ model in Chapter 2 as well. All the model parameters, that is, the fixed cost K, the holding cost rate h, the backorder cost rate b, and the distribution of demand remain unchanged over time. As a consequence, the optimal policy is stationary, too; i.e., the optimal values of s and S remain unchanged over time.

We will begin with a short review of dynamic programming when an average cost criterion is used to establish an optimal policy. Following this review, we will formulate the inventory management problem and present an algorithm which can be used to compute the optimal policy.

10.3.1 A Primer on Dynamic Programming with an Average Cost Criterion

Let us begin by establishing the meaning of the long-run average cost as an objective function. Recall the dynamic programming formulation stated for the finite-horizon problem (10.1). Suppose we are given a state x at the beginning of the horizon, and suppose we divide the optimal cost $v_1(x_1)$ by the length of the horizon T, and then increase the length of the horizon. The long-run average cost $v^a(x_1)$ is equal to this optimal average cost as the length of the horizon approaches infinity.

10.3.2 Formulation and Background Results

Let x_t and y_t be the levels of net inventory at the beginning of period t before and after the order placement decisions. If an order is placed, $y_t = S$; otherwise, $y_t = x_t$. Since we work in an infinite-horizon setting, exactly which period we are in is inconsequential. In a finite-horizon setting, the end-of-horizon cost structure is different from that of other periods, and this difference influences periods that are closer to the end of the horizon more than periods that are farther away. As a result, even if all the cost parameters are time-independent, the optimal policy (s_t, S_t) is time-dependent. On the other hand, the end-of-horizon effect is not present in an infinite-horizon setting. Thus, when the cost structure and the demand process are stationary, any two periods are indistinguishable. Therefore, we will not use the subscript t in x_t and y_t henceforth.

Recall that an order is placed at the beginning of a period if $x \leq s$. We assume that the demand takes only integer values in a period. Let p_j be the probability that the demand equals j in a period, where j lies in the set of non-negative integers, $\{0, 1, 2, \ldots, \}$. Let $V^1(y)$ be the one-period expected cost of holding inventory and backordering when the net inventory after order placement is y. Therefore, $V^1(y)$ is equal to

$$V^1(y) = hE\left[(y-D)^+\right] + bE\left[(D-y)^+\right]$$

$$= h\sum_{j=0}^{y} p_j(y-j) + b\sum_{j=y+1}^{\infty} p_j(j-y).$$

Note that $V^1(y)$ depends on the choice of the values for s and S since these values influence the value of y for a given x. If $x \leq s$, then $y = S$; otherwise, $y = x$.

Let v^{a*} be the minimal average cost over all the possible choices of s and S. Let v_R^a be the average cost when a policy R is followed. A policy R is characterized by a pair of numbers s_R and S_R such that $S_R > s_R$. The dynamic programming formulation is

$$g(y) = V^1(y) - v^{a*} + \sum_{j=0}^{\infty} p_j g(y-j) \tag{10.2}$$

$$g(S) = 0. \tag{10.3}$$

The evolution of net inventory over time can be understood in terms of a stochastic process, which in this case is a renewal process. Whenever the net inventory hits s or below, an order is placed to raise it to the level S. Thus, the state S is reached over and over again. Whenever the state S is achieved, a renewal takes place.

Next, we want to obtain two intermediate quantities that will be used to compute the average cost v_R^a and $g(x)$. Let y be the net inventory after the ordering decision is made in the current period. Then $y > s$. The two intermediate quantities are the expected time until the placement of the next order and the cost incurred until this happens. The expected time until the placement of next order is denoted by $t(y)$ and is equal to the expected time it will take for the net inventory to become less than or equal to s as a consequence of demand. The next order will be placed, at the earliest, in the next period. Further, suppose demand in the current period is equal to j and suppose $y - j > s$. Then no order will be placed in the next period and the expected time from the next period until the order placement will be equal to $t(y-j)$. The probability of this scenario is p_j. For each value of j, different scenarios may be created as listed in the following table:

Demand	Corresponding time	Probability
0	$t(y)$	p_0
1	$t(y-1)$	p_1
2	$t(y-2)$	p_2
$\vdots$	$\vdots$	$\vdots$
$y-s-2$	$t(s+2)$	p_{y-s-2}
$y-s-1$	$t(s+1)$	p_{y-s-1}

Thus, the total expected time until the next order is equal to

$$t(y) = 1 + p_0 t(y) + p_1 t(y-1) + \cdots + p_{y-s-2} t(s+2) + p_{y-s-1} t(s+1)$$

$$t(y) = 1 + \sum_{j=0}^{y-s-1} p_j t(y-j). \tag{10.4}$$

Observe that $t(y)$ is independent of s; it just depends on $y - s$. For example, when $s = 5$ we can compute $t(9)$ and when $s = 8$ we can compute $t(12)$. Since $9 - 5 = 12 - 8$, $t(9) = t(12)$. This observation will be useful in the implementation of the algorithm later.

Before we compute the expected cost until the placement of the next order, let us discuss the computation of $t(y)$. Given the recursive nature of the above equation, which is called a renewal equation, calculating $t(y)$ is straightforward. We illustrate the computation in the following example.

Suppose $s = 5$ and demand is discretely uniformly distributed over 0 through 5. Let us compute the expected time until the placement of the next order when $y = 10$.

The probability of demand in a period being equal to $0, 1, 2, 3, 4$, and 5 is equal to $1/6$. Since $s + 1 = 6$, using (10.4),

$$t(10) = 1 + p_0 t(10) + p_1 t(9) + p_2 t(8) + p_3 t(7) + p_4 t(6). \qquad (10.5)$$

But we do not know the values of $t(6), t(7), t(8)$, and $t(9)$. Suppose we had to find the expected time until the next order placement for $y = 6, t(6)$. Then (10.4) would be

$$t(6) = 1 + p_0 t_6 = 1 + \frac{1}{6} t(6),$$

which is easily solvable and we get $t(6) = 6/5$. Next, we use (10.4) to find $t(7)$

$$t(7) = 1 + p_0 t(7) + p_1 t(6) = 1 + \frac{1}{6} t(7) + \frac{1}{6} \left(\frac{6}{5} \right),$$

which can be solved to get $t(7) = 36/25 = 1.44$. In a similar way, we can find $t(8)$ as

$$t(8) = 1 + \frac{1}{6} t(8) + \frac{1}{6} t(7) + \frac{1}{6} t(6) = 1 + \frac{1}{6} t(8) + \frac{11}{25},$$

which leads to $t(8) = 1.73$. Finally, we find

$$t(9) = 1 + \frac{1}{6} t(9) + \frac{1}{6} t(8) + \frac{1}{6} t(7) + \frac{1}{6} t(6) = 1 + \frac{1}{6} t(9) + \frac{91}{125},$$

which leads to $t(9) = 2.07$. Substituting for $t(6), t(7), t(8)$, and $t(9)$ in (10.5), we get

$$t(10) = 1 + \frac{1}{6} (t(10) + 2.07 + 1.73 + 1.44 + 1.2),$$

and so $t(10) = 2.49$. To summarize, for any value of y, first compute $t(s+1)$. Use it to compute $t(s+2)$ and use that along with $t(s+1)$ to compute $t(s+3)$. Continue these recursive computations until you obtain $t(y)$.

Let the expected cost until the next placement of orders be denoted by $\mathcal{C}_s(y)$. The procedure for obtaining an expression for $\mathcal{C}_s(y)$ is similar to that for $t(y)$. The cost incurred in the current period is $V^1(y)$. Suppose demand in the following period is for j units where $y - j > s$. Then the next order will not be placed in the following period.

The expected cost from the next period until the placement of the next order is $\mathcal{C}_s(y-j)$, and the probability of this scenario is p_j. Once again, a table can be constructed. The total expected cost until the next order placement is thus equal to

$$\mathcal{C}_s(y) = V^1(y) + p_0\mathcal{C}_s(y) + p_1\mathcal{C}_s(y-1) + \cdots + p_{y-s-1}\mathcal{C}_s(s+1)$$

$$= V^1(y) + \sum_{j=0}^{y-s-1} p_j\mathcal{C}_s(y-j). \tag{10.6}$$

Once again, the recursive nature of the above expression makes the computation of $\mathcal{C}_s(y)$ easier. We illustrate the computation in the following example.

Suppose the holding cost is $h = 1$, the backorder cost is $b = 10$, and the purchasing cost per unit is $c = 3$. When $s = 4$ and demand is uniformly distributed from 0 through 7, let us compute the expected cost until the next order is placed if the current inventory level y is 8.

Note that demand takes integer values $0, 1, \ldots, 7$, each with probability $1/8$. Using (10.6),

$$\mathcal{C}_s(8) = V^1(8) + p_0\mathcal{C}_s(8) + p_1\mathcal{C}_s(7) + p_2\mathcal{C}_s(6) + p_3\mathcal{C}_s(5); \tag{10.7}$$

but we do not know the values of $\mathcal{C}_s(7)$, $\mathcal{C}_s(6)$, and $\mathcal{C}_s(5)$. We will first compute these values recursively. Using (10.6),

$$\mathcal{C}_s(5) = V^1(5) + p_0\mathcal{C}_s(5).$$

Here $V^1(5) = hE\left[(5-D)^+\right] + bE\left[(D-5)^+\right]$. If demand equals 0, 1, 2, 3, or 4, there will be a positive amount of inventory remaining at the end of the period. If demand equals 6 or 7, there will be a backlog. Therefore, the expected end-of-the-period holding cost is equal to

$$hE[(5-D)^+] = 1\left(\frac{1}{8}(5-4) + \frac{1}{8}(5-3) + \frac{1}{8}(5-2) + \frac{1}{8}(5-1) + \frac{1}{8}(5-0)\right) = 1.875.$$

Similarly, the expected end-of-period backorder cost is equal to

$$bE[(D-5)^+] = 10\left(\frac{1}{8}(6-5) + \frac{1}{8}(7-5)\right) = 3.75,$$

and so $V^1(5) = 5.625$. Therefore, $\mathcal{C}_s(5) = 5.625/\left(\frac{7}{8}\right) = 6.43$. Next, we use (10.6) to find

$$\mathcal{C}_s(6) = V^1(6) + p_0\mathcal{C}_s(6) + p_1\mathcal{C}_s(5).$$

As before, $V^1(6)$ can be computed as

$$V^1(6) = hE[(6-D)^+] + bE[(D-6)^+]$$
$$= 1\left(\frac{1}{8}(6-5) + \frac{1}{8}(6-4) + \frac{1}{8}(6-3) + \frac{1}{8}(6-2) + \frac{1}{8}(6-1) + \frac{1}{8}(6-0)\right)$$
$$+ 10\left(\frac{1}{8}(7-6)\right) = 3.875.$$

Thus

$$\mathcal{C}_s(6) = 3.875 + \frac{1}{8}\mathcal{C}_s(6) + \frac{1}{8}(6.43),$$

which can be solved to get $\mathcal{C}_s(6) = 5.35$. Next,

$$\mathcal{C}_s(7) = V^1(7) + p_0\mathcal{C}_s(7) + p_1\mathcal{C}_s(6) + p_2\mathcal{C}_s(5).$$

$V^1(7)$ can be computed as

$$V^1(7) = hE[(7-D)^+] + bE[(D-7)^+] = 3.5,$$

and so

$$\mathcal{C}_s(7) = 3.5 + \frac{1}{8}\mathcal{C}_s(7) + \frac{1}{8}(5.35) + \frac{1}{8}(6.43),$$

which can be solved to get $\mathcal{C}_s(7) = 5.68$. Finally,

$$V^1(8) = hE[(8-D)^+] + bE[(D-8)^+] = 4.5,$$

and $C_s(8) = 7.64$.

Computing the values of $t(y)$ and $\mathcal{C}_s(y)$ for $y > s$ allows us to write down an expression for $g(x)$ and v_R^a:

$$v_R^a = \frac{\mathcal{C}_s(S) + K}{t(S)} \tag{10.8}$$
$$g_R(x) = \mathcal{C}_s(x) + K - v_R^a t(x), \quad x > s, \tag{10.9}$$
$$= K, \quad x \le s \tag{10.10}$$

Let us now turn our attention to stating an algorithm for finding optimal values for s and S that is based on these observations.

10.3.3 Algorithm

Step 1: First, compute the lower and upper bounds for the optimal values for s and S. These bounds are denoted by LB, MB, and UB and are such that s is in $\{LB, LB+1, \ldots, MB-1\}$ and S is in $\{MB, MB+1, \ldots, UB\}$. We compute these bounds in the following manner:

1. MB is the smallest integer minimizing $V^1(y)$.
2. UB is the smallest integer greater than or equal to MB such that $V^1(UB+1) \geq V^1(MB) + K$.
3. LB is the smallest integer such that $V^1(LB+1) \leq V^1(MB) + K$.

Choose any values s and S such that s is in $\{LB, LB+1, \ldots, MB-1\}$ and S is in $\{MB, MB+1, \ldots, UB\}$.

Step 2: Compute the function $t(y)$ for $y = s+1, s+2, \ldots, s+UB-LB$ using (10.4).

Step 3: Compute $\mathcal{C}_s(y)$ for $y = s+1, s+2, \ldots, UB$ using (10.6).

Step 4: Compute v_R^a using (10.8) and $g_R(x)$ using (10.9) and (10.10).

Step 5: Find S_1 in $\{MB, MB+1, \ldots, UB-1, UB\}$ and use it to compute

$$g_R(S_1) = \min_{MB \leq y \leq UB} g_R(y).$$

Step 6: Now, find all y in $\{s+1, s+2, \ldots, MB-1\}$ satisfying the inequality

$$K + g_R(S_1) < g_R(y).$$

If more than one value of y satisfies the above inequality, set s_1 to be largest such y and go to Step 7. Otherwise, search for all the values of y in $\{LB, LB+1, \ldots, s-1\}$ that satisfy $V^1(y) < v_R^a$. If there is more than one such value of y, set s_1 to be the smallest of them, and go to Step 7. If we cannot find a suitable value of y, then set $s_1 = s$ and go to Step 7.

Step 7: If $s_1 = s$ and $S_1 = S$, (s,S) is optimal. Otherwise, go to Step 2.

Clearly this is not as simple an algorithm to execute and understand as the ones we developed earlier. The reason this algorithm is more complicated is due to the non-convexity of the objective function. We will now illustrate the execution of the algorithm. As you work through this example, observe how various expressions lack monotonicity as well as convexity.

Suppose $K = 7, h = 1, b = 10$ and demand in each period is uniformly distributed over $0, 1, \ldots, 7$. Let us find the optimal policy.

Iteration 1

Step 1: To obtain the minimum of $V^1(y) = hE(y-D)^+ + bE(D-y)^+$, we compute the values of $V^1(y)$ for $y = 2, 3, 4, \ldots, 15, 16$, which are given in the table below.

y	$V^1(y)$	y	$V^1(y)$	y	$V^1(y)$
2	19.125	7	3.50	12	8.50
3	13.250	8	4.50	13	9.50
4	8.750	9	5.50	14	10.50
5	5.625	10	6.50	15	11.50
6	3.875	11	7.50	16	12.50

Clearly, the minimum occurs at $y = 7$. Hence, $MB = 7$. To compute UB, we know that UB is the smallest integer greater than or equal to MB such that $V^1(UB+1) \geq V^1(MB) + K = 10.5$. Thus, $UB + 1 = 14$ and $UB = 13$. Similarly, LB is the smallest integer such that $V^1(LB+1) \leq V^1(MB) + K$, or $LB = 3$. Thus, s lies in $\{3, 4, 5, 6\}$ and S lies in $\{7, 8, 9, 10, 11, 12, 13, 14\}$. We (arbitrarily) take $s = 5$ and $S = 12$. Note that $S - s = 7$ and that $\sqrt{\frac{(2)(3.5)(7)}{1}} = Q_E = 7$. Hence we use the EOQ lot size to initialize the search.

Step 2: The values for $t(y)$ for $y = s + 1 = 6, 7, \ldots, s + UB - LB = 15$ are given in the following table:

y	$t(y)$	y	$t(y)$
6	1.1429	11	2.2282
7	1.3061	12	2.5465
8	1.4927	13	2.9103
9	1.7060	14	3.326
10	1.9497	15	3.8012

Step 3: Given the values of $V^1(y)$, we next compute the values of $\mathcal{C}_s(y)$ for $y = s + 1 = 6, 7, \ldots, UB = 13$.

y	$\mathcal{C}_s(y)$	y	$\mathcal{C}_s(y)$
6	4.4286	10	10.8569
7	4.6327	11	13.5508
8	6.4373	12	16.6294
9	8.4998	13	20.1479

Step 4: Compute v_R^a and $g_R(x)$ as follows:

$$v_R^a = \frac{\mathcal{C}_s(S) + K}{t(S)}$$

$$= \frac{16.6294 + 7}{2.5465} = 9.279179$$

and

y	$g_R(y)$	y	$g_R(y)$	y	$g_R(y)$
2	7	6	.8237	10	$-.2344$
3	7	7	$-.4871$	11	$-.125$
4	7	8	$-.4138$	12	0
5	7	9	$-.3301$	13	.1429

Step 5: The minimum of $g_R(y)$ over the range MB through UB occurs at $y = MB = 7$. Thus $S_1 = 7$.

Step 6: Only one value of y lies strictly between s and MB, $y = 6$. The sum $K + g_R(S_1) = 6.5129$, which is greater than $g_R(6) = .82$. Hence, no y exists such that $K + g_R(S_1) < g_R(y)$. On the other hand, there exist values of y such that $V^1(y) < v_R^a$ for $y \le s$; these values are $y = 4$ and 5. We take s_1 to be the smaller of the two, that is, $s_1 = 4$.

Step 7: Since the pair (s_1, S_1) is not equal to the pair (s, S), we proceed to the next iteration. Set $s = 4$ and $S = 7$.

Iteration 2

We do not have to perform Steps 1 and 2 here. Step 1 is the same for every iteration. In Step 2, we compute $t(y)$. The value of $t(y)$ is independent of the value of s; it just depends on $y - s$. As a result, $t(s+1) = t(6)$ in the previous iteration is equal to $t(s+1) = t(5)$ in the current iteration. Therefore, the new table for $t(y)$ is as follows:

y	$t(y)$	y	$t(y)$
5	1.1429	10	2.2282
6	1.3061	11	2.5465
7	1.4927	12	2.9103
8	1.7060	13	3.326
9	1.9497	14	3.8012

Step 3: Compute values of $\mathcal{C}_s(y)$ for $y = s+1 = 5, 6, \ldots, UB = 13$.

y	$\mathcal{C}_s(y)$	y	$\mathcal{C}_s(y)$	y	$\mathcal{C}_s(y)$
5	6.4286	8	7.6368	11	15.3413
6	5.3470	9	9.8706	12	18.6757
7	5.6822	10	12.4236	13	22.4865

Step 4: Compute v_R^a and $g_R(y)$:

$$v_R^a = \frac{\mathcal{C}_s(S) + K}{t(S)} = \frac{5.6822 + 7}{1.4927} = 8.49609$$

and

y	$g_R(y)$	y	$g_R(y)$	y	$g_R(y)$
2	7	6	1.25	10	.49271
3	7	7	0	11	.70596
4	7	8	.14286	12	.94966
5	3.7187	9	.306122	13	1.22819

Step 5: The minimum of $g_R(y)$ over the range MB through UB again occurs at $y = 7$, and so $S_1 = 7$.

Step 6: The values of y that are between s and S are $y = 5$ and $y = 6$. But $K + g_R(S_1) = 7$, which is greater than $g_R(5)$ and $g_R(6)$. Thus, when $y = 5$ or $y = 6$, $K + g_R(S_1) > g_R(y)$. Also, there is no value of y for which $V^1(y) < v_R^a$, $y \leq s$. Let $s_1 = s = 4$.

Step 7: Since $s_1 = s$ and $S_1 = S$, we have reached optimality. The optimal solution is $s^* = 4$ and $S^* = 7$.

10.4 Proof of Theorem 10.1

In our proof we assume that the salvage value is equal to the purchasing cost, that is, $w = c$. Recall that in Section 10.1 we defined the functions $V_t^1(x_t, y_t)$ and $v_t(x_t)$. We use these functions to define $g_t(x_t)$ and $G_t(y_t)$ as follows. Let

$$g_t(x_t) = cx_t + v_t(x_t)$$

and

$$G_t(y_t) = cy_t + hE[(y_t - D_t)^+] + bE[(D_t - y_t)^+] + E[v_{t+1}(y_t - D_t)].$$

Since $E[v_{t+1}(y_t - D_t)] = E[g_{t+1}(y_t - D_t)] - E[c(y_t - D_t)]$,

$$G_t(y_t) = cy_t + hE[(y_t - D_t)^+] + bE[(D_t - y_t)^+] + E[g_{t+1}(y_t - D_t)] - E[c(y_t - D_t)]$$
$$= \begin{cases} cE(D_t) + hE[(y_t - D_t)^+] + bE[(D_t - y_t)^+] + E[g_{t+1}(y_t - D_t)], & t < T, \\ cE(D_t) + hE[(y_t - D_t)^+] + bE[(D_t - y_t)^+], & t = T. \end{cases}$$

Also let

$$G^1(y_t) = hE[(y_t - D_t)^+] + bE[(D_t - y_t)^+].$$

Let y_t^* be the minimizer of $G_t(y_t)$, that is, $y_t^* = \arg\min G_t(y_t)$ and y_t^0 be the minimizer of $G^1(y_t)$. The proof we present closely follows the approach taken by Huh and

Janakiraman [177]. This type of analysis was first considered by Veinott [355]. The proof has four major parts.

Part 1: For any $y_t^2 < y_t^1 \leq y_t^0$ or $y_t^0 \leq y_t^1 < y_t^2$, we have $G_t(y_t^1) - G_t(y_t^2) \leq K$.

In this step, we show that if an inventory system in period t (characterized by y_t^1) is operating closer to the one-period cost's minimizer compared to, say, another inventory system which is characterized by y_t^2, that is, either $y_t^2 < y_t^1 \leq y_t^0$ or $y_t^0 \leq y_t^1 \leq y_t^2$, then the second system's cost advantage is no more than K.

First, let $t = T$. Note that $G_T(y_T)$ is convex in y_t, and since $G_T(y_T)$ and $G^1(y_T)$ differ by just a constant, both have the same minimizer and $y_t^0 = y_T^*$. Therefore, $G_T(y_T^1) \leq G_T(y_T^2)$ if y_T^1 is closer to y_T^*.

Now, suppose $t < T$. Owing to the convexity of $G^1(y_t)$, and since y_t^1 is closer to y_t^0 compared to y_t^2, $G^1(y_t^1) \leq G^1(y_t^2)$. Now, demand D_t occurs. Suppose we know about the ordering information of system 2 in period $t + 1$ but do not know about the ordering information of system 1 in that period. In other words, suppose we know the inventory level of system 2 after order delivery y_{t+1}^2 but do not know the inventory level of system 1 after order delivery, y_{t+1}^1.

If $y_t^0 \leq y_t^1 < y_t^2$, then $x_{t+1}^1 = y_t^1 - D_t < y_t^2 - D_t = x_{t+1}^2 \leq y_{t+1}^2$ since both systems see the same demand. In short, $x_{t+1}^1 < y_{t+1}^2$. On the other hand, if $y_t^2 < y_t^1 \leq y_t^0$, then $x_{t+1}^1 \leq y_t^0$. Regardless of whether $y_t^2 < y_t^1 \leq y_t^0$ or $y_t^0 \leq y_t^1 < y_t^2$, there are only two possible relationships between x_{t+1}^1 and y_{t+1}^2: $x_{t+1}^1 < y_{t+1}^2$ or $y_{t+1}^2 < x_{t+1}^1 \leq y_t^0$. We analyze each of these cases in turn.

Suppose $x_{t+1}^1 < y_{t+1}^2$. In this case, if system 1 places an order equal to $y_{t+1}^2 - x_{t+1}^1$ in period $t + 1$, then the inventory level of system 1 after placing the order is $x_{t+1}^1 + (y_{t+1}^2 - x_{t+1}^1) = y_{t+1}^2$. Further, suppose system 1 imitates system 2's ordering policy after period $t + 1$. Clearly, after period $t + 1$, the costs are equal. In period $t + 1$, we do not know if system 2 places an order. But system 1 does place an order and incurs a cost equal to K. Also, the holding and backordering costs in period $t + 1$ are equal for both the systems since $y_{t+1}^1 = y_{t+1}^2$. So the maximum possible difference in the costs of the two systems over periods $t + 1, t + 2, \ldots, T$ is K. In monetary terms in period t, this difference is equal to K. In other words, $G_t(y_t^1) - G_t(y_t^2) \leq K$.

Suppose $y_{t+1}^2 < x_{t+1}^1 \leq y_t^0$. In this case, suppose that system 1 does not place an order until period $t' > t + 1$, where period t' is the first period in which the beginning net inventory of system 1 falls below the post-ordering net inventory of system 2. That is, $x_{t'}^1 < y_{t'}^2$. In period t', system 1 places an order to raise the inventory equal to $y_{t'}^2$. Thus, system 1 places an order quantity equal to $y_{t'}^2 - x_{t'}^1$ and incurs a fixed ordering cost equal to K. After period t', system 1 mimics system 2 and thus has the same cost. The cost differs between system 1 and 2 in periods $t + 1, t + 2, \ldots, t'$. In periods $t + 1, t + 2, \ldots, t' - 1$, system 1 has a post-order inventory closer to y_t^0 compared to system 2. Thus, in each of these periods considered individually, the expected holding and backorder costs of system 1 are lower than the expected costs for system 2. In period

t', system 1 places an order and system 2 may or may not have placed an order. Thus, system 1's cost is at most K higher than system 2's cost in period t'. Combining the costs of systems 1 and 2, we see that in the worst case, system 1 incurs the same cost as does system 2 in periods $t+1, t+2, \ldots, t'-1$ but K more in period t'. In terms of the dollar value in period t, this is equal to K. Consequently, $G_t(y_t^1) - G_t(y_t^2) \leq K$.

Next, we show that if the starting net inventory in period t, x_t, is higher than either y_t^* or y_t^0, placing an order is more costly than not placing an order.

Part 2: If $x_t \geq \min\{y_t^*, y_t^0\}$, then it is optimal not to order in period t, that is, $y_t = x_t$.

Suppose an order is placed that increases the inventory to some level y_t. We will show that the cost will be higher than if no order were placed. We use the result from Part 1. To apply this result, we have to know exactly where x_t is relative to y_t^0; we know only that it is more than $\min\{y_t^*, y_t^0\}$. This creates two cases.

Suppose $x_t \geq y_t^0$. Since y_t is strictly greater than x_t, y_t is strictly greater than y_t^0 as well. In Part 1, we substitute $y_t^1 = x_t$ and $y_t^2 = y_t$, and we get

$$G_t(x_t) - G_t(y_t) \leq K.$$

This means that the savings $G_t(x_t) - G_t(y_t)$ due to having a higher inventory level of y_t are less than the fixed cost of order placement, and placing an order is not justified.

Now suppose $y_t^* < x_t \leq y_t^0$. Once again, we use Part 1. Substitute $y_t^1 = x_t$ and $y_t^2 = y_t^*$. We get

$$G_t(x_t) - G_t(y_t^*) \leq K.$$

Since y_t^* is the global minimum, $G_t(y_t^*) \leq G_t(y_t)$. This implies that

$$G_t(x_t) - G_t(y_t) \leq G_t(x_t) - G_t(y_t^*) \leq K.$$

Once again, this shows that placing an order is not justified since the savings due to a higher inventory level are not sufficient to justify incurring the ordering cost.

Part 3: $G_t(y_t)$ is non-increasing in the interval $(-\infty, \min\{y_t^*, y_t^0\})$.

The proof for this step is similar to the proof for Part 1. Once again, we consider two systems that have post-order inventory levels in period t equal to y_t^1 and y_t^2, respectively, such that $y_t^2 \leq y_t^1 \leq y_t^0$. We want to show that $G_t(y_t^1) \leq G_t(y_t^2)$.

In period t, since y_t^1 is closer to y_t^0 compared to y_t^2, $G^1(y_t^1) \leq G^1(y_t^2)$. After period t, we want system 1 to mimic system 2's operation. That is, if system 2 places an order, then system 1 will also do so. If system 2 does not place an order, neither does system 1. Suppose t' is used to denote the first period after period t in which system 2 places an order. Both systems see identical demands in periods $\{t+1, t+2, \ldots, t'-1\}$. Consequently,

$$y_t^0 \geq y_{t+1}^1 = x_{t+1}^1 = y_t^1 - D_t \geq y_t^2 - D_t = x_{t+1}^2 = y_{t+1}^2$$
$$y_t^0 \geq y_{t+2}^1 = x_{t+2}^1 = y_{t+1}^1 - D_{t+1} \geq y_{t+1}^2 - D_{t+1} = x_{t+2}^2 = y_{t+2}^2$$

$$\vdots$$

$$y_t^0 \geq y_{t'-1}^1 = x_{t'-1}^1 = y_{t'-2}^1 - D_{t'-2} \geq y_{t'-2}^2 - D_{t'-2} = x_{t'-1}^2 = y_{t'-1}^2.$$

Thus, in periods $t+1, t+2, \ldots, t'-1$, the post-order inventory in system 1 is always closer to y_t^0 compared to the post-order inventory in system 2. As a result, system 1 has lower expected holding and backorder costs than does system 2 for periods $t+1, t+2, \ldots, t'-1$. In period t', both system 1 and system 2 place an order and incur the ordering cost of K. Now, suppose system 2's post-order inventory level is $y_{t'}^2$. System 1 places an order such that its post order inventory $y_{t'}^1$ is also equal to $y_{t'}^2$. After period t', system 1 just imitates system 2. Since the expected costs in any period depend only on the post-order inventory levels, systems 1 and 2 have the same cost in periods $t', t'+1, \ldots, T$ (that is, $G_t(y_{t'}^1) = G_t(y_{t'}^2)$). Therefore,

$$
\begin{aligned}
G_t(y_t^1) - G_t(y_t^2) &= \left(G^1(y_t^1) + G^1(y_{t+1}^1) + \cdots + G^1(y_{t'-1}^1) \right) \\
&\quad - \left(G^1(y_t^2) + G^1(y_{t+1}^2) + \cdots + G^1(y_{t'-1}^2) \right) \\
&\leq 0.
\end{aligned}
$$

If system 2 does not place any order after period t, then system 1's expected costs will be lower than system 2's expected costs in every period through the end of the horizon.

Part 4: There exist s_t and S_t for any t such that it is optimal to order if and only if $x_t \leq s_t$ and the order-up-to level is S_t.

If an order is placed, the order-up-to level S_t has to be equal to y_t^* because that has the least cost. Now, when should we place an order? From Part 2, we know that an order is placed only if $x_t \leq \min\{y_t^*, y_t^0\}$. For an order to be placed, the savings due to having a higher inventory level have to be at least K. That is, $G_t(x_t) - G_t(y_t^*) \geq K$. Since $G_t(y_t)$ is non-increasing in y_t, if we can find one value of x_t (let that be x_t^1) for which an order must be placed, then for every $x_t < x_t^1$ there will be sufficient cost savings to justify placing an order. Indeed, s_t is the largest value of x_t for which an order is placed.

Suppose demand is described by a continuous random variable. Then s_t is the largest point where $s_t < S_t$ such that

$$G_t(s_t) - G_t(S_t) = K.$$

In the discrete demand random variable case, s_t is the largest integer satisfying $G_t(s_t) \geq G_t(S_t) + K$ and $s_t < S_t$.

10.5   A Heuristic Method for Calculating *s* and *S*

The algorithm presented in Section 10.3 is relatively complex. Hence, in many practical situations heuristics are employed to obtain solutions. We will discuss one such heuristic, which is attributed to Wagner [358].

Suppose that demand is independent and identically distributed from period to period. Assume the planning horizon is infinite in length and that the cost structure is unchanging over time. The costs considered in the model are fixed ordering costs, denoted by K; holding costs, denoted by h ($ per period per unit held in stock at the period's end); and backorder costs, denoted by b ($ per unit short at the end of a period). Positive procurement lead times are permitted, where τ represents the known, constant lead time, measured in periods. Finally, we assume demand over a lead time is normally distributed. Let μ and σ represent the per-period expected demand and standard deviation of demand, respectively.

The dynamics of the system are illustrated in the following figure. The lead time is assumed to be one period in length in the example. At the beginning of a period we check the value of the inventory position. If it is less than or equal to s, we place an order, thereby increasing its value to S. If it exceeds s, then no order is placed. In the figure we place orders only in periods 0, 4 and 7.

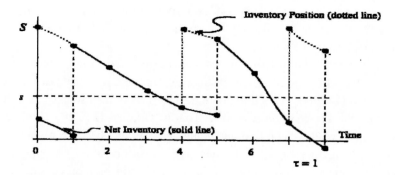

Fig. 10.1. Dynamics of the system.

The value of s must be chosen to account for demand over $\tau + 1$ periods and not just τ periods. For example, in period 2, in the figure, we do not place an order. The on-hand stock at that time must be adequate to cost-effectively meet demand not only in period 2 but also in period 3. This is the case because the next opportunity to place an order occurs at the beginning of period 3. If an order were placed at that time, those units of inventory would not be available for sale until the beginning of period 4. Thus, in general, we must consider demand over $\tau + 1$ periods when establishing the value of s.

As a consequence, we can write

$$s = (\tau + 1)\mu + u\sqrt{\tau + 1}\,\sigma,$$

where u is the number of standard deviations of demand over $\tau + 1$ time periods selected to provide safety stock. Note that $u\sqrt{\tau + 1}\,\sigma$ is not equal to the safety stock. For it to be so, the inventory position would have to equal s every time an order is placed, which, in general, is not the case. Furthermore, $u\sqrt{\tau + 1}\,\sigma$ is an upper bound on the safety stock.

Let

$$\Phi(u) = \int_{-\infty}^{u} \frac{1}{\sqrt{2\pi}} e^{-\frac{1}{2}x^2}\, dx,$$

the standard normal cumulative distribution function, and

$$G(u) = \int_{u}^{\infty} (x - u)\frac{1}{\sqrt{2\pi}} e^{-\frac{1}{2}x^2}\, dx,$$

the standard normal loss function.

The heuristic we are about to state has close ties to ideas presented in earlier chapters. First, we will see that the EOQ plays a prominent role. Second, we will see that the newsvendor model solution is also present.

Heuristic

Step 1: Compute $Q_E = \sqrt{\frac{2\mu K}{h}}$.

Step 2: Find u such that $G(u) = \frac{hQ_E}{b\sqrt{\tau + 1}\,\sigma}$.

Step 3: If $Q_E > 1.5\mu$, let

$$s = (\tau + 1)\mu + u\sqrt{\tau + 1}\,\sigma \text{ and}$$
$$S = s + Q_E;\text{ otherwise, go to Step 4.}$$

Step 4: Find v such that

$$\Phi(v) = \frac{b}{b + h}.$$

Furthermore, let

$$w = \min(u, v).$$

Set $s = (\tau + 1)\mu + w\sqrt{\tau + 1}\,\sigma$ and $S = (\tau + 1)\mu + \min\left[u\sqrt{\tau + 1}\,\sigma + Q, v\sqrt{\tau + 1}\,\sigma\right]$.

Observe that in Step 4, v is chosen to satisfy the newsvendor optimality condition. In Step 2, $\sqrt{\tau + 1}\,\sigma G(u)$ is equal to the expected number of backorders per cycle. Note also that if the value of Q_E is large so that no more than a single order is outstanding at the end of a period and the extent of the overshoot of s is usually small just prior to placing an order, then $1 - \frac{\sqrt{\tau+1}\,\sigma G(u)}{Q_E}$ is approximately equal to the fill rate. Finally, observe that when $w = v$, then an order will be placed in every period experiencing positive demand.

Let us now see how this heuristic operates. In our first example, suppose $h = 1, b = 20, K = 32, \tau = 0, \mu = 9$ and $\sigma^2 = 9$.

Step 1: $Q_E = \sqrt{\frac{2 \cdot 9 \cdot 32}{1}} = 24$.

Step 2: Find u such that $G(u) = \frac{2}{5}$. Hence $u = 0$.

Step 3: Since $Q_E > 1.5\mu = 13.5$,

$$s = 9 + 0.0 = 9$$

and

$$S = s + Q_E = 33.$$

Next, suppose $\mu = \sigma^2 = 36$. All the other parameter values in the preceding example remain unaltered. Then

Step 1: $Q_E = \sqrt{\frac{2 \cdot 36 \cdot 32}{1}} = 48$.

Step 2: Find u such that $G(u) = \frac{2}{5}$, or $u = 0$.

Step 3: Since $Q_E = 48 < 36 \cdot 1.5 = 54$, go to Step 4.

Step 4: Find v such that $\Phi(v) = .95$. Then $v = 1.645$.

$$w = \min(0, 1.645) = 0 \text{ so } s = 36 + 0.0 = 36$$

and

$$S = 36 + \min\{48, 9.87\} \approx 46.$$

10.6   Exercises

10.1. In Step 3 of the approximation algorithm in Section 10.2.1, show that an order is placed in period t only if the net inventory at the beginning of period t, x_t, is less than the demand in period t, d_t.

10.2. Consider the inventory management at a retailer of an electronic product whose product-life is estimated to be two years. For simplicity we divide this horizon into four decision-making epochs. The lead time is zero period in length. We are now in period 2 and the demand in this period is going to be 100 units. An order was placed in period 1 after which the inventory level became 250. The demand in period 1 was 200. Demands in periods 3 and 4 are independent and have the following distributions, respectively:

$$D_3 = 150 \text{ with probability } 0.75$$
$$= 50 \text{ with probability } 0.25$$
$$D_4 = 100 \text{ with probability } 0.75$$
$$= 50 \text{ with probability } 0.25.$$

Further, let the holding cost h be \$2 per unit per period and the backordering cost b be \$15 per unit per period. Should an order be placed if the order placement cost is \$400? If so, determine the order quantity.

10.3. In the previous problem, will your decision change if the order placement cost were \$200? Once again, determine the order quantity if necessary.

10.4. Solve Problems 10.2 and 10.3 by taking the demand to be normally distributed where the mean and standard deviation for periods 3 and 4 are the same as in Problem 10.2. That is, the mean for period 3 is $\mu_3 = 125$, and the standard deviation for period 3 is $\sigma_3 = 43.30$. The mean for period 4 is $\mu_4 = 87.5$, and the standard deviation for period 4 is $\sigma_4 = 21.65$.

10.5. Suppose a firm manages its inventory using a periodic review system. The fixed cost of placing an order is \$12 and the holding cost is \$1 per unit per period. Furthermore, the shortage cost per unit per period is \$5. The procurement lead time is zero.

(a) The demand in each period is 2 units with certainty. Use the algorithm described in Section 10.3 to compute the optimal policy

(b) Now suppose that the demand in each period is either one or two units with probability .2 and .8, respectively. Find the optimal values for s and S.

10.6. When determining whether or not to place an order in a periodic review management environment, a manager checks the inventory position at the beginning of each period. If it is a value of s or lower, an order is placed to increase its value to S. Thus, the inventory position random variable can assume only values $s+1, s+2, \ldots, S$ following the point in time after an ordering decision is made.

First, show how to represent the transitions of the inventory position random variable from period to period as a Markov chain. Assume that the demand process and costs are stationary. Lead times can be positive. Note the structure of the transition matrix.

Second, using the ideas presented in Section 10.2 for computing the function $t(y)$, show how the stationary probability distribution for the inventory position random variable just following the placing of an order can be calculated.

Third, using this distribution, find the probability distribution of the net inventory random variable at the end of a period.

Finally, find the probability distribution of the "overshoot" of s. That is, when an order is placed, what is the size of the order when following a given (s, S) policy?

10.7. Llenroc Electronics consumes a certain type of component at an average of 100 units per week. Usage varies from week to week. Data indicate that weekly demand is normally distributed with a standard deviation of 30 units. Usage is independent from week to week. Llenroc manages the component using an (s,S) policy. Each week the company reviews its inventory position and places an order if the inventory position is s or lower. The lead time for replenishment of stock from its supplier is three weeks. There is a fixed ordering cost of $900. The cost to carry a unit of the component is estimated to be $2 per week. Additionally, there is a shortage cost of $50 per unit per week due to the incremental shipping cost when units are not available when desired.

Use the method described in Section 10.5 to find s and S.

References

1. Afentakis, P., Gavish, B., Karmarkar, U.: Computationally efficient optimal solutions to the lot-sizing problem in multistage assembly systems. Management Science **30**, 222–239 (1984)

2. Aggarwal, A., Park, J.: Improved algorithms for economic lot-size problems. Operations Research **41**, 549–571 (1993)

3. Aggarwal, P.K., Moinzadeh, K.: Order expedition in multi-echelon production/distribution systems. IIE Transactions **26**(2), 86–96 (1994)

4. Aggarwal, S.: A review of current inventory theory and its applications. International Journal of Production Research **12**, 443–472 (1974)

5. Agrawal, V., Cohen, M.A., Zheng, Y.S.: Service parts logistics: A benchmark analysis. IIE Transactions, Special Issue on Supply Chain Co-ordination and Integration **29**(8), 627–639 (1997)

6. Tripp et al., R.: A decision support system for assessing and controlling the effectiveness of multi-echelon logistics actions. Interfaces **21**(4), 11–25 (1991)

7. Albright, S.C.: An approximation to the stationary distribution of a multi-echelon repairable-item inventory system. Naval Research Logistics **36**, 179–195 (1989)

8. Albright, S.C., Soni, A.: Markovian multi-echelon repairable inventory system. Naval Research Logistics Quarterly **35**, 49–61 (1988)

9. Alfredsson, P., Verrijdt, J.: Modeling emergency supply flexibility in a two-echelon inventory system. Management Science **45**(10), 1416–1431 (1999)

10. Allen, S.G.: Redistribution of total stock over several user locations. Naval Research Logistics Quarterly **5**, 51–59 (1958)

11. Allen, S.G.: A redistribution model with set-up charge. Management Science **8**(1), 99–108 (1961)

12. Allen, S.G.: Computation for the redistribution model with set-up charge. Management Science **8**(4), 482–489 (1962)

13. Allen, S.G., D'Esopo, D.A.: An ordering policy for repairable stock items. Operations Research **16**(3), 669–674 (1968)

14. Allen, S.G., D'Esopo, D.A.: An ordering policy for stock items when delivery can be expedited. Operations Research **16**(4), 880–883 (1968)

15. Anily, S., Federgruen, A.: One warehouse multiple retailer systems with vehicle routing costs. Management Science **36**, 92–114 (1990)

16. Archibald, B., Silver, E.: (s, S) policies under continuous review and discrete compound Poisson demands. Management Science **24**, 899–908 (1978)

17. Archibald, T., E., S.S.A., L.C., T.: An optimal policy for a two depot inventory problem with stock transfer. Management Science **43**(2), 173–183 (1997)

18. Aronis, K., Magou, I., Dekker, R., Tagaras, G.: Inventory control of spare parts using a Bayesian approach: A case study. European Journal of Operational Research **154**, 730–739 (2004)

19. Arrow, K., Harris, T., Marschak, J.: Optimal inventory policy. Econometrica **19**, 250–272 (1951)

20. Arrow, K., Karlin, S., Scarf, H.: Studies in Applied Probability and Management Science. Stanford University, Stanford, CA (1962)

21. Arrow, K.J., Karlin, S., Scarf, H. (eds.): Studies in the Mathematical Theory of Inventory and Production. Stanford University Press, Stanford, California (1958)

22. Askin, R.: A procedure for production lot sizing with probabilistic dynamic demand. AIIE Transactions **13**, 132–137 (1981)

23. Atkins, D., Sun, D.: 98%-effective lot sizing for series inventory systems with backlogging. Operations Research **43**, 335–345 (1995)

24. Aviv, Y., Federgruen, A.: Stochastic inventory models with limited production capacity and periodically varying parameters. Probability in the Engineering and Informational Sciences **11**, 107–135 (1997)

25. Axsäter, S.: Worst case performance for lot sizing heuristics. European Journal of Operations Research **9**, 339–343 (1982)

26. Axsäter, S.: Modeling emergency lateral transshipments in inventory systems. Management Science **36**(11), 1329–1338 (1990)

27. Axsäter, S.: Simple solution procedures for a class of two-echelon inventory problems. Operations Research **38**(1), 64–69 (1990)

28. Axsäter, S.: Exact and approximate evaluation of batch-ordering policies for two-level inventory systems. Operations Research **41**, 777–785 (1993)

29. Axsäter, S.: Inventory Control, first edn. Kluwer Academic Publishers (2000)

30. Axsäter, S.: Evaluation of unidirectional lateral transshipments and substitutions in inventory systems. European Journal of Operational Research **149**(2), 438–447 (2003)

31. Axsäter, S.: A new decision rule for lateral transshipments in inventory systems. Management Science **49**(9), 1168–1179 (2003)

32. Axsäter, S.: Note: Optimal policies for serial inventory systems under fill rate constraints. Management Science **49**(2), 247–253 (2003)

33. Axsäter, S., Rosling, K.: Installation vs. echelon stock policies for multilevel inventory control. Management Science **39**, 1274–1280 (1993)

34. Axsäter, S., Schneeweiss, C., Silver, E.: Multi-stage production planning and inventory control. Springer-Verlag, Berlin (1986)

35. Azoury, K.: Bayes solution to dynamic inventory models under unknown demand distribution. Management Science **31**, 1150–11,600 (1985)

36. Bagchi, U.: Modeling lead time demand for lumpy demand and variable lead time. Naval Research Logistics Quarterly **34**, 687–704 (1987)

37. Bagchi, U., Hayya, J., Chu, C.: The effect of leadtime variability: The case of independent demand. Journal of Operations Management **6**, 159–177 (1986)

38. Baker, K.: The inventory-queueing analogy and a Markovian production and inventory model. Opsearch **10**, 24–37 (1973)

39. Baker, K., Dixon, P., Magazine, M., Silver, E.: An algorithm for the dynamic lot size problem with time-varying production capacity constraints. Management Science **24**, 1710–1720 (1978)

40. Balana, A.R., Gross, D., Soland, R.M.: Optimal provisioning for single-echelon repairable item inventory control in a time varying environment. IIE Transactions **15**, 344–352 (1983)

41. Banerjee, A., Burton, J., Banerjee, S.: A simulation study of lateral shipments in single supplier, multiple buyers supply chain networks. International Journal of Production Economics **81-82**, 103–114 (2003)

42. Barankin, E.W.: A delivery-lag inventory model with an emergency provision. Naval Research Logistics Quarterly **8**, 285–311 (1961)

43. Beckmann, M.: An inventory model for arbitrary interval and quantity distributions of demand. Management Science **8**, 35–57 (1961)

44. Bessler, S., Veinott Jr, A.F.: Optimal policy for a dynamic multi-echelon inventory problem. Naval Research Logistics Quarterly **13**, 355–389 (1966)

45. Billington, P., McClain, J., Thomas, L.: Heuristics for multilevel lot-sizing with a bottleneck. Management Science **32**, 989–1006 (1986)

46. Bitran, G., Magnanti, T., Yanasse, H.: Approximation methods for the uncapacitated dynamic lot size problem. Management Science **30**, 1121–1140 (1984)

47. Bitran, G., Yanasse, H.: Computational complexity of the capacitated lot size problem. Management Science **28**, 1174–1186 (1982)

48. Bollapragada, S., Morton, T.: Myopic heuristics for the random yield problem (1994). Working paper. Carnegie-Mellon University, Pittsburgh

49. Bollapragada, S., Morton, T.: A simple heuristic for computing non-stationary (s, S) policies (1994). Working paper. Carnegie-Mellon University, Pittsburgh

50. Boyaci, T., Gallego, G.: Serial production/distribution systems under service constraints. Manufacturing and Service Operations Management **3**, 43–50 (2001)

51. Bradley, J., Robinson, L.: Improved base-stock approximations for independent stochastic lead times with order crossover. Manufacturing and Service Operations Management **7**, 319–329 (2005)

52. Brooks, R., Geoffrion, A.: Finding Everett's Lagrange multipliers by linear programming. Operations Research **14**(6), 1149–1153 (1966)

53. Brown, R.G.: Decision Rules for Inventory Management. Holt, Rinehart, and Winston, New York (1967)

54. Browne, S., Zipkin, P.: Inventory models with continuous, stochastic demands. Annals of Applied Probability **1**, 419–435 (1991)

55. Burns, L., Hall, R., Blumenfeld, D., Daganzo, C.: Distribution strategies that minimize transportation and inventory costs. Operations Research **33**, 469–490 (1985)

56. Buyukkurt, M.D., Parlar, M.: A comparison of allocation policies in a two-echelon repairable item inventory model. International Journal of Production Economics **29**, 291 (1993)

57. Buzacott, J., Shanthikumar, G.: Stochastic Models of Manufacturing Systems. Prentice-Hall, Englewood Cliffs, NJ (1992)

58. Cachon, G.: Competitive supply chain inventory management (1999). In Tayur et al.

59. Caggiano, K., Jackson, P., Muckstadt, J., Rappold, J.A.: Optimizing service parts inventory in a multiechelon, multi-item supply chain with time-based customer service-level agreements. Operations Research **55**, 303–318 (2007)

60. Caglar D., C.L., Simchi-Levi, D.: Two-echelon spare parts inventory system subject to a service constraint. IIE Transactions **36**(7), 655–666 (2004)

61. Carrillo, M.J.: Generalizations of Palm's theorem and dyna-METRIC's demand and pipeline variability. Report R-3698-AF, RAND Corporation, Santa Monica, California (1989)

62. Cetinkaya, S., Parlar, M.: Note: Optimality conditions for an (s, s) policy with proportional and lump-sum penalty costs. Management Science **48**, 1635–1639 (2002)

63. Chan, L., Muriel, A., Shen, Z.J., Simchi-Levi, D., Teo, C.P.: Effective zero-inventory-ordering policies for the single-warehouse multiretailer problem with piecewise linear cost structures. Management Science **48**, 1446–1460 (2002)

64. Chen, F., Song, J.: Optimal policies for multi-echelon inventory problems with Markov-modulated demand. Operations Research **49**(2), 226–234 (2001)

65. Chen, F., Zheng, Y.: Evaluating echelon stock (r, nq) policies in serial production/inventory systems with stochastic demand. Management Science **40**, 1262–1275 (1994)

66. Chen, F., Zheng, Y.: Lower bounds for multi-echelon stochastic inventory systems. Management Science **40**(11), 1426–1443 (1994)

67. Chen, F., Zheng, Y.: One-warehouse, multi-retailer systems with centralized stock information. Operations Research **45**, 275–287 (1997)

68. Chen, F., Zheng, Y.: Near-optimal echelon-stock (r, nq) policies in multi-stage serial systems. Operations Research **46**, 592–602 (1998)

69. Chen, X., Sim, M., Simchi-Levi, D., Sun, P.: Risk aversion in inventory management. Operations Research **55**, 828–842 (2007)

70. Clark, A., Scarf, H.: Optimal policies for a multi-echelon inventory problem. Management Science **6**, 476–490 (1960)

71. Clark, A.J.: An informal survey of multi-echelon inventory theory. Naval Research Logistics Quarterly **19**(4), 621–650 (1972)

72. Clark, A.J., Scarf, H.E.: Approximate solutions to a simple multi-echelon inventory problem. In: K.J. Arrow, S. Karlin, H. Scarf (eds.) Studies in Applied Probability and Management Science, chap. 5. Stanford University Press, Stanford, California (1962)

73. Cohen, M.A., Ernst, R.: Operations related groups (ORGs): A clustering procedure for production/inventory systems. Journal of Operations Management **9**(4), 574–598 (1990)

74. Cohen, M.A., Kleindorfer, P.R., Lee, H.L.: Optimal stocking policies for low usage items in multi-echelon inventory systems. Naval Research Logistics Quarterly **33**(1), 17–38 (1986)

75. Cohen, M.A., Kleindorfer, P.R., Lee, H.L.: Service constrained (s, S) inventory systems with priority demand classes and lost sales. Management Science **34**(4), 482–499 (1988)

76. Cohen, M.A., Kleindorfer, P.R., Lee, H.L.: Near optimal service constrained stocking policies for spare parts. Operations Research **37**(1), 104–117 (1989)

77. Cohen, M.A., Lee, H.L.: Strategic analysis of integrated production-distribution systems: Models and methods. Operations Research **36**(2), 216–228 (1988)

78. Crawford, G.: Variability in the demands for aircraft spare parts. Report R-3318-AF, RAND Corporation, Santa Monica, California (1988)

79. Crawford, G.B.: Palm's theorem for nonstationary processes. Report R-2750-RC, RAND Corporation, Santa Monica, CA (1981)

80. Crowston, W., Wagner, M.: Dynamic lot size models for multistage assembly systems. Management Science **20**, 14–21 (1973)

81. Dada, M.: A two-echelon inventory system with priority shipments. Management Science **38**(8), 1140–1153 (1992)

82. Dallery, Y., Gershwin, S.: Manufacturing flow line systems: A review of models and analytical results. Queueing systems **12**, 3–94 (1992)

83. Das, C.: Supply and redistribution rules for two-location inventory systems: One-period analysis. Management Science **21**(7), 765–776 (1975)

84. Demmy, W.S., Presutti, V.J.: Multi-echelon inventory theory in the air force logistics command. In: L.B. Schwarz (ed.) Multi-Level Production/Inventory Control Systems: Theory and Practice, *Studies in the Management Sciences*, vol. 16, pp. 279–297. North Holland Publishing, Amsterdam (1981)

85. Dhakar, T.S., Schmidt, C.P., Miller, D.M.: Base stock level determination for high cost low demand critical repairable spares. Computers and Operations Research **21**, 411–420 (1994)

86. Diaby, M., Bahl, H., Karwan, M., Zionts, S.: A Lagrangian relaxation approach for very-large-scale capacitated lot-sizing. Management Science **38**, 1329–1340 (1992)

87. Dvoretzky, A., Kiefer, J., Wolfowitz, J.: The inventory problem. Econometrica **20**, 187–222 (1952)

88. Ebeling, C.E.: Optimal stock levels and service channel allocations in a multi-item repairable asset inventory system. IIE Transactions **23**(2), 115–120 (1991)

89. Ehrhardt, R.: (s,S) policies for a dynamic inventory model with stochastic leadtimes. Operations Research **32**, 121–132 (1984)

90. Ehrhardt, R.: Easily computed approximations for (s,S) inventory system operating characteristics. Naval Research Logistics Quarterly **32**, 347–359 (1985)

91. Ehrhardt, R., Mosier, C.: A revision of the power approximation for computing (s,S) policies. Management Science **30**, 618–622 (1984)

92. Ehrhardt, R., Taube, L.: An inventory model with random replenishment quantities. International Journal of Production Research **25**, 1795–1804 (1987)

93. El-Najdawi, M.: A compact heuristic for common cycle lot-size scheduling in multi-stage, multi-product production processes. International Journal of Production Economics **27**, 29–42 (1992)

94. El-Najdawi, M., Kleindorfer, P.: Common cycle lot-size scheduling for multi-product, multi-stage production. Management Science **39**, 872–885 (1993)

95. Eppen, G., Gould, F., Pashigian, B.: Extensions of the planning horizon theorem in the dynamic lot size model. Management Science **15**, 268–277 (1969)

96. Eppen, G., Martin, R.: Solving multi-item capacitated lot-size problems using variable redefinition. Operations Research **35**, 832–848 (1987)

97. Eppen, G.D.: Effects of centralization on expected costs in a multi-location newsboy problem. Management Science **25**(5), 498–501 (1979)

98. Erkip, N., Hausman, W., Nahmias, S.: Optimal centralized ordering policies in multi-echelon inventory systems with correlated demands. Management Science **36**(3), 381–392 (1990)

99. Erlenkotter, D.: An early classic misplaced: Ford W. Harris' economic order quantity model of 1915. Management Science **35**, 898–900 (1989)

100. Erlenkotter, D.: Ford Whitman Harris and the economic order quantity model. Operations Research **38**, 937–946 (1990)

101. Evans, J.: An efficient implementation of the Wagner-Whitin algorithm for dynamic lot-sizing. Journal of Operations Management **5**, 229–235 (1985)

102. Everett III, H.: Generalized Lagrange multiplier method for solving problems of optimal allocation of resources. Operations Research **11**(3), 399–417 (1965)

103. Evers, P.T.: Hidden benefits of emergency transshipments. Journal of Business Logistics **18**(2), 55–77 (1997)

104. Evers, P.T.: Filling customer orders from multiple locations: A comparison of pooling methods. Journal of Business Logistics **20**(1), 121–140 (1999)

105. Faaland, B., Schmitt, T.: Cost-based scheduling of workers and equipment in a fabrication and assembly shop. Operations Research **41**, 253–268 (1993)

106. Federgruen, A.: Centralized planning models for multi-echelon inventory systems under uncertainty (1993). Chapter 3 in Graves et al.

107. Federgruen, A., Meissner, J., Tzur, M.: Progressive interval heuristics for multi-item capacitated lot-sizing problems. Operations Research **55**, 490–502 (2007)

108. Federgruen, A., Queyranne, M., Zheng, Y.: Simple power of two policies are close to optimal in a general class of production/distribution networks with general joint setup costs. Mathematics of Operations Research **17**, 951–963 (1992)

109. Federgruen, A., Tzur, M.: A simple forward algorithm to solve general dynamic lot sizing models with n periods in $O(n \log n)$ or $O(n)$ time. Management Science **37**, 909–925 (1991)

110. Federgruen, A., Tzur, M.: The dynamic lot sizing models with backlogging: A simple $O(n \log n)$ algorithm and minimal forecast horizon procedure. Naval Research Logistics **40**, 459–478 (1993)

111. Federgruen, A., Tzur, M.: Minimal forecast horizons and a new planning procedure for the general dynamic lot sizing model: Nervousness revisited. Operations Research **42**, 456–468 (1994)

112. Federgruen, A., Zheng, Y.: An efficient algorithm for computing an optimal (r, Q) policy in continuous review stochastic inventory systems. Operations Research **40**, 808–813 (1992)

113. Federgruen, A., Zheng, Y.: The joint replenishment problem with general joint cost structures. Operations Research **40**, 384–403 (1992)

114. Federgruen, A., Zheng, Y.: Efficient algorithms for finding optimal power-of-two policies for production/distribution systems with general joint setup costs. Operations Research **43**, 458–470 (1995)

115. Federgruen, A., Zipkin, P.: Allocation policies and cost approximations for multilocation inventory systems. Naval Research Logistics Quarterly **31**, 97–129 (1984)

116. Federgruen, A., Zipkin, P.: Approximations of dynamic, multilocation production and inventory problems. Management Science **30**(1), 69–84 (1984)

117. Federgruen, A., Zipkin, P.: An efficient algorithm for computing optimal (s, s) policies. Operations Research **32**(**6**), 1268–1285 (1984)

118. Federgruen, A., Zipkin, P.: An inventory model with limited production capacity and uncertain demands i: The average-cost criterion. Mathematics of Operations Research **11**(2), 193–207 (1986)

119. Federgruen, A., Zipkin, P.: An inventory model with limited production capacity and uncertain demands ii: The average-cost criterion. Mathematics of Operations Research **11**(2), 208–215 (1986)

120. Feeney, G.J., Sherbrooke, C.C.: The $(s - 1, s)$ inventory policy under compound Poisson demand. Management Science **12**(5), 391–411 (1966)

121. Fisher, W.W., Brennan, J.J.: The performance of cannibalization policies in a maintenance system with spares, repair, and resource constraints. Naval Research Logistics Quarterly **33**, 1–15 (1986)

122. Florian, M., Klein, M.: Deterministic production planning with concave costs and capacity constraints. Management Science **18**, 12–20 (1971)

123. Fox, B.L., Landi, D.M.: Searching for the multiplier in one-constraint optimization problems. Operations Research **18**(2), 253–262 (1970)

124. Fukuda, Y.: Optimal disposal policies. Naval Research Logistics Quarterly **8**, 221–227 (1961)

125. Gallego, G.: New bounds and heuristics for (Q, r) policies. Management Science **44**, 219–233 (1998)

126. Gallego, G., Moon, I.: The distribution-free newsboy problem: Review and extensions. Journal of the Operational Research Society **44**, 825–834 (1993)

127. Gallego, G., Ozer, O.: Integrating replenishment decisions with advance demand information. Management Science **47**, 1344–1360 (2001)

128. Gallego, G., Ozer, O.: Optimal replenishment policies for multiechelon inventory problems under advance demand information. Manufacturing and Service Operations Management **5**, 157–175 (2003)

129. Gallego, G., Ozer, O., Zipkin, P.: Bounds, heuristics and approximations for distribution systems. Operations Research **55**, 503–517 (2007)

130. Gallego, G., Zipkin, P.: Stock positioning and performance estimation in serial production-transportation systems. Manufacturing & Service Operations Management **1**, 77–88 (1999)

131. Galliher, H., Morse, P., Simond, M.: Dynamics of two classes of continuous review inventory systems. Operations Research **7**, 362–384 (1959)

132. Glasserman, P.: Bounds and asymptotics for planning critical safety stocks. Operations Research **45**(2), 244–257 (1997)

133. Glasserman, P., Tayur, S.: The stability of a capacitated, multi-echelon production-inventory system under a base-stock policy. Operations Research **42**(5), 913–925 (1994)

134. Glasserman, P., Tayur, S.: Sensitivity analysis for base-stock levels in multiechelon production-inventory systems. Management Science **41**(2), 263–281 (1995)

135. Glasserman, P., Tayur, S.: A simple approximation for a multistage capacitated production-inventory system. Naval Research Logistics **43**(1), 41–58 (1996)

136. Godfrey, G., Powell, W.: An adaptive, distribution-free algorithm for the newsvendor problem with censored demands, with applications to inventory and distribution. Management Science **47**, 1101–1112 (2001)

137. Graves, S.: Multi-stage lot-sizing: An iterative procedure (1981). Chapter 5 in Schwartz

138. Graves, S.: A multi-echelon inventory model for a repairable item with one-for-one replenishment. Management Science **31**(10), 1247–1256 (1985)

139. Graves, S.: A single-item inventory model for a nonstationary demand process. Manufacturing & Service Operations Management **1**, 50–61 (1999)

140. Graves, S., Kan, R., Zipkin, P.: Logistics of production and inventory. Handbooks in Operations Research and Management Science **4** (1993). Elsevier (North Holland), Amsterdam

141. Gross, D.: Centralized inventory control in multilocation supply systems. In: H.E. Scarf, et al. (eds.) Multistage Inventory Models and Techniques, chap. 3. Stanford University Press, Stanford, California (1963)

142. Gross, D.: On the ample service assumption of Palm's theorem in inventory modeling. Management Science **28**(9), 1065–1079 (1982)

143. Gross, D., Harris, C.M.: On one-for-one ordering inventory polices with state-dependent lead-times. Operations Research **19**(3), 735–760 (1971)

144. Gross, D., Harris, C.M.: Fundamentals of Queuing Theory (2nd ed.). John Wiley and Sons, New York (1985)

145. Gross, D., Ince, J.F.: Spares provisioning for a heterogeneous population. Technical Report T-376, Program in Logistics, The George Washington University (1978)

146. Gross, D., Kahn, H.D., Marsh, J.: Queuing models for spares provisioning. Naval Research Logistics Quarterly **24**(4), 521–536 (1977)

147. Gross, D., Kioussis, L.C., Miller, D.R.: A network decomposition approach for approximating the steady-state behavior of Markovian multi-echelon reparable item inventory system. Management Science **33**(11), 1453–1468 (1987)

148. Gross, D., Kioussis, L.C., Miller, D.R.: Transient behavior of large Markovian multi-echelon repairable item inventory systems using a truncated state space approach. Naval Research Logistics Quarterly **34**, 173–198 (1987)

149. Gross, D., Miller, D.R.: Multi-echelon repairable-item provisioning in a time varying environment using the randomization technique. Naval Research Logistics Quarterly **31**, 347–361 (1984)

150. Gross, D., Miller, D.R., Soland, R.M.: A closed queuing network model for multi-echelon repairable item provisioning. IIE Transactions **15**, 344–352 (1983)

151. Gross, D., Soriano, A.: On the economic application of airlift to product distribution and its impact on inventory levels. Naval Research Logistics Quarterly **19**, 501–507 (1972)

152. Gupta, A.: Approximate solution of a single-base multi-indentured repairable-item inventory system. Journal of the Operational Research Society **44**, 701–710 (1993)

153. Gupta, A., Albright, S.C.: Steady-state approximations for a multi-echelon multi-indentured repairable-item inventory system. European Journal of Operational Research **62**, 340–353 (1992)

154. Gurbuz, M., Moinzadeh, K., Zhou, Y.P.: Coordinated replenishment strategies in inventory/distribution systems. Management Science **53**, 293–307 (2007)

155. Haber, S.E., Sitgreaves, R.: An optimal inventory model for the intermediate echelon when repair is possible. Management Science **21**(6), 638–648 (1975)

156. Haber, S.E., Sitgreaves, R., Solomon, H.: A demand prediction technique for items in military inventory systems. Naval Research Logistics Quarterly **16**, 297–308 (1975)

157. Hadley, G., Whitin, T.M.: Analysis of Inventory Systems. Prentice-Hall, Englewood Cliffs, N.J. (1963)

158. Hadley, G., Whitin, T.M.: An inventory transportation model with N locations. In: Scarf, Gilford, Shelly (eds.) Multistage Inventory Models and Techniques, chap. 5. Stanford University Press, Stanford, California (1963)

159. Hamann, T., Proth, J.: Inventory control of repairable tools with incomplete information. International Journal of Production Economics **31**, 543–550 (1993)

160. Hanssmann, F.: Operations Research in Production and Inventory Control. Wiley, New York (1962)

161. Hariharan, R., Zipkin, P.: Customer-order information, leadtimes, and inventories. Management Science **41**(10), 1599–1607 (1995)

162. Harris, F.: How many parts to make at once. Factory: The Magazine of Management **10**, 135–136 (1913)

163. Hausman, W.H.: Communication: On optimal repair kits under a job completion criterion. Management Science **28**(11), 1350–1351 (1982)

164. Hausman, W.H., Scudder, G.D.: Priority scheduling rules for repairable inventory systems. Management Science **28**(11), 1215–1232 (1982)

165. Henig, M., Gerchak, Y.: The structure of periodic review policies in the presence of random yield. Operations Research **38**, 634–643 (1990)

166. Herer, Y., Roundy, R.: Heuristics for a one-warehouse multiretailer distribution problem with performance bounds. Operations Research **45**, 102–115 (1997)

167. Herron, D.: A comparison of techniques for multi-item inventory analysis. Production and Inventory Management pp. 103–115 (1978)

168. Heyman, D.P.: Return policies for an inventory system with positive and negative demands. Naval Research Logistics Quarterly **25**(4), 581–596 (1978)

169. Higa, I., A, G., Machado, A.: Waiting time in an $(S-1,S)$ inventory system. Operations Research **23**, 674–680 (1975)

170. Hoadley, B., Heyman, D.P.: A two-echelon inventory model with purchases, dispositions, shipments, returns, and transshipments. Naval Research Logistics Quarterly **24**, 1–19 (1977)

171. van Hoesel, S., Romeijn, H., Morales, D., Wagelmans, A.: Integrated lot sizing in serial supply chains with production capacities. Management Science **51**, 1706–1719 (2005)

172. Hopp, W., Spearman, M., Zhang, R.Q.: Easily implementable inventory control policies. Operations Research **45**(3), 327–340 (1997)

173. van Houtum, G.J., Scheller-Wolf, A., Yi, J.: Optimal control of serial inventory systems with fixed replenishment intervals. Operations Research **55**, 674–687 (2007)

174. Howard, J.V.: Service exchange systems–the stock control of repairable items. Journal of the Operational Research Society **35**, 235–245 (1984)

175. Hsu, V., Lowe, T.: Dynamic economic lot size models with period-pair-dependent backorder and inventory costs. Operations Research **49**, 316–321 (2001)

176. Huggins, E., Olsen, T.: Supply chain management with guaranteed delivery. Management Science **49**(9), 1154–1167 (2003)

177. Huh, W., Janakiraman, G.: Optimality results in inventory-pricing control: An alternate approach (2006). To Appear in Operations Research as a Technical Note

178. Huh, W., Janakiraman, G.: Optimality of (s,S) policies. Management Science **54**, 139–150 (2008)

179. Iglehart, D.: Dynamic programming and stationary analysis in inventory problems (1963)

180. Iglehart, D.: Optimality of (s,S) policies in the infinite horizon dynamic inventory problem. Management Science **9**, 259–267 (1963)

181. Iglehart, D., Karlin, S.: Optimal policy for dynamic inventory process with nonstationary stochastic demands (1962)

182. Jackson, P., Maxwell, W., Muckstadt, J.: The joint replenishment problem with powers of two restrictions. AIIE Transactions **17**, 25–32 (1985)

183. Jackson, P., Maxwell, W., Muckstadt, J.: Determining optimal reorder intervals in capacitated production distribution systems. Management Science **34**, 938–958 (1988)

184. Jackson, P.L.: Stock allocation in a two-echelon distribution system or 'what to do until your ship comes in'. Management Science **34**(7), 880–895 (1988)

185. Jackson, P.L., Muckstadt, J.A.: Risk pooling in a two-period, two-echelon inventory stocking and allocation problem. Naval Research Logistics **36**(1), 1–26 (1989)

186. Janakiraman, G., Muckstadt, J.A.: A decomposition approach for a class of capacitated serial systems,. Tech. rep.

187. Janakiraman, G., Muckstadt, J.A.: Extending the 'single unit - single customer' approach to capacitated systems. Technical Report 1360, School of Operations Research, Cornell University, Ithaca, NY (2003). Under revision for Operations Research

188. Janakiraman, G., Muckstadt, J.A.: Optimality of multi-tier base-stock policies for a class of capacitated serial systems. Technical Report 1361, School of Operations Research, Cornell University, Ithaca, NY (2003). Submitted to Operations Research

189. Janakiraman, G., Seshadri, S., Shanthikumar, J.: A comparison of the optimal costs of two canonical inventory systems. Operations Research **55**, 866–875 (2007)

190. Johansen, S., Thorstenson, A.: Optimal and approximate (Q, r) inventory policies with lost sales and gamma-distributed lead time. International Journal of Production Economics **30**, 179–194 (1993)

191. Johansen, S., Thorstenson, A.: Optimal (r, Q) inventory policies with Poisson demands and lost sales: Discounted and undiscounted cases. International Journal of Production Economics **46**, 359–371 (1996)

192. Johnson, G., Thompson, H.: Optimality of myopic inventory policies for certain dependent demand processes. Management Science **21**, 1303–1307 (1975)

193. Johnson, M., Lee, H., Davis, T., Hall, R.: Expressions for item fill rates in periodic inventory systems. Naval Research Logistics **42**, 57–80 (1995)

194. Jones, P., Inman, R.: When is the economic lot scheduling problem easy? IIE Transactions **21**, 11–20 (1989)

195. Jung, W.: Recoverable inventory systems with time-varying demand. Production and Inventory Management Journal **34**, 77–81 (1993)

196. Kalchschmidt, M., Zotteri, G., Verganti, R.: Inventory management in a multi-echelon spare parts supply chain. International Journal of Production Economics **81-82**, 397–414 (2003)

197. Kaplan, A.J.: Incorporating redundancy considerations into stockage models. Naval Research Logistics **36**, 625–638 (1989)

198. Kaplan, R.: A dynamic inventory model with stochastic lead times. Management Science **16**(7), 491–507 (1970)

199. Karlin, S.: Optimal inventory policy for the Arrow-Marschak dynamic model (1958). Chapter 9 in Arrow et al.

200. Karlin, S.: Dynamic inventory policy with varying stochastic demands. Management Science **6**, 231–258 (1960)

201. Karlin, S., Scarf, H.: Inventory models of the Arrow-Harris-Marschak type with time lag. In: K.J. Arrow, S. Karlin, H. Scarf (eds.) Studies in the Mathematical Theory of Inventory and Production, chap. 10. Stanford University Press, Stanford, California (1958)

202. Karmarkar, U.: Policy structure in multi-state production/inventory problems: An application of convex analysis (1981). Chapter 16 in Schwarz

203. Karmarkar, U.: The multilocation multiperiod inventory problem: Bounds and approximations. Management Science **33**, 86–94 (1987)

204. Karmarkar, U., Kekre, S., Kekre, S.: The dynamic lot-sizing problem with startup and reservation costs. Operations Research **35**, 389–398 (1987)

205. Karmarkar, U.S., Patel, N.R.: The one-period, N-location distribution problem. Naval Research Logistics Quarterly **24**, 559–575 (1977)

206. Kim, J., Shin, K., Yu, H.: Optimal algorithm to determine the spare inventory level for a repairable-item inventory system. Computers and Operations Research **23**, 289–297 (1996)

207. Kimball, G.: General principles of inventory control. Journal of Manufacturing and Operations Management **1**, 119–130 (1988)

208. Krishnan, K.S., Rao, V.R.K.: Inventory control in N warehouses. Journal of Industrial Engineering **16**, 212–215 (1965)

209. Kruse, W.K.: Waiting time in an $(S-1,S)$ inventory system with arbitrarily distributed lead-times. Operations Research **28**(2), 348–352 (1980)

210. Kruse, W.K.: Waiting time in a continuous review (s,S) inventory system with constant lead times. Operations Research **29**(1), 202–207 (1981)

211. Kukreja, A., Schmidt, C., Miller, D.: Stocking decisions for low-usage items in a multilocation inventory system. Management Science **47**, 1371–1383 (2001)

212. Kunnumkal, S., Topaloglu, H.: Using stochastic approximation methods to compute optimal base-stock levels in inventory control problems. Operations Research pp. Published online before print Jan 14, 2008

213. Lee, H., Nahmias, S.: Single-product, single-location models (1993). Chapter 1 in Graves et al.

214. Lee, H.L.: A multi-echelon inventory model for repairable items with emergency lateral transshipments. Management Science **33**(10), 1302–1316 (1987)

215. Lee, H.L., Moinzadeh, K.: Operating characteristics of a two-echelon system for repairable and consumable items under batch operating policy. Naval Research Logistics **34**, 365–380 (1987)

216. Lee, H.L., Moinzadeh, K.: A repairable item inventory system with diagnostic and repair service. European Journal of Operational Research **40**, 210–221 (1989)

217. Levi, R., Pal, M., Roundy, R., Shmoys, D.: Approximation algorithms for stochastic inventory control models. Mathematics of Operations Research **32**, 284–302 (2007)

218. Lim, W.S., Ou, J., Teo, C.P.: Inventory cost effect of consolidating several one-warehouse multiretailer systems. Operations Research **51**, 668–672 (2003)

219. Lippman, S., McCardle, K.: The competitive newsboy. Operations Research **45**, 54–65 (1977)

220. Little, J.: A proof of the queueing formula $l = \lambda w$. Operations Research **9**, 383–387 (1961)

221. Love, S.: A facilities in series model with nested schedules. Management Science **18**, 327–338 (1972)

222. Lovejoy, W.: Stopped myopic policies in some inventory models with generalized demand processes. Management Science **38**, 688–707 (1992)

223. Lundin, R., Morton, T.: Planning horizons for the dynamic lot size model. Operations Research **23**, 711–734 (1975)

224. Maes, J., Wassenhove, L.V.: Multi-item capacitated dynamic lot-sizing heuristics: A general review. Journal of the Operational Research Society **39**, 991–1004 (1988)

225. Mamer, J.W., Smith, S.A.: Optimizing field repair kits based on job completion rate. Management Science **28**(11), 1328–1333 (1982)

226. Matta, K.F.: A simulation model for repairable items/spare parts inventory systems. Computers and Operations Research **12**, 395–409 (1985)

227. Maxwell, W., Muckstadt, J.: Establishing consistent and realistic reorder intervals in production-distribution systems. Operations Research **33**, 1316–1341 (1985)

228. Miller, B.L.: Dispatching from depot repair in a recoverable item inventory system: On the optimality of a heuristic rule. Management Science **21**(3), 316–325 (1974)

229. Miller, B.L., Modarres-Yazdi, M.: The distribution of recoverable inventory items from a repair center when the number of consumption centers is large. Naval Research Logistics Quarterly **25**(4), 597–604 (1978)

230. Mitchell, J.: 98%-effective lot-sizing for one warehouse, multi-retailer inventory systems with backlogging. Operations Research **35**, 399–404 (1987)

231. Moinzadeh, K., Aggarwal, P.K.: An information based multiechelon inventory system with emergency orders. Operations Research **45**, 694–701 (1997)

232. Moinzadeh, K., Lee, H.L.: Batch size and stocking levels in multi-echelon repairable systems. Management Science **32**(12), 1567–1581 (1986)

233. Moinzadeh, K., Nahmias, S.: A continuous review model for an inventory system with two supply modes. Management Science **34**(6), 761–773 (1988)

234. Moinzadeh, K., Schmidt, C.P.: An $(S-1,S)$ inventory system with emergency orders. Operations Research **39**(2), 308–321 (1991)

235. Monahan, J.: A quantity discount pricing model to increase vendor profits. Management Science **30**, 720–726 (1984)

236. Morse, P.: Queues, Inventories and Maintenance. Wiley, New York (1958)

237. Morton, T.: The near-myopic nature of the lagged-proportional-cost inventory problem with lost sales. Operations Research **19**, 1708–1716 (1971)

238. Morton, T., Pentico, D.: The finite-horizon non-stationary stochastic inventory problem. Near-myopic bounds, heuristics, testing. Management Science **41**, 334–343 (1995)

239. Muckstadt, J.: Analysis and Algorithms for Service Parts Supply Chains, first edn. Springer (2005)

240. Muckstadt, J., Roundy, R.: Analysis of multistage production systems (1993). Chapter 3 in Graves et al.

241. Muckstadt, J.A.: A model for a multi-item, multi-echelon, multi-indenture inventory system. Management Science **20**(4), 472–481 (1973)

242. Muckstadt, J.A.: Navmet: a four-echelon model for determining the optimal quality and distribution of navy spare aircraft engines. Technical Report 263, School of Operations Research and Industrial Engineering, Cornell University, Ithaca, NY (1976)

243. Muckstadt, J.A.: Some approximations in multi-item, multi-echelon inventory systems for recoverable items. Naval Research Logistics Quarterly **25**(3), 377–394 (1978)

244. Muckstadt, J.A.: A three-echelon, multi-item model for recoverable items. Naval Research Logistics Quarterly **26**(2), 199–221 (1979)

245. Muckstadt, J.A.: Comparative adequacy of steady-state versus dynamic models for calculating stockage requirements. Report R-2636-AF, RAND Corporation, Santa Monica, California (1980)

246. Muckstadt, J.A.: A multi-echelon model for indentured, consumable items. Technical Report 548, School of Operations Research, Cornell University, Ithaca, NY (1982)

247. Muckstadt, J.A., Issac, M.H.: An analysis of a single item inventory system with returns. Naval Research Logistics Quarterly **28**, 237–254 (1981)

248. Muckstadt, J.A., Thomas, L.J.: Are multi-echelon inventory methods worth implementing in systems with low-demand-rate items? Management Science **26**(5), 483–494 (1980)

249. Muharremoglu, A., Tsitsiklis, J.: Echelon base stock policies in uncapacitated serial inventory systems (2001). Http://web.mit.edu/jnt/www/publ.html

250. Muharremoglu, A., Tsitsiklis, J.: A single-unit decomposition approach to multi-echelon inventory systems (2007). July 2001 (revised September 2007)Operations Research, forthcoming

251. Naddor, E.: Optimal and heuristic decisions in single and multi-item inventory systems. Management Science **21**, 1234–1249 (1975)

252. Nahmias, S.: Simple approximations for a variety of dynamic leadtime lost-sales inventory models. Operations Research **27**(5), 904–924 (1979)

253. Nahmias, S.: Managing reparable item inventory systems: A review. In: L.B. Schwarz (ed.) Multi-Level Production/Inventory Control Systems: Theory and Practice, *Studies in the Management Sciences*, vol. 16, pp. 253–278. North Holland Publishing, Amsterdam (1981)

254. Nahmias, S.: Perishable inventory theory: A review. Operations Research **30**, 680–708 (1982)

255. Nahmias, S.: Demand estimation in lost sales inventory systems. Naval Research Logistics **41**, 739–757 (1994)

256. Nahmias, S.: Production and Operations Analysis. McGraw-Hill/Irwin, fifth edition (2004)

257. Nahmias, S., Rivera, H.: A deterministic model for a repairable item inventory system with a finite repair rate. International Journal of Production Research **17**, 215–221 (1976)

258. Needham, P.M.: The influence of individual cost factors on the use of emergency transshipments. Transportation Research, Part E, Logistics and Transportation Review **34**, 149–161 (1998)

259. O'Malley, T.J.: The aircraft availability model: Conceptual framework and mathematics. Report, Logistics Management Institute, Washington, D.C. (1983)

260. Orlicky, J.: Material Requirement Planning. McGraw-Hill, New York, NY (1975)

261. Ozer, O.: Replenishment strategies for distribution systems under advanced demand information. Management Science **49**(3), 255–272 (2003)

262. Ozer, O., Wei, W.: Inventory control with limited capacity and advance demand information. Operations Research **52**, 988–1000 (2004)

263. Palm, C.: Analysis of the Erlang traffic formulae for busy-signal arrangements. Ericsson Technics **5**, 39–58 (1938)

264. Pena Perez, A., Zipkin, P.: Dynamic scheduling rules for a multiproduct make-to-stock queue. Operations Research **45**, 919–930 (1997)

265. Platt, D., Robinson, L., Freund, R.: Tractable (Q,R) heuristic models for constrained service levels. Management Science **43**, 951–965 (1997)

266. Porteus, E.: Investing in reduced setups in the EOQ model. Management Science **31**, 998–1010 (1985)

267. Porteus, E.: Investing in new parameter values in the discounted EOQ model. Naval Research Logistics Quarterly **33**, 39–48 (1986)

268. Porteus, E.: Optimal lot sizing, process quality improvement and setup cost reduction. Operations Research **34**, 137–144 (1986)

269. Porteus, E.: Foundations of Stochastic Inventory Theory, first edn. Stanford University Press (2002)

270. Porteus, E., Landsdowne, Z.: Optimal design of a multi-item, multi-location, multi-repair type repair and supply system. Naval Research Logistics Quarterly **21**(2), 213–238 (1974)

271. Presutti, V.J., Trepp, R.C.: More ado about economic order quantities (EOQ). Naval Research Logistics Quarterly **17**, 243–251 (1970)

272. Pyke, D.F.: Priority repair and dispatch policies for repairable-item logistics systems. Naval Research Logistics **37**, 1–30 (1990)

273. Ramaswami, V., Neuts, M.: Some explicit formulas and computational methods for infinite-server queues with phase-type arrivals. Journal of Applied Probability **17**, 498–514 (1980)

274. Rao, A.: A survey of MRP II software suppliers' trends in support of JIT. Production and Inventory Management **30**, (3rd quarter) 14–17 (1989)

275. Rao, A.: Convexity and sensitivity properties of the (R, T) inventory control policy for stochastic demand models (1994). Working paper, Cornell University, Ithaca, NY

276. Rao, U.: Properties of the periodic review (r, t) inventory control policy for stationary, stochastic demand. Manufacturing and Service Operations Management **5**, 37–53 (2003)

277. Rappold, J.A., Muckstadt, J.A.: A computationally efficient approach for determining inventory levels in a capacitated multi-echelon production-distribution system. Naval Research Logistics **47**(5), 377–398 (2000)

278. Resh, M., Friedman, M., Barbosa, L.: On a general solution of the deterministic lot size problem with time-proportional demand. Operations Research **24**, 718–725 (1976)

279. Richards, F.R.: Comments on the distribution of inventory position in a continuous-review (s, S) inventory system. Operations Research **23**, 366–371 (1975)

280. Richards, F.R.: A stochastic model of a repairable-item inventory system with attrition and random lead times. Operations Research **24**(1), 118–130 (1976)

281. Robinson, L., Bradley, J., Thomas, L.: Consequences of order crossover under order-up-to inventory policies. Manufacturing and Service Operations Management **3**, 175–188 (2001)

282. Robinson, L.W.: Optimal and approximate policies in multiperiod multilocation inventory models with transshipment. Operations Research **38**(2), 278–295 (1990)

283. Rose, M.: The $(s-1, s)$ inventory model with arbitrary backordered demand and constant delivery times. Operations Research **20**(5), 1020–1032 (1972)

284. Rosling, K.: Optimal lot-sizing for dynamic assembly systems (1986). Chapter 7 in Axsäter et al.

285. Rosling, K.: Optimal inventory policies for assembly systems under random demands. Operations Research **37**, 565–579 (1989)

286. Rosling, K.: Inventory cost rate functions with nonlinear shortage costs. Operations Research **50**, 1007–1017 (2002)

287. Roundy, R.: 98% effective integer-ratio lot-sizing for one warehouse multi-retailer systems. Management Science **31**, 1416–1430 (1985)

288. Roundy, R.: 98% effective lot-sizing rule for a multi-product, multi-stage production inventory system. Mathematics of Operations Research **11**, 699–727 (1986)

289. Roundy, R.: Efficient, effective lot sizing for multistage production systems. Operations Research **41**, 371–385 (1989)

290. Roundy, R.: Rounding off to powers of two in continuous relaxations of capacitated lot sizing problems. Management Science **35**, 1433–1442 (1989)

291. Roundy, R.O., Muckstadt, J.A.: Heuristic computation of periodic-review base stock inventory policies. Management Science **46**(1), 104–109 (2000)

292. Rustenburg, J., van Houtum, G., Zijm, W.: Spare parts management for technical systems: resupply of spare parts under limited budgets. IIE Transactions **32**(10), 1013–1026 (2000)

293. Sahin, I.: On the stationary analysis of continuous review (s, S) inventory systems with constant lead times. Operations Research **27**, 717–729 (1979)

294. Sahin, I.: On the objective function behavior in (s, S) inventory models. Operations Research **30**, 709–725 (1982)

295. Sahin, I.: On the continuous review (s, S) inventory model under compound renewal demand and random lead times. Journal of Applied Probability **20**, 213–219 (1983)

296. Scarf, H.: A min-max solution of an inventory problem (1958). Chapter 12 in Arrow et al.

297. Scarf, H.: Stationary operating characteristics of an inventory model with time lag (1958). Chapter 16 in Arrow et al.

298. Scarf, H.: The optimality of (s,S) policies in the dynamic inventory problem (1960). Chapter 13 in Arrow et al.

299. Scarf, H., Guilford, D., Shelly, M.: Multistage Inventory Models and Techniques. Stanford University, Stanford, CA (1963)

300. Schaefer, M.K.: A multi-item maintenance center inventory model for low-demand repairable items. Management Science **29**(9), 1062–1068 (1983)

301. Schaefer, M.K.: Replenishment policies for inventories of recoverable items with attrition. Omega **17**, 281–287 (1989). Oxford

302. Schmidt, C., Nahmias, S.: Optimal policy for a two-stage assembly system under random demand. Operations Research **33**, 1130–1145 (1985)

303. Schrady, D.A.: A deterministic inventory model for repairable items. Naval Research Logistics Quarterly **14**(3), 391–398 (1967)

304. Schultz, C.R.: On the optimality of the $(s-1,s)$ policy. Naval Research Logistics **37**, 715–723 (1990)

305. Schwartz, L.B. (ed.): Multi-Level Production/Inventory Systems: Theory and Practice, *Studies in the Management Sciences*, vol. 16. North Holland Publishing, New York (1981)

306. Scudder, G.D.: Priority scheduling and spares stocking policies for a repair shop: The multiple failure case. Management Science **30**(6), 739–749 (1984)

307. Scudder, G.D., Hausman, W.: Spares stocking policies for repairable items with dependent repair times. Naval Research Logistics Quarterly **29**, 303–322 (1982)

308. Sethi, S., Cheng, F.: Optimality of (s,S) polices in inventory models with Markovian demand. Operations Research **45**, 931–939 (1997)

309. Shang, K., Song, J.S.: Newsvendor bounds and heuristic for optimal policies in serial supply chains. Management Science **49**, 618–638 (2003)

310. Shang, K., Song, J.S.: A closed-form approximation for serial inventory systems and its application to system design. Manufacturing and Service Operations Management **8**, 394–406 (2006)

311. Shang, K., Song, J.S.: Serial supply chains with economies of scale: Bounds and approximations. Operations Research **55**, 843–853 (2007)

312. Shanker, K.: Exact analysis of a two-echelon inventory system for recoverable items under batch inspection policy. Naval Research Logistics Quarterly **28**, 579–601 (1981)

313. Sherbrooke, C.C.: Generalization of a queuing theorem of Palm to finite populations. Management Science **12**(11), 907–908 (1966)

314. Sherbrooke, C.C.: METRIC: A multi-echelon technique for recoverable item control. Operations Research **16**(1), 122–141 (1968)

315. Sherbrooke, C.C.: An evaluator for the number of operationally ready aircraft in a multilevel supply system. Operations Research **19**(3), 618–635 (1971)

316. Sherbrooke, C.C.: Waiting time in an $(s-1,s)$ inventory system-constant service time case. Operations Research **23**(4), 819–820 (1975)

317. Sherbrooke, C.C.: VARI-METRIC: Improved approximations for multi-indenture, multi-echelon availability models. Operations Research **34**(2), 311–319 (1986)

318. Sherbrooke, C.C.: Multi-echelon inventory systems with lateral supply. Naval Research Logistics **39**, 29–40 (1992)

319. Sherbrooke, C.C.: Optimal Inventory Modeling of Systems: Multi-Echelon Techniques, second edition. Springer, New York (2004)

320. Silver, E.A.: Inventory allocation among an assembly and its repairable subassemblies. Naval Research Logistics Quarterly **19**, 261–280 (1972)

321. Silver, E.A., Meal, H.: A heuristic for selecting lot size quantities for the case of a deterministic time-varying demand rate and discrete opportunities for replenishment. Production and Inventory Management pp. (2nd quarter) 64–74 (1973)

322. Silver, E.A., Peterson, R.: Decision Systems for Inventory Management and Production Planning. Wiley, New York, NY (1985)

323. Simon, R.M.: Stationary properties of a two-echelon inventory model for low demand items. Operations Research **19**(3), 761–773 (1971)

324. Simon, R.M., D'Esopo, D.A.: Comments on a paper by S.G. Allen and D.A. D'Esopo: 'an ordering policy for repairable stock items'. Operations Research **19**(4), 986–988 (1971)

325. Simpson Jr, K.E.: A theory of allocation of stocks to warehouses. Operations Research **7**(6), 797–805 (1959)

326. Simpson, V.P.: Optimum solution structure for a repairable inventory problem. Operational Research **26**(2), 270–281 (1978)

327. Singhal, V.: Inventories, risk and the value of the firm. Journal of Manufacturing and Operations Management **1**, 4–43 (1988)

328. Sivazlian, B.: A continuous review (s,S) inventory system with arbitrary interarrival distribution between unit demand. Operations Research **22**, 65–71 (1974)

329. Slay, F.M.: Lateral resupply in a multi-echelon inventory system. Report AF501-2, Logistics Management Institute, Washington, D.C. (1986)

330. Smith, C.H., Schaefer, M.K.: Optimal inventories for repairable redundant systems with aging components. Journal of Operations Management **5**, 339–349 (1985)

331. Smith, S.A.: Optimal inventories for an $(S-1,S)$ system with no backorders. Management Science **23**(5), 522–528 (1977)

332. Smith, S.A., Chambers, J.C., Shlifer, E.: Optimal inventories based on job completion rate for repairs requiring multiple items. Management Science **26**(8), 849–852 (1980)

333. Sobel, M.: Fill rates of single-stage and multistage supply systems. Manufacturing and Service Operations Management **6**(1), 41–52 (2004)

334. Sobel, M., Heyman, D.: Stochastic Models in Operations Research, Vol. I: Stochastic Processes and Operating Characteristics. Dover Publications (2003)

335. Sobel, M., Heyman, D.: Stochastic Models in Operations Research, Vol. II: Stochastic Optimization. Dover Publications (2003)

336. Song, J., Zipkin, P.: Evaluation of base-stock policies in multiechelon inventory systems with state-dependent demands. Naval Research Logistics **39**, 715–728 (1992)

337. Song, J., Zipkin, P.: Inventory control in a fluctuating demand environment. Operations Research **41**, 351–370 (1993)

338. Song, J., Zipkin, P.: Inventory control with information about supply conditions. Management Science **42**, 1409–1419 (1993)

339. Song, J., Zipkin, P.: Managing inventory with the prospect of obsolescence. Operations Research **44**, 215–222 (1993)

340. Starr, M., Miller, D.: Inventory Control: Theory and Practice. Prentice-Hall, Englewood Cliffs, NJ (1962)

341. Svoronos, A., Zipkin, P.: Estimating the performance of multi-level inventory systems. Operations Research **36**(1), 57–72 (1988)

342. Svoronos, A., Zipkin, P.: Evaluation of one-for-one replacement policies for multiechelon inventory systems. Management Science **37**(1), 68–83 (1991)

343. Tagaras, G.: Effects of pooling on the optimization and service levels of two-location inventory systems. IIE Transactions **21**(3), 250–258 (1989)

344. Tagaras, G., Cohen, M.A.: Pooling in two-location inventory systems with non-negligible replenishment lead times. Management Science **38**(8), 1067–1083 (1992)

345. Tagaras, G., Vlachos, D.: A periodic review inventory system with emergency replenishments. Management Science **47**(3), 415–429 (2001)

346. Tayur, S.: Computing the optimal policy in capacitated inventory models. Stochastic Models **9**(4) (1993)

347. Tayur, S., Ganeshan, R., Magazine, M.: Quantitative Models for Supply Chain Management. Kluwer, Norwell, MA (1999)

348. Thomas, D.: Measuring item fill-rate performance in a finite horizon. Manufacturing and Service Operations Management **7**, 74–80 (2005)

349. Thonemann, U., Brown, A., Hausman, W.: Easy quantification of improved spare parts inventory policies. Management Science **48**, 1213–1225 (2002)

350. Treharne, J., Sox, C.: Adaptive inventory control for nonstationary demand and partial information. Management Science **48**, 607–624 (2002)

351. Veinott, A.: On the optimality of (s,S) inventory policies: New conditions and a new proof. Siam Journal on Applied Mathematics **14**, 1067–1083 (1966)

352. Veinott, A.: The status of mathematical inventory theory. Management Science **12**, 745–777 (1966)

353. Veinott, A.: Optimal policy for a multi-product, dynamic, nonstationary inventory problem. Management Science **12**, 206–222 (1995)

354. Veinott, A., Wagner, H.: Computing optimal (s,S) inventory policies. Management Science **11**, 525–552 (1965)

355. Veinott Jr., A.F.: On the optimality of (s,S) inventory policies: New conditions and a new proof. SIAM Journal on Applied Mathematics **14(5)**, 1067–1083 (1966)

356. Verridjt, J., Adan, I., de Kok, T.: A trade-off between emergency repair and inventory investment. IIE Transactions **30**, 119–132 (1998)

357. Wagelmans, A., Hoesel, S.V., Kolen, A.: Economic lot sizing: An $O(n \log n)$ algorithm that runs in linear time in the Wagner-Whitin case. Operations Research **40**, S145–S156 (1992)

358. Wagner, H.: Principles of Operations Research (2nd ed.). Prentice Hall, Englewood Cliffs, NJ (1975)

359. Wagner, H., O'Hagan, M., Lundh, B.: An empirical study of exact and approximately optimal inventory policies. Management Science **11**, 690–723 (1965)

360. Wagner, H., Whitin, T.: Dynamic version of the economic lot size model. Management Science **5**, 89–96 (1958)

361. Wang, Y., Cohen, M.A., Zheng, Y.S.: A two-echelon repairable inventory system with stocking-center-dependent depot replenishment lead times. Management Science **46**(11), 1441–1453 (2000)

362. Whitin, T.: The Theory of Inventory Management. Princeton University, Princeton, NJ (1953)

363. Wolff, R.: Poisson arrivals see time averages. Operations Research **30**, 223–231 (1982)

364. X., Fan, F., Liu, X., Xie, J.: Storage-space capacitated inventory system with (r, q) policies. Operations Research **55**, 854–865 (2007)

365. Xu, K., Evers, P., Fu, M.: Estimating customer service in a two-location continuous review inventory model with emergency transshipments. European Journal of Operational Research **145**(3), 569–584 (2003)

366. Yanagi, S., Sasaki, M.: Reliability analysis for a two-echelon repair system considering lateral resupply, return policy and transportation times. Computers and Industrial Engineering **27**(1-4), 493–497 (1994)

367. Zacks, S.: A two-echelon multi-station inventory model for navy applications. Naval Research Logistics Quarterly **17**(1), 79–85 (1970)

368. Zangwill, W.: A deterministic multiproduct, multifacility production and inventory model. Operations Research **14**, 486–507 (1966)

369. Zangwill, W.: Minimum concave cost flows in certain networks. Management Science **14**, 429–450 (1968)

370. Zangwill, W.: A backlogging model and a multi-echelon model of a dynamic economic lot size production system. Management Science **15**, 506–527 (1969)

371. Zhang, H.: A note on the convexity of service-level measures of the (r, q) system. Management Science **44**, 431–432 (1998)

372. Zheng, Y.: A simple proof for the optimality of (s, S) policies for infinite-horizon inventory problem. Journal of Applied Probability **28**, 802–810 (1991)

373. Zheng, Y.: On properties of stochastic inventory systems. Management Science **38**, 87–103 (1992)

374. Zheng, Y., Chen, F.: Inventory policies with quantized ordering. Naval Research Logistics **39**, 285–305 (1992)

375. Zheng, Y., Federgruen, A.: Finding optimal (s, S) policies is about as simple as evaluating a single policy. Operations Research **39**, 654–665 (1991)

376. Zijm, W., Avsar, Z.: Capacitated two-indenture models for repairable item systems. International Journal of Production Economics **81-82**, 573–588 (2003)

377. Zipkin, P.: Inventory service-level measures: Convexity and approximation. Management Science **32** (1986)

378. Zipkin, P.: Stochastic lead times in continuous-time inventory models. Naval Research Logistics Quarterly **33** (1986)

379. Zipkin, P.: Critical number policies for inventory models with periodic data. Management Science **35**(1), 71–80 (1989)

380. Zipkin, P.: Foundations of Inventory Management, first edn. Irwin-McGraw Hill (2002)

Index